MATERIALS SCIENCE AND TECHNOLOGIES

HYSTERESIS

TYPES, APPLICATIONS AND BEHAVIOR PATTERNS IN COMPLEX SYSTEMS

MATERIALS SCIENCE AND TECHNOLOGIES

Additional books in this series can be found on Nova's website
under the Series tab.

Additional e-books in this series can be found on Nova's website
under the e-book tab.

MATERIALS SCIENCE AND TECHNOLOGIES

HYSTERESIS

TYPES, APPLICATIONS AND BEHAVIOR PATTERNS IN COMPLEX SYSTEMS

JOSÉ CARLOS DIAS
EDITOR

nova publishers
New York

For permission to use material from this book please contact us:
Telephone 631-231-7269; Fax 631-231-8175
Web Site: http://www.novapublishers.com

Library of Congress Cataloging-in-Publication Data

Hysteresis : types, applications, and behavior patterns in complex systems / editor, Josi Carlos Dias (ISCTE-IUL Business School Lisbon, Portugal).
pages cm. -- (Materials science and technologies)
Includes bibliographical references and index.
ISBN 978-1-63321-336-4 (hardcover)
1. System theory. 2. Hysteresis. I. Dias, Josi Carlos.
Q295.H97 2014
003--dc23
2014022552

Published by Nova Science Publishers, Inc. † New York

CONTENTS

Preface		**vii**
Chapter 1	Hysteresis Effects and First Passage Time Densities under Alternative Modeling Architecture Assumptions *José Carlos Dias and Manuela Larguinho*	**1**
Chapter 2	Hysteresis Effects in Economics: Different Methods for Describing Economic Path-Dependence *Ansgar Belke, Matthias Göcke and Laura Werner*	**19**
Chapter 3	Hysteretic Properties of Magnetic Composite Microwires *A. Chizhik, A. Stupakiewicz, V. Zablotskii, A. Zhukov, A. Maziewski and J. Gonzalez*	**43**
Chapter 4	Theoretical and Experimental Study of Magnetic Hysteresis Cycle *R. Masrour, O. Mounkachi, H. El Moussaoui, M. Hamedoun, A. Benyoussef and E. K. Hlil*	**59**
Chapter 5	Hysteretic Behavior in Some Natural Processes *Jenő Takács and Akos Nemcsics*	**89**
Chapter 6	Hysteresis Phenomena in Aerodynamics *G. Alonso and J. Meseguer*	**107**
Chapter 7	Hysteresis Modulation in Structural Control Systems and Semi-Active Implementation *Marco Domaneschi*	**143**
Chapter 8	Percolation Analysis of Tendencies of Relative Permeability Hysteresis Behavior *V. V. Kadet and A. M. Galechyan*	**165**
Chapter 9	Threshold/Flat-Band Voltage Hysteresis of Non-Volatile Semiconductor Memory Devices *Zs. J. Horváth*	**187**

Chapter 10 Behavior Patterns, Origin of Problems and Solutions Regarding Hysteresis Phenomena in Complex Battery Systems **215**
Mohammad Rezwan Khan, Jorge V. Barreras, Ana I. Stan, M. Swierczynski, Søren J. Andreasen and Søren K. Kær

Chapter 11 Biological Hysteresis Illustrated from Different Study Cases: Control Factors, Response Parameters and Functional Meanings **227**
Nabil Semmar and Asma Hammami-Semmar

Chapter 12 Neuromechanical Models Revealing Hysteresis Mechanism in Gait Transitions of Animals **285**
Shinya Aoi

Index **301**

PREFACE

Complex hysteresis problems are present in many branches of science ranging from magnetism, physics, and other streams of natural sciences to other fields unrelated to natural sciences such as finance, economy, or fiscal policy. This book makes an extensive treatment of many hysteresis problems that have been studied in the academic literature.

The book is composed by 12 chapters. A very short description of each one is presented below.

The contribution of Dias and Larguinho reviews the so-called real options literature on hysteresis economic problems under both interest rate and output price uncertainty.

The work of Belke, Göcke, and Werner present on overview of the implementation of hysteresis in economics with an emphasis on: (1) the differentiation between micro and macroeconomic hysteresis; and (2) different methods applied in econometrics in order to capture economic path-dependency empirically.

The chapter of Chizhik, Stupakiewicz, Zablotskii, Zhukov, Maziewski, and Gonzalez studies magnetic hysteresis in metallic microwires using the magneto-optical Kerr effect using both magnetometry and microscopy.

The work of Masrour, Mounkachi, El moussaoui, Hamedoun, and Benyoussef develop a theoretical and experimental study of magnetism hysteresis cycle and illustrate it through many practical applications.

The contribution of Takács and Nemcsics shows how one can use the a hyperbolic model to successfully analyze complicated hysteretic behaviors appearing in physics such as in spin-valve technology, semiconductor, and surface physics.

The contribution of Alonso and Meseguer is related with the hysteresis phenomena that appear in both aeronautical and civil engineering aerodynamics.

The work of Domaneschi employs hysteresis models for simulating control devices in complex structural systems and presents several numerical applications.

The work of Kadet and Galechyan present a model of relative permeability hysteresis for drainage and imbibition based on percolation theory.

The work of Horváth analyze the types and structure of different non-volatile semiconductor memory devices and the principle of their operation.

The chapter of Khan, Barreras, Stan, Swierczynski, Andreasen, and *Kær* study hysteresis effects commonly found in complex battery systems.

The contribution of Semmar and Hammami-Semmar presents several case studies to illustrate biological hysteresis concerning intrinsic parameters governed by different external forces and showing different vaiability meaning.

The chapter of Aoi examines the hysteresis mechanism of the gait transitions in quadruped and insect locomotion from a dynamic point of view.

José Carlos Dias
PhD in Finance
Director of the MSc Finance
ISCTE-IUL Business School
Lisbon, Portugal
Jose.Carlos.Dias@iscte.pt
March 2014

In: Hysteresis
Editor: Jos Carlos Dias

ISBN: 978-1-63321-336-4

Chapter 1

HYSTERESIS EFFECTS AND FIRST PASSAGE TIME DENSITIES UNDER ALTERNATIVE MODELING ARCHITECTURE ASSUMPTIONS

***José Carlos Dias*[1,*] *and Manuela Larguinho*[2,†]**
[1]BRU-UNIDE and ISCTE-IUL Business School, Lisboa, Portugal
[2]Instituto Politécnico de Coimbra, ISCAC
Quinta Agrícola, Bencanta, Coimbra, Portugal

Abstract

Models of hysteresis are useful for explaining optimal entry and exit decisions of a firm under uncertain economic environments. This chapter reviews the existent real options literature on hysteresis modeling problems with a special focus on interest rate uncertainty and output price uncertainty. Subsequently, the chapter sheds some light on the computation of the *ex ante* probability of entry within a specified time horizon for an idle firm, and on the computation of the *ex ante* probability of exit within a specified time horizon for an operating firm.

Keywords: real options, hysteresis, uncertainty, hitting times

1. Introduction

Most of the real options literature assumes the underlying asset price follows a geometric Brownian motion (hereafter, GBM) due to its analytical tractability. While such modeling framework can be a reasonable approximation for the stochastic behavior in many economic applications, the accumulated empirical evidence suggests that such distributional assumption is not able to properly capture the behavior patterns commonly found in many complex capital budgeting decisions. Moreover, a special focus is typically given on output price uncertainty assuming discount rates remain constant. For many decisions, however, revenue

*E-mail address: jose.carlos.dias@iscte.pt
†E-mail address: mlarguinho@iscac.pt

or cost streams may be relatively static and investment is driven by interest uncertainty, e.g. the decision to invest in durable machinery and equipment.

In this chapter, we review some of the most recent attempts of the real options literature to analyze the economic hysteresis problem under both interest rate and output price uncertainty. In section 2 we revisit the [4] work on hysteresis effects under interest rate uncertainty assuming interest rates follow the [2] term structure interest rate model. We then extend the analysis for analyzing the *ex ante* probability of entry (exit) within a specified time horizon for an idle (operating) firm.

In section 3 we review the classic hysteresis model of [5] where uncertainty stems from the output price of the firm, assumed to follow a GBM assumption given its modeling tractability. We then consider the effect of mean-reversion on the investment and divestment trigger points considering the modeling framework proposed by [12]. By doing so, we are able to present a slightly different (but equivalent) solution for analyzing entry and exit decisions under the inhomogeneous geometric Brownian motion (hereafter, IGBM) model of [12]. This allow us also to shed some light on the impact of mean-reversion on the investment and divestment thresholds as well as to numerically analyze the *ex ante* probability of entry (exit) within a specified time horizon for an idle (active) firm. Finally, in section 4 we present the concluding remarks.

2. Hysteresis Effects under Interest Rate Uncertainty

2.1. The Firm's Policy

In this section we will consider decisions on reversible investments under interest rate uncertainty. Following [4], we will analyze the case of a firm that can invest $\overline{I}$ (the investment cost) and receive a perpetuity (a project) with constant cash flow rate. Since the interest rate, r, used to discount the perpetual flows is driven by the CIR process of [2], the perpetuity value is stochastic. In an unfavorable economic situation, the firm can divest (receiving the divestment proceeds, $\underline{I}$), and so recoup a fraction α of the invested capital, with $\alpha = \underline{I}/\overline{I}$ $(0 \leq \alpha \leq 1)$. Investment in this perpetual project will be triggered when interest rates are low (high perpetuity value) and in particular at a critical level $\underline{r}$.

The value function for the idle project (without flow) but with option to open is denoted by $F_0(r)$, and the value function for the active project (with perpetuity) which has the option to shut is denoted by $F_1(r) + P(r)$. Investment will be triggered as the interest rate r falls to the lower threshold $\underline{r}$ whilst divestment will be triggered as rates rise to the upper trigger $\overline{r}$ $(\overline{r} > \underline{r})$. Thus, the decision to switch from the idle state to the operating state, and vice-versa, can be described by the following two *value-matching* conditions

$$F_0(\underline{r}) + \overline{I} = F_1(\underline{r}) + P(\underline{r})$$
$$F_0(\overline{r}) + \underline{I} = F_1(\overline{r}) + P(\overline{r}).$$

Thus, an idle firm invests when rates fall to $\underline{r}$ and an operating firm will disinvest once the interest rate rise to $\overline{r}$. Let us denote by V the value premium of active to idle firms, this depend on current rates r:

$$V(r) = F_1(r) + P(r) - F_0(r), \tag{1}$$

and the value matching conditions can be coupled with two *smooth-pasting* (first order) conditions

$$V(\underline{r})=\overline{I},\ V'(\underline{r})=0,\ V(\overline{r})=\underline{I},\ V'(\overline{r})=0. \tag{2}$$

The range $(\underline{r},\ \overline{r})$ is the hysteretic band of the problem since idle firms do not invest and operating firms do not suspend within this intermediate level of interest rate.

2.2. The Interest Rate Dynamics

We will assume that the interest rate, r, under the risk-neutral measure $\mathbb{Q}$, follows the CIR process which is given by the stochastic differential equation (sde):

$$dr_t=(\kappa+\lambda)\left[\frac{\kappa\theta}{(\kappa+\lambda)}-r_t\right]dt+\sigma\sqrt{r_t}dW_t^{\mathbb{Q}}, \tag{3}$$

where $\kappa^*=\kappa+\lambda$ is the speed of adjustment (reversion rate or reverting rate), $\theta^*=\kappa\theta/k^*$ is the long-run mean of the instantaneous interest rate (asymptotic interest rate or reverting level), σ is the volatility of the process, λ is the market price of risk parameter, and $W_t^{\mathbb{Q}}$ is a standard Brownian motion under $\mathbb{Q}$. It is well known that the $\kappa\theta$ term plays a key role under this diffusion and has important implications for capture of the interest rate process r at a value of zero. The condition $2\kappa\theta\geq\sigma^2$ ensures that the interest rate remains positive.[1] In a CIR economy, the price of a general interest rate claim $F(r,t)$ with cash flow rate $C(r,t)$ satisfies the following partial differential equation

$$\frac{1}{2}\sigma^2r\frac{\partial^2F(r,t)}{\partial r^2}+\kappa(\theta-r)\frac{\partial F(r,t)}{\partial r}+\frac{\partial F(r,t)}{\partial t}-\lambda r\frac{\partial F(r,t)}{\partial r}-rF(r,t)+C(r,t)=0. \tag{4}$$

Let us denote by $P(r)$ the value of a perpetuity, that pays coupons at a constant unit rate $C(r,t)=1$, which satisfies the following ordinary differential equation (ode):[2]

$$\frac{1}{2}\sigma^2r\frac{d^2P(r)}{dr^2}+\kappa(\theta-r)\frac{dP(r)}{dr}-\lambda r\frac{dP(r)}{dr}-rP(r)+1=0. \tag{5}$$

The solution of ode (5) is the value of a perpetuity given by

$$P(r)=E_{t_0}^{Q}\left[\int_{t_0}^{\infty}e^{-\int_{t_0}^{t}r_sds}dt\right]=\int_{t_0}^{\infty}Z(r,t_0,t)dt, \tag{6}$$

where $Z(r,t,T)$ is the price of a zero coupon bond under CIR diffusion. To evaluate the value of this perpetuity, we will use a method proposed by [3], which has a better performance in terms of accuracy, as been shown by [10].

To determinate the perpetual option to invest in or divest from a project, we consider a particular situation of the time homogeneous part of (5), which gives the following ode, where $C(r)=0$ since no cash flows are present in the options

$$\frac{1}{2}\sigma^2r\frac{d^2F(r)}{dr^2}+\kappa(\theta-r)\frac{dF(r)}{dr}-\lambda r\frac{dF(r)}{dr}-rF(r)=0. \tag{7}$$

---

[1] See [6] for a complete description of the boundary conditions.

[2] We must note that since $\lim_{t\to\infty}\frac{\partial P(r,t)}{\partial t}=0$, the value of a perpetuity is not time dependent.

[4] have shown that the general solution of the ode (7) is given by

$$F_i(r) = C_1 e^{v_i r} M(a_i, b, z_i) + C_2 e^{v_i r} U(a_i, b, z_i), \text{ with} i \in \{0, 1\}, \tag{8}$$

with

$$\gamma = \left[(\kappa+\lambda)^2 + 2\sigma^2\right]^{1/2},$$

$$a_{0,1} = \frac{\kappa\theta}{\sigma^2}\left(1 \mp \frac{\kappa+\lambda}{\gamma}\right),$$

$$b = \frac{2\kappa\theta}{\sigma^2},$$

$$z_{0,1} = \pm\frac{2\gamma r}{\sigma^2},$$

$$v_{0,1} = \frac{\kappa+\lambda\mp\gamma}{\sigma^2},$$

and where C_1 and C_2 are constants to be determined from boundary conditions, and $M(a,b,z)$ is the Kummer's confluent hypergeometric function as given in [1, Equation 13.1.2], and finally $U(a,b,z)$ is the Tricomi's confluent hypergeometric function as presented in [1, Equation 13.1.3].

Given the near-zero and asymptotic behaviour of the functions $M(a,b,z)$ and $U(a,b,z)$ and the necessary boundary conditions, [4] found that the expected net present value in the idle state with the option to open and the option to switch out of the perpetuity are respectively given by:

$$F_0(r) = C_2 e^{v_0 r} U(a_0, b, z_0), \tag{10}$$

$$F_1(r) = C_1 e^{v_1 r} M(a_1, b, z_1), \tag{11}$$

which, using (1) and (2), lead to two value matching conditions

$$F_1(\underline{r}) + P(\underline{r}) - F_0(\underline{r}) = \overline{I} \text{ and } F_1(\overline{r}) + P(\overline{r}) - F_0(\overline{r}) = \underline{I}. \tag{12}$$

The first order (smooth pasting) conditions are given by

$$\frac{dF_0(\underline{r})}{d\underline{r}} = \frac{dF_1(\underline{r})}{d\underline{r}} + \frac{dP(\underline{r})}{d\underline{r}} \text{ and } \frac{dF_0(\overline{r})}{d\overline{r}} = \frac{dF_1(\overline{r})}{d\overline{r}} + \frac{dP(\overline{r})}{d\overline{r}}. \tag{13}$$

Thus, the solution of the two sided control problem rests on the determination of the two embedded constants C_1, C_2, and two thresholds $\underline{r}$, $\overline{r}$.

2.3. First Passage Time Distributions

In this subsection we will compute and analyze the entry (exit) probability for an idle (active) firm to enter (exit) in (from) the market within a specified horizon T. These probabilities will be a measure of the fraction of idle (active) firms, under competitive equilibrium, that will enter (exit) the market by horizon T.

For any optimal policy pair $(\underline{r}, \overline{r})$, the probability that an idle firm will enter by time T is equal to $P(\tau_L \leq T)$, where $\tau_L := \inf\{t \geq t_0 : r_t = \underline{r}\}$ is the first hitting time of the interest rate to the investment threshold $\underline{r}$. Similarly, the probability that an active firm will exit the market by time T is equal to $P(\tau_U \leq T)$ with $\tau_U := \inf\{t \geq t_0 : r_t = \overline{r}\}$ is the first hitting time of the interest rate to the divestment threshold $\overline{r}$.

Since for the CIR process there are no simple closed-form solutions to compute these entry and exit probabilities, it will be necessary to compute first an approximate solution of the CIR interest rates, and then to simulate these probabilities. So, to compute an approximate solution we will use the Euler-Maruyama (EM) method, as described in [7]. Following this method, the CIR interest rate are calculated, at dates $0 = t_0 < t_1 < \cdots < t_N = T$, as

$$r_{i+1} = r_i(1 - k(t_{i+1} - t_i)) + k\theta(t_{i+1} - t_i) + \sigma\sqrt{|r_i(t_{i+1} - t_i)|}Z_i, \tag{14}$$

where $Z_{i+1} \sim \mathcal{N}(0,1)$ random variables.

[7] argument that this discretization scheme provides qualitatively correct approximations to the first and second moments. In the usual Euler method we have the problem that the discrete process for the interest rate r become negative with non-zero probability, making the computation of $\sqrt{r_i}$ impossible. The EM scheme solves this problem.

Estimates of the probabilities of entry and exit are calculated via

$$P(\tau_L \leq T) = \frac{\underline{M}}{M}, \tag{15}$$

and

$$P(\tau_U \leq T) = \frac{\overline{M}}{M}, \tag{16}$$

where $\underline{M} = \{r_{i+1} \in M : r_{i+1} \leq \underline{r}\}$ and $\overline{M} = \{r_{i+1} \in M : r_{i+1} \geq \overline{r}\}$, and M is a large number of possible future interest rates, r, at dates $0 = t_0 < t_1 < \cdots < t_N = T$ generated via the EM method.

2.4. Computational Results

Table 1 replicates the results for the upper and lower interest rate thresholds for the entry and exit problem in a CIR economy as presented in [4]. For this table we have used the following parameters values: $k = 0.2330$, $\theta = 0.0808$, $\sigma \in \{0.0854, 0.30\}$, and $\lambda = 0$. We also have considered an investment cost of $\overline{I} = 10$, and $\alpha \in \{0.00, 0.25, 0.50, 0.75, 1.00\}$. When $\alpha = 0.00$, an operating firm never shuts it projects. This table reveals also that higher levels of flexibility reduce, as expected, the band of inaction. We also present the case of perfect reversibility, $\alpha = 1.00$, to illustrate the limiting case.

Table 2 presents the entry and exit probabilities under the CIR diffusion modeling assumption, with the same parameter values as in Table 1. To compute these probabilities we have used the aforementioned Euler and Maruyama scheme.

Table 2 shows that there is a direct relation between entry/exit probabilities and the recouped fraction α of the invested capital. From this table, we also can conclude that the investment/divestment decision have a direct relation with the volatility of the interest rate.

Table 1. Upper and lower interest rate thresholds for the entry and exit problem in a CIR economy

	$\sigma = 0.0854$		$\sigma = 0.30$	
α	$\underline{r}$	$\overline{r}$	$\underline{r}$	$\overline{r}$
0.00	0.0723	$+\infty$	0.0238	$+\infty$
0.25	0.0723	0.7098	0.0238	0.9328
0.50	0.0723	0.3969	0.0244	0.5505
0.75	0.0723	0.2375	0.0288	0.3414
1.00	0.10	0.10	0.10	0.10

This table values the upper and lower interest rate thresholds for the entry and exit problem under the CIR diffusion model. σ is the volatility of the interest rate, $\overline{I} = 10$ is the investment cost, $\alpha \in \{0.00, 0.25, 0.50, 0.75, 1.00\}$ is the ratio of divestment proceeds in the investment cost, $k = 0.2339$ is the intensity with which the interest rate is drawn back towards a long-run mean, $\theta = 0.0808$ is the long-run mean of the instantaneous interest rate, and $\lambda = 0.00$ is the market risk parameter.

3. Hysteresis Effects under Output Price Uncertainty

3.1. The Firm's Policy

In this section we will consider decisions on reversible investments under output price uncertainty. We will analyze how output price uncertainty and costly reversibility affects the optimal entry and exit policy of a competitive price-taking firm. We will consider the case of a firm that can invest $\overline{I}$ (again, the investment cost) to entry in the market. As usual, the entry mode is in the form of a unit of investment, namely, a single project of a given size. While in the market, it can produce a unit flow of output at the variable cost C. In an unfavorable economic situation, the firm can divest, but must pay a lump-sum cost $\underline{I}$ to do so, and must pay the entry cost $\overline{I}$ again should it decide to reenter at some future time. We assume that the magnitudes $\overline{I}$, $\underline{I}$, C, and the riskless interest rate, r, are constant and non-stochastic. Moreover, uncertainty stems from the output equilibrium price P which is assumed to be exogenous to the firm. Throughout this section we will study the cases where the output price follows a geometric Brownian motion (GBM) and an inhomogeneous geometric Brownian motion (IGBM).

Let us define $V_0(P)$ as the expected net present value of the firm of starting with an initial output price P in the idle state and following dynamic optimal entry-exit policies. Similarly, let us define by $V_1(P)$ the value of the active firm. The firm will entry in the market when the output price P rises to the upper threshold $\overline{P}$, whilst the firm will exit from the market when the output price falls down to the lower threshold $\underline{P}$ $(\overline{P} > \underline{P})$. So, the optimal entry and exit policy is determined through these two thresholds at which a firm optimally switches from the idle to the operating state and vice versa. At the optimal entry threshold $\overline{P}$, the idle firm exercises its entry option by paying $\overline{I}$ in order to receive an "underlying asset" of value $V_1(\overline{P})$, which includes both an option to exit and a flow reward component. Similarly, at the optimal exit threshold $\underline{P}$, the active firm exercises its exit option by paying $\underline{I}$ in favour of regaining an "underlying asset" worth $V_0(\underline{P})$ (i.e. an option to enter the market again).

The entry and exit thresholds are determined numerically through a set of two *value-*

Table 2. Probability of entry for an idle firm and the probability of exit for an active firm within a specified horizon, T, for the CIR processes, computed via Monte Carlo simulation

	$\sigma = 0.0854$		$\sigma = 0.30$	
α	Entry probability	Exit probability	Entry probability	Exit probability
0.25	0.4100	0.0000	0.7223	0.1060
0.50	0.6351	0.0006	0.8147	0.2773
0.75	0.7877	0.0532	0.8847	0.4753

This table values the probabilities of entry in the market during a time horizon T for the interest rates under the CIR model. In each simulation, 100,000 paths and 1,000 time steps are used. σ is the volatility of the interest rate, $\overline{I} = 10$ is the investment cost, $\alpha \in \{0.25, 0.50, 0.75\}$ is the ratio of divestment proceeds in the investment cost, $k = 0.2339$ is the intensity with which the interest rate is drawn back towards a long-run mean, $\theta = 0.0808$ is the long-run mean of the instantaneous interest rate, and $\lambda = 0.00$ is the market risk parameter. In all cases, $r_0^* = (\overline{r} + \underline{r})/2$ and the time horizon is $T = 10$.

matching conditions

$$V_0(\overline{P}) + \overline{I} = V_1(\overline{P}) \tag{17}$$
$$V_0(\underline{P}) - \underline{I} = V_1(\underline{P}), \tag{18}$$

and two *smooth-pasting* conditions

$$V_0'(\overline{P}) = V_1'(\overline{P}) \tag{19}$$
$$V_0'(\underline{P}) = V_1'(\underline{P}). \tag{20}$$

The value-matching conditions ensure that the gain in value from exercising the option is exactly equal to the cost of doing so. The smooth-pasting conditions require that the first derivative of the firm value function must take the same value before and after the option (to enter or exit) has been exercised.

3.2. Hysteresis under the GBM Model

Throughout this subsection, the output price P is assumed, under the physical measure $\mathbb{P}$, to follow a GBM process which is given by the stochastic differential equation (sde):

$$dP_t = \mu P_t dt + \sigma P_t dW_t^{\mathbb{P}}, \tag{21}$$

where $W_t^{\mathbb{P}}$ is a standard Brownian motion under $\mathbb{P}$, and μ and σ denote, respectively, the constant growth rate and the constant volatility of the market price P.

The value function $V_0(P)$ for an idle firm satisfies the differential equation

$$\frac{1}{2}\sigma^2 P^2 V_0''(P) + (\mu - \lambda\rho\sigma) P V_0'(P) - r V_0(P) = 0, \tag{22}$$

where $\lambda = (\mathbb{E}[R_m] - r)/\sigma(R_m)$ is the market price of risk (with $\mathbb{E}[R_m]$ and $\sigma(R_m)$ being the expected return and standard deviation, respectively, of the market portfolio return R_m), and ρ is the correlation between the output price P and the market portfolio. Both λ and ρ are assumed constant. The corresponding function $V_1(P)$ for an active firm satisfies

$$\frac{1}{2}\sigma^2 P^2 V_1''(P) + (\mu - \lambda\rho\sigma) P V_1'(P) - rV_1(P) + P - C = 0. \tag{23}$$

As shown by [5], the general solutions of (22) and (23) are easy to obtain. Both are linear and have the same homogeneous part. These solutions are known as complementary solutions, which can be found together. What determines the difference between the two contingent solutions and the options they represent are their boundary conditions. So, to solve these equations subject to appropriate boundary conditions, [5] has found a solution of the form

$$V(P) = AP^{\xi},$$

where A and ξ are constants to be determined. Substituting $V(P)$, $V'(P)$ and $V''(P)$ into the homogeneous equation

$$\frac{1}{2}\sigma^2 P^2 V''(P) + (\mu - \lambda\rho\sigma) P V'(P) - rV(P) = 0, \tag{24}$$

gives the following quadratic equation

$$\frac{1}{2}\sigma^2 \xi(\xi - 1) + (\mu - \lambda\rho\sigma)\xi - r = 0. \tag{25}$$

The solution of this equation has two roots given by

$$\xi_1 = \frac{1}{2} - \frac{(\mu - \lambda\rho\sigma)}{\sigma^2} + \sqrt{\left(\frac{(\mu - \lambda\rho\sigma)}{\sigma^2} - \frac{1}{2}\right)^2 + \frac{2r}{\sigma^2}} > 1,$$

$$\xi_2 = \frac{1}{2} - \frac{(\mu - \lambda\rho\sigma)}{\sigma^2} - \sqrt{\left(\frac{(\mu - \lambda\rho\sigma)}{\sigma^2} - \frac{1}{2}\right)^2 + \frac{2r}{\sigma^2}} < 0.$$

Therefore, the general solution of the homogeneous differential equation is given by

$$V(P) = A_{\phi} P^{\xi_1} + B_{\phi} P^{\xi_2}, \tag{27}$$

with $\phi \in \{0, 1\}$, and where 0 and 1 stand for the idle and active firm, respectively. Then, the solutions to (22) and (23) subject to the boundaries conditions

$$\lim_{P \to 0^+} V_0(P) = 0 \tag{28}$$

and

$$\lim_{P \to +\infty} V_1(P) = \mathbb{E}\left[\int_0^{+\infty} (P_s - C)\, e^{-rs} ds | P_0 = P\right] \tag{29}$$

are

$$V_0(P) = A_0 P^{\xi_1} \tag{30}$$

and

$$V_1(P) = B_1 P^{\xi_2} + \frac{P}{r+\lambda\rho\sigma-\mu} - \frac{C}{r}, \tag{31}$$

with A_0 and B_1 being constants to be determined.

$V_0(P)$ corresponds to the value function of an idle firm over the range of prices $(0,\overline{P})$. In other words, it is the option value to become active in the market at the trigger price $\overline{P}$. Similarly, $V_1(P)$ corresponds to the value function of an active firm over the range of prices $(\underline{P},+\infty)$. Note that the first component on the right-hand side of equation (31) is the value of the option to exit at the trigger price $\underline{P}$ while the sum of the second and third components represent the expected value of operating in the market forever.

To apply the smooth-pasting conditions (19) and (20), one needs to compute the first derivatives of the value functions V_0 and V_1. Using straightforward calculus, it follows that:

$$V_0'(P) = A_0 \xi_1 P^{\xi_1 - 1}, \tag{32}$$

and

$$V_1'(P) = B_1 \xi_2 P^{\xi_2 - 1} + \frac{1}{r+\lambda\rho\sigma-\mu}. \tag{33}$$

We can express the non-linear system of equations (17) - (20) in matrix form. So it can be shown that the optimal solution $\mathbf{X} = [\overline{P}, \underline{P}, A_0, B_1]'$ is uniquely determined by solving the system **F(X) = 0**, where

$$\mathbf{F}(\mathbf{X}) = \begin{bmatrix} -A_0\overline{P}^{\xi_1} + B_1\overline{P}^{\xi_2} + \varphi\overline{P} - \overline{X} \\ -A_0\underline{P}^{\xi_1} + B_1\underline{P}^{\xi_2} + \varphi\underline{P} - \underline{X} \\ -A_0\xi_1\overline{P}^{\xi_1} + B_1\xi_2\overline{P}^{\xi_2} + \varphi\overline{P} \\ -A_0\xi_1\underline{P}^{\xi_1} + B_1\xi_2\underline{P}^{\xi_2} + \varphi\underline{P} \end{bmatrix} \tag{34}$$

with

$$\varphi = (r+\lambda\rho\sigma-\mu)^{-1}, \tag{35a}$$

$$\overline{X} = \frac{C}{r} + \overline{I}, \tag{35b}$$

$$\underline{X} = \frac{C}{r} - \underline{I}. \tag{35c}$$

Following [11], and using the last two equations, we can define the value-matching and smooth-pasting conditions in another matrix form with $[A_0, B_1, \overline{X}, \underline{X}]$ as a function of $[\overline{P}, \underline{P}]$

$$\begin{bmatrix} 1 & 0 & \overline{P}^{\xi_1} & -\overline{P}^{\xi_2} \\ 0 & 1 & \underline{P}^{\xi_1} & -\underline{P}^{\xi_2} \\ 0 & 0 & \xi_1\overline{P}^{\xi_1} & -\xi_2\overline{P}^{\xi_2} \\ 0 & 0 & \xi_1\underline{P}^{\xi_1} & -\xi_2\underline{P}^{\xi_2} \end{bmatrix} \begin{bmatrix} \overline{X} \\ \underline{X} \\ A_0 \\ B_1 \end{bmatrix} = \varphi \begin{bmatrix} \overline{P} \\ \underline{P} \\ \overline{P} \\ \underline{P} \end{bmatrix}.$$

Inverting the matrix to recover the vector $\left[A_0, B_1, \overline{X}, \underline{X}\right]$, the matrix product is most easily evaluated as function of a fraction $\gamma \equiv \underline{P}/\overline{P}$ (a time independent ratio of the lower and upper switching thresholds)

$$\begin{bmatrix} \overline{X} \\ \underline{X} \\ A_0 \\ B_1 \end{bmatrix} = \frac{\varphi \overline{\mathbf{P}}}{\xi_1 \xi_2 (\gamma^{\xi_2} - \gamma^{\xi_1})} \begin{bmatrix} \xi_1\xi_2(\gamma^{\xi_2} - \gamma^{\xi_1}) - \xi_2\gamma^{\xi_2} + \xi_1\gamma^{\xi_1} - (\xi_1 - \xi_2)\gamma \\ \xi_1\xi_2(\gamma^{\xi_2+1} - \gamma^{\xi_1+1}) + \xi_2\gamma^{\xi_1+1} - \xi_1\gamma^{\xi_2+1} - (\xi_1 - \xi_2)\gamma^{\xi_1+\xi_2} \\ \frac{\xi_2\varphi^{-\xi_1}(\gamma^{\xi_2} - \gamma)}{\overline{P}^{\xi_1}} \\ \frac{\xi_1\varphi^{-\xi_1}(\gamma^{\xi_1} - \gamma)}{\overline{P}^{\xi_2}} \end{bmatrix}$$

Define the variable $\pi = \underline{X}/\overline{X}$ (the ratio of switching costs). This variable can be expressed as a function of γ by dividing the first two rows of the last matrix

$$\pi(\gamma) = \frac{\xi_1\xi_2(\gamma^{\xi_2+1} - \gamma^{\xi_1+1}) + \xi_2\gamma^{\xi_1+1} - \xi_1\gamma^{\xi_2+1} - (\xi_1 - \xi_2)\gamma^{\xi_1+\xi_2}}{\xi_1\xi_2(\gamma^{\xi_2} - \gamma^{\xi_1}) - \xi_2\gamma^{\xi_2} + \xi_1\gamma^{\xi_1} - (\xi_1 - \xi_2)\gamma} \tag{36}$$

This is the equation that determines the entire hysteresis system. Since the $\pi(\gamma)$ is monotonic and increasing in γ, it admits inverse function. Then for every choice of π, once the optimal γ is numerically retrieved from (36), the thresholds and option constants can be recovered by substituting in the following equations

$$\begin{bmatrix} \overline{P} \\ \underline{P} \end{bmatrix} = \xi_1\xi_2(\gamma^{\xi_2} - \gamma^{\xi_1}) \begin{bmatrix} \frac{\overline{X}}{\xi_1\xi_2(\gamma^{\xi_2} - \gamma^{\xi_1}) - \xi_2\gamma^{\xi_2} + \xi_1\gamma^{\xi_1} - (\xi_1 - \xi_2)\gamma} \\ \frac{\underline{X}}{\xi_1\xi_2(\gamma^{\xi_2} - \gamma^{\xi_1}) + \xi_2\gamma^{\xi_1} - \xi_1\gamma^{\xi_2} - (\xi_1 - \xi_2)\gamma^{\xi_1+\xi_2-1}} \end{bmatrix}, \tag{37}$$

$$\begin{bmatrix} A_0 \\ B_1 \end{bmatrix} = \frac{\varphi \overline{\mathbf{P}}}{\xi_1\xi_2(\gamma^{\xi_2} - \gamma^{\xi_1})} \begin{bmatrix} \frac{\xi_2\varphi^{-\xi_1}(\gamma^{\xi-2} - \gamma)}{\overline{P}^{\xi_1}} \\ \frac{\xi_1\varphi^{-\xi_1}(\gamma^{\xi-1} - \gamma)}{\overline{P}^{\xi_2}} \end{bmatrix}. \tag{38}$$

3.3. Hysteresis under the IGBM Model

Along this subsection, the output price P is assumed, under the physical measure $\mathbb{P}$, to follow a IGBM process which is given by the stochastic differential equation (sde):

$$dP_t = k(\theta - P_t)dt + \sigma P_t dW_t^{\mathbb{P}}, \tag{39}$$

where $W_t^{\mathbb{P}}$ is a standard Brownian motion under $\mathbb{P}$, and κ, θ, and σ denote, respectively, the (constant) speed of reversion, the (constant) long-run mean price level, and the (constant) volatility parameter.

The value function $V_0(P)$ for an idle firm satisfies the differential equation

$$\frac{1}{2}\sigma^2 P^2 V_0''(P) + \big(k\theta - (k + \lambda\rho\sigma)P\big)V_0'(P) - rV_0(P) = 0, \tag{40}$$

with λ being (again) the market price of risk, and ρ the correlation between the output price P and the market portfolio. As usual, both λ and ρ are assumed to be constant. The corresponding function $V_1(P)$ for an active firm satisfies

$$\frac{1}{2}\sigma^2 P^2 V_1''(P) + \big(k\theta - (k+\lambda\rho\sigma)P\big)V_1'(P) - rV_1(P) + P - C = 0. \tag{41}$$

To find the solutions to these two equations, we start by solving the homogeneous ode

$$PV''(P) + \left(\frac{2k\theta}{\sigma^2 P} - \frac{2(k+\lambda\rho\sigma)}{\sigma^2}\right)V'(P) - \frac{2r}{\sigma^2 P}V(P) = 0. \tag{42}$$

Similarly to [5], we define a new function $h(P)$ by

$$V(P) = P^{\xi}h(P), \tag{43}$$

where ξ is a constant that will be chosen in order to make $h(P)$ satisfy a differential equation with a known solution. Then substituting $V(P)$, $V'(P)$, and $V''(P)$ into (42) and rearranging gives the following equation

$$\begin{aligned} &P^{\xi-1}h(P)\left[\xi(\xi-1) - \tfrac{2(k+\lambda\rho\sigma)\xi}{\sigma^2} - \tfrac{2r}{\sigma^2}\right] + \\ &+P^{\xi}\left[Ph''(P) + \left[\left(2\xi - \tfrac{2(k+\lambda\rho\sigma)}{\sigma^2}\right) + \tfrac{2k\theta}{\sigma^2 P}\right]h'(P) + \tfrac{2k\theta\xi}{\sigma^2 P^2}h(P)\right] = 0. \end{aligned} \tag{44}$$

Equation (44) must hold for any value of P, so the bracketed terms in both the first and second lines of equation (44) must equal zero. Let us first choose ξ to set the bracketed terms of the first line of equation (44) equal to zero:

$$\phi(\xi) = \xi(\xi-1) - \frac{2(k+\lambda\rho\sigma)\xi}{\sigma^2} - \frac{2r}{\sigma^2} = 0. \tag{45}$$

This quadratic equation has two solutions for ξ, since $\phi''(\xi) = 2 > 0$ and $\phi(0) = -2r/\sigma^2 < 0$ and $\phi(1) = -2(k+\lambda\rho\sigma+r)/\sigma^2 < 0$; one root must be less than zero, ξ_1, and the other must be greater than one, ξ_0. Written out explicitly,

$$\xi_1 = \frac{2(k+\lambda\rho\sigma)+\sigma^2 - \sqrt{8r\sigma^2 + (-2k-2\lambda\rho\sigma-\sigma^2)^2}}{2\sigma^2} \tag{46a}$$

$$\xi_0 = \frac{2(k+\lambda\rho\sigma)+\sigma^2 + \sqrt{8r\sigma^2 + (-2k-2\lambda\rho\sigma-\sigma^2)^2}}{2\sigma^2}. \tag{46b}$$

To find a general solution to the bracketed term in the second line of (44) we make the change of variable $z = 2k\theta/(\sigma^2 P)$ and we consider a new function $h(P) = g(z)$, so that $h'(P) = -\frac{z^2\sigma^2 g'(z)}{2k\theta}$ and $h''(P) = \frac{2\sigma^4 z^3 g'(z)}{(2k\theta)^2} + \frac{\sigma^4 z^4 g''(z)}{(2k\theta)^2}$. By making this substitution we transform this equation into

$$zg''(z) + (b-z)g'(z) - ag(z) = 0, \tag{47}$$

with $a = -\xi$ and $b = 2 - 2\xi + 2(k+\lambda\rho\sigma)/\sigma^2$. Equation (47) is the Kummer's equation, as defined in [1, Equation 13.1.1], which has a general solution given by

$$g(x) = A_\phi M(a,b,x) + B_\phi U(a,b,x).$$

If we multiply the general solution of Kummer's equation by P^{ξ} and reverse the change of variables, we get the general solution for $V(P)$, which is solution of the ode (42)

$$V(P) = A_{\phi} P^{\xi} M\left(a, b, \frac{2k\theta}{\sigma^2 P}\right) + B_{\phi} P^{\xi} U\left(a, b, \frac{2k\theta}{\sigma^2 P}\right). \tag{48}$$

Then the solutions to (40) and (41) subject to the boundaries conditions (28) and (29) are

$$V_0(P) = A_0 P^{\xi_0} U\left(-\xi_0, 2 - 2\xi_0 + \frac{2(k+\lambda\rho\sigma)}{\sigma^2}, \frac{2k\theta}{\sigma^2 P}\right) \tag{49}$$

and

$$\begin{aligned} V_1(P) &= B_1 P^{\xi_1} M\left(-\xi_1, 2 - 2\xi_1 + \frac{2(k+\lambda\rho\sigma)}{\sigma^2}, \frac{2k\theta}{\sigma^2 P}\right) \\ &+ \frac{P}{r+\lambda\rho\sigma} + \frac{k\theta}{r(k+\lambda\rho\sigma)} - \frac{k\theta}{(r+k+\lambda\rho\sigma)(k+\lambda\rho\sigma)} - \frac{C}{r}, \end{aligned} \tag{50}$$

respectively, with A_0 and B_1 being two constants to be determined. $M(a,b,x)$ and $U(a,b,x)$ are the Kummer's functions, as defined by [1, Equations 13.1.2 and 13.1.3].

$V_0(P)$ corresponds to the value function of an idle firm over the range of prices $(0, \overline{P})$. In other words, it is the option value to become active in the market at the trigger price $\overline{P}$. Similarly, $V_1(P)$ corresponds to the value function of an active firm over the range of prices $(\underline{P}, +\infty)$. Note that the first component on the right-hand side of equation (50) is the value of the option to exit at the trigger price $\underline{P}$ while the sum of the second and third components represent the expected value of operating in the market forever.

To apply the smooth-pasting conditions (19) and (20), one needs to compute the first derivatives of the value functions V_0 and V_1. Using straightforward calculus, it follows that:

$$\begin{aligned} V_0'(P) &= -A_0 \xi_0 P^{\xi_0 - 1} \left[U\left(-\xi_0, 2 - 2\xi_0 + \frac{2(k+\lambda\rho\sigma)}{\sigma^2}, \frac{2k\theta}{\sigma^2 P}\right) \right. \\ &\left. - \frac{2k\theta}{\sigma^2 P} U\left(1 - \xi_0, 3 - 2\xi_0 + \frac{2(k+\lambda\rho\sigma)}{\sigma^2}, \frac{2k\theta}{\sigma^2 P}\right) \right], \end{aligned} \tag{51}$$

and

$$\begin{aligned} V_1'(P) &= -B_1 \xi_1 P^{\xi_1 - 1} \left[M\left(-\xi_1, 2 - 2\xi_1 + \frac{2(k+\lambda\rho\sigma)}{\sigma^2}, \frac{2k\theta}{\sigma^2 P}\right) \right. \\ &\left. + \frac{k\theta}{P(\sigma^2 - \xi_1\sigma^2 + (k+\lambda\rho\sigma))} M\left(1 - \xi_1, 3 - 2\xi_1 + \frac{2(k+\lambda\rho\sigma)}{\sigma^2}, \frac{2k\theta}{\sigma^2 P}\right) \right]. \end{aligned} \tag{52}$$

We can express the non-linear system of equations (17) - (20) in matrix form. So it can be shown that the optimal solution $\mathbf{X} = [\overline{P}, \underline{P}, A_0, B_1]'$ is uniquely determined by solving the system **F(X) = 0**, where

$$\mathbf{F}(\mathbf{X}) = \begin{bmatrix} -A_0 U(\overline{P}) + B_1 M(\overline{P}) + \varphi \overline{P} + \overline{X} \\ -A_0 U(\underline{P}) + B_1 M(\underline{P}) + \varphi \underline{P} + \underline{X} \\ -A_0 V(\overline{P}) + B_1 W(\overline{P}) + \varphi \overline{P} \\ -A_0 V(\underline{P}) + B_1 W(\underline{P}) + \varphi \underline{P} \end{bmatrix} \tag{53}$$

with

$$\varphi = (r+k+\lambda\rho\sigma)^{-1}, \tag{54a}$$

$$a_1 = -\xi_1, \tag{54b}$$

$$a_2 = -\xi_0, \tag{54c}$$

$$b_1 = 2+2a_1+\frac{2(k+\lambda\rho\sigma)}{\sigma^2}, \tag{54d}$$

$$b_2 = 2+2a_2+\frac{2(k+\lambda\rho\sigma)}{\sigma^2}, \tag{54e}$$

$$M(y) = y^{-a_1}M(a_1,b_1,z(y)), \tag{54f}$$

$$U(y) = y^{-a_2}U(a_2,b_2,z(y)), \tag{54g}$$

$$W(y) = a_1\left[M(y)+\frac{z(y)}{b_1}y^{-a_1}M(a_1+1,b_1+1,z(y))\right], \tag{54h}$$

$$V(y) = a_2\left[U(y)-z(y)y^{-a_2}U(a_2+1,b_2+1,z(y))\right], \tag{54i}$$

and

$$\overline{X} = \frac{k\theta}{r(k+\lambda\rho\sigma)} - \frac{k\theta}{(r+k+\lambda\rho\sigma)(k+\lambda\rho\sigma)} - \frac{C}{r} - \overline{I}, \tag{55a}$$

$$\underline{X} = \frac{k\theta}{r(k+\lambda\rho\sigma)} - \frac{k\theta}{(r+k+\lambda\rho\sigma)(k+\lambda\rho\sigma)} - \frac{C}{r} + \underline{I}. \tag{55b}$$

We note that [12] provides an alternative solution to equation (40), where such solution depends on two Kummer's function $M(a,b,x)$. Our alternative and equivalent solution depends only on one Tricomi's function $U(a,b,x)$, thus making the smooth-pasting condition easiest to apply and turning the economic hysteresis problem simpler to analyze and understand.

3.4. First Passage Time Distributions

In this subsection, we will compute and analyze the entry (exit) probability for an idle (active) firm to enter (exit) in (from) the market within a specified time horizon T. These probabilities will be a measure of the fraction of idle (active) firms, under competitive equilibrium, that will enter (exit) the market by time T.

For any optimal policy pair $(\underline{P},\overline{P})$, the probability that an idle firm will enter by time T is equal to $P(\tau_U \leq T)$, where $\tau_U := \inf\{t \geq t_0 : P_t = \overline{P}\}$ is the first hitting time of the output price to the investment threshold $\overline{P}$. Similarly, the probability that an active firm will exit the market by time T is equal to $P(\tau_L \leq T)$ with $\tau_L := \inf\{t \geq t_0 : P_t = \underline{P}\}$ is the first hitting time of the output price to the divestment threshold $\underline{P}$.

To compute the entry and exit probabilities for the GBM process, we will use the closed-form solutions as given by [9, Equations 3.3.2 and 3.3.3], respectively, that is

$$P(\tau_U \leq T) = N\left(\frac{\ln(\frac{P_0^*}{\overline{P}}) + (\mu - \frac{\sigma^2}{2})T}{\sigma\sqrt{T}}\right) + \left(\frac{\overline{P}}{P_0^*}\right)^{\frac{2\mu}{\sigma^2}-1} N\left(\frac{\ln(\frac{P_0^*}{\overline{P}}) - (\mu - \frac{\sigma^2}{2})T}{\sigma\sqrt{T}}\right), \quad (56)$$

$$P(\tau_L \leq T) = N\left(\frac{\ln(\frac{\underline{P}}{P_0^*}) - (\mu - \frac{\sigma^2}{2})T}{\sigma\sqrt{T}}\right) + \left(\frac{\underline{P}}{P_0^*}\right)^{\frac{2\mu}{\sigma^2}-1} N\left(\frac{\ln(\frac{\underline{P}}{P_0^*}) + (\mu - \frac{\sigma^2}{2})T}{\sigma\sqrt{T}}\right), \quad (57)$$

where $N(.)$ is the normal standard cumulative distribution function, and $P_0^* = \frac{\overline{P}+\underline{P}}{2}$ is the current level of the price process. We further note that equation (57) corrects a typo in [12, Equation 27].

Since for the IGBM process, there are no simple closed-form solutions to compute the entry and exit probabilities, it will be necessary to compute first an approximate solution of IGBM output prices, and then to simulate these probabilities. So, to compute an approximate solution we will use the *Pathwise Adapted Linearization (PAL)* scheme, suggested by [8]. As refereed by the authors this is a stable method, since θ and/or k tend to be not too large, the scheme's proximity to exactness translates into a remarkable accuracy when used in applications. Following this method, the IGBM output prices are calculated, at dates $0 = t_0 < t_1 < \cdots < t_N = T$, as

$$P_{i+1} = P_i e^{-\delta_i(t_{i+1}-t_i)} + k\theta\left(\frac{1 - e^{-\delta_i(t_{i+1}-t_i)}}{\delta_i}\right), \quad (58)$$

where $i = 0, 1, ..., N-1$ and $Z_{i+1} \sim i.i.d.\mathcal{N}(0,1)$ random variables, and with

$$\delta_i = k + \frac{1}{2}\delta^2 - \delta\beta_i, \quad (59)$$

and

$$\beta_i = \frac{Z_{i+1} - Z_i}{t_{i+1} - t_i}. \quad (60)$$

Estimates of the probabilities of entry and exit are calculated via

$$\Pr(\tau_U \leq T) = \frac{M_U}{M} \quad (61)$$

and

$$\Pr(\tau_L \leq T) = \frac{M_L}{M} \quad (62)$$

where $M_U = \{P_{i+1} \in M : P_{i+1} \geq \overline{P}\}$ and $M_L = \{P_{i+1} \in M : P_{i+1} \leq \underline{P}\}$, $i = 0, 1, ..., N-1$ are the number of simulated paths with at least one price P_{i+1} exceeding or falling below the entry and exit price thresholds, respectively.

3.5. Computational Results

Table 3 presents the entry and exit probabilities for the GBM diffusion model using the parameter values $\sigma \in \{0.10, 0.15, 0.20\}$, $r = 0.04$, $\lambda = 0.4$, $\rho = 1$, and $T = 10$ years. The parameter μ will be adjusted so that the difference $(\mu - \lambda\rho\sigma)$ remains constant and equal to 2 (so, for $\sigma = 0.10, 0.15$, and 0.20 we will have $\mu = 0.06, 0.08$, and 0.10, respectively). In addition, we use an investment cost $\overline{I} = 3$, a divestment cost $\underline{I} = 2$, and a variable cost $C = 2$. To compute these probabilities we have used the expressions (56) and (57).

Table 3 shows that there is a direct relation between the volatility parameter and the probability of exit, while for the probability of entry there is no evidence of direct or indirect relation.

Table 3. Probability of entry for an idle firm and the probability of exit for an active firm within a specified horizon, T, for the GBM process

Volatility	Entry probability	Exit probability
$\sigma = 0.10$	0.5888	0.1602
$\sigma = 0.15$	0.5818	0.2942
$\sigma = 0.20$	0.5820	0.3945

This table values the probabilities of entry and exit in the market during the time horizon T under the GBM diffusion model. σ_0 is the volatility of the market price, and μ is the growth rate of the market price, which is adjusted for each σ value ($\mu \in \{0.06, 0.08, 0.10\}$, respectively). In all cases $P_0^* = (\overline{P} + \underline{P})/2$, the time horizon is $T = 10$ years, the variable flow cost is $C = 2$, $r = 0.04$ is the riskless interest rate, $\lambda = 0.4$ is the market price of the risk, $\rho = 1$ is the correlation between output price and the market portfolio, and the entry and exit sunk costs are $\overline{I} = 3$ and $\underline{I} = 2$, respectively.

Table 4 presents the entry and exit probabilities for the IGBM diffusion model using the parameter values $k \in \{0.00, 0.05, 0.10\}$, $\theta \in \{0.40, 1.00, 1.40\}$, $\sigma \in \{0.10, 0.15, 0.25\}$, $r = 0.04$, $\lambda = 0.4$, $\rho \in \{0, 1\}$, and $T = 10$ years. In addition, we use an investment cost $\overline{I} = 3$, divestment cost $\underline{I} = 2$, and variable cost $C = 1$.

Table 4 shows that when the long-run output price level, θ, is low or moderate, there is an inverse relation between entry probability and the speed of mean-reversion, k, while the relation is direct for θ high. We also conclude that the probability of entry increases with the volatility parameter, and decreases when the correlation between the equilibrium output price and market portfolio, ρ, increases.

Table 4 shows that when the long-run output price level, θ, is low or moderate, there is a direct relation between exit probability and the speed of mean-reversion, k, while the relation is inverse for high θ. We also conclude that the probability of exit increases with the volatility parameter (with the exception when $\rho = 0$ and $k = 0.10$). The exit probability decreases when the correlation between the equilibrium output price and market portfolio, ρ, increases (with the exception when $\theta = 0.40$).

Table 4. Probability of entry for an idle firm and probability of exit for an active firm within a specified time horizon, T, for IGBM processes, computed via Monte Carlo simulation

	$\theta = 0.40$			$\theta = 1.00$			$\theta = 1.40$		
	$\sigma_0 = 0.10$	$\sigma_0 = 0.15$	$\sigma_0 = 0.25$	$\sigma_0 = 0.10$	$\sigma_0 = 0.15$	$\sigma_0 = 0.25$	$\sigma_0 = 0.10$	$\sigma_0 = 0.15$	$\sigma_0 = 0.25$
Panel A: Probability of entry									
$\rho = 1$									
$k = 0.00$				0.2551	0.3446	0.4363			
$k = 0.05$	0.0295	0.1194	0.2745	0.1012	0.2015	0.3333	0.2195	0.2905	0.3816
$k = 0.10$	0.0010	0.0274	0.1544	0.0233	0.1002	0.2425	0.2359	0.2482	0.3338
$\rho = 0$									
$k = 0.00$				0.2829	0.3611	0.4416			
$k = 0.05$	0.0488	0.1531	0.2966	0.1729	0.2724	0.3750	0.3383	0.3897	0.4437
$k = 0.10$	0.0022	0.0401	0.1792	0.0826	0.1871	0.3148	0.4524	0.4616	0.4741
Panel B: Probability of exit									
$\rho = 1$									
$k = 0.00$				0.1561	0.2542	0.4049			
$k = 0.05$	0.3621	0.4306	0.5340	0.1455	0.2585	0.4228	0.0491	0.1535	0.3442
$k = 0.10$	0.6604	0.6423	0.6711	0.1431	0.2723	0.4549	0.0002	0.0532	0.2709
$\rho = 0$									
$k = 0.00$				0.1977	0.2860	0.3875			
$k = 0.05$	0.4677	0.4933	0.5497	0.1152	0.2134	0.3750	0.0012	0.0406	0.2249
$k = 0.10$	0.7448	0.7046	0.6936	0.0421	0.1350	0.3173	0.0000	0.0000	0.0048

This table values the probabilities of entry in the market during a time horizon T for the IGBM. In each simulation, 100,000 paths and 1,000 time steps are used. θ is the long-run output price level, k is the speed of mean reversion, σ is the volatility of the output price process, and ρ is the correlation between the equilibrium output price and the market portfolio. In all cases, $P_0^* = (\overline{P} + \underline{P})/2$, the time horizon is $T = 10$ years, the market price of risk is $\lambda = 0.4$, the variable flow cost is $C = 1$, and the entry and exit sunk costs are $\overline{I} = 3$ and $\underline{I} = 2$, respectively.

4. Conclusion

In this chapter, we review the hysteresis economic problem under both interest rate uncertainty and output price uncertainty. Our results demonstrate that the incorrect specification of the underlying state variable for valuing investments on real assets can expose a firm to significant errors of analysis which may lead to non-optimal investment and divestment decisions.

References

[1] Abramowitz, M., Stegun, I.A.: *Handbook of Mathematical Functions*. Dover, New York (1972)

[2] Cox, J.C., Ingersoll, J.E, Ross, S.A.: A theory of the term structure of interest rate. *Econometrica* 53(2), 385–408 (1985)

[3] Delbaen, F.: Consols in the CIR model. *Mathematical Finance* 3(2), 125–134 (1993)

[4] Dias, J.C., Shackleton, M.B.: Hysteresis effects under CIR Interest rates. *European Journal of Operational Research* 211, 594–600 (2011)

[5] Dixit, Avinash K.: Entry and Exit Decisions under Uncertainty. *Journal of Political Economy* 97 (2), 620–638 (1989)

[6] Feller, W.: Two singular diffusion problems. *Annals of Mathematics* 54, 173–182 (1951)

[7] Higham, D.J., Mao, X..:Convergence of Monte Carlo simulations involving the mean-reverting square root process. *Journal of Computational Finance* 8(3), 35–61 (2006)

[8] Kahl, C., Jäckel, P..: Fast Strong Approximation Monte Carlo Schemes for Stochastic Volatility Models. *Quantitative Finance* 6(6), 513–536 (2006)

[9] Jeanblanc, M., Yor, M., Chesney, M.: M*athematical Methods for Financial Markets*. Springer, New York (2009)

[10] Larguinho, M., Dias, J.C., Braumann, C.A..: A Note on (Dis)Investment Options and Perpetuities under CIR Interest Rates. Recent Developments in Modeling and Applications in Statistics, *Studies in Theoretical and Applied Statistics*, Oliveira, P. E., Temido, M. G., Henriques, C., and Vichi, M. (eds.), Springer, Berlin, Part 5, 203–211, Springer, Berlin. (2013)

[11] Shackleton, M.B., Tsekrekos, A.E., Wojakowski, R.: Strategic entry and market leadership in a two-player real options game. *Journal of Banking & Finance* 28, 179–201 (2004)

[12] Tsekrekos, Andrianos E.: The Effect of Mean Reversion on Entry and Exit Decisions Under Uncertainty. *Journal of Economic Dynamics and Control* 34 (4), 725–742 (2010)

In: Hysteresis
Editor: José Carlos Dias

ISBN: 978-1-63321-336-4

Chapter 2

HYSTERESIS EFFECTS IN ECONOMICS: DIFFERENT METHODS FOR DESCRIBING ECONOMIC PATH-DEPENDENCE

Ansgar Belke[1], Matthias Göcke[2] and Laura Werner[2]
[1]University of Duisburg-Essen, Germany
[2]Justus Liebig University of Giessen, Gießen, Hesse, Germany

ABSTRACT

Relations between economic variables are often characterized by a situation where initial conditions and the past realizations of economic variables matter. I.e. past (transient) exogenous disturbances and past states of the economic system do have an influence on the current economic relations. Typical examples are the dynamics of (un)employment in business cycles and the dynamics of the nexus of exchange rate and exports. Since the standard characteristics of hysteresis apply – i.e. permanent effects of a temporary stimulus, resulting in path-dependent multiple equilibria – these economic phenomena are correctly titled as “hysteresis”. Analogous to magnetism, the pattern of hysteresis depends on the scope: Based on sunk-adjustment costs (e.g. entry costs of starting export activity on foreign markets) *micro*economic behavior (e.g. of *single firms* on export markets) shows a discontinuous switching-pattern (being active on a foreign market or not) as described by a non-ideal relay, analogous to the magnetism of a single iron crystal. Correspondingly, *macro*economic dynamics of *aggregate* economic variables (e.g. the exports volume of a *whole country,* based on an aggregation over firms with heterogeneous cost structures) show a pattern similar to the well known hysteresis-loop of an entire piece of iron. The aggregate macroeconomic loop is characterized by a smooth/continuous transition between different “branches” of the loop, occurring with changes in the direction of the (e.g. exchange rate) movement.

Empirical research in economics is using different methods in order to capture path-dependent effects. First econometric approaches tried to describe these effects by simple time-series processes with unit- (or zero)-root dynamics. However, since unit-root-dynamics are not related to genuine multiple equilibria but on the order of integration of time series, these first attempts were expanded by more sophisticated time-series models integrating structural breaks, threshold-cointegration or non-linear autoregressive distributed lag-models. Another branch of empirical studies tries to keep closer to the original concept of the macro-loop, trying to apply an explicit *Mayergoyz-Preisach* aggregation procedure for heterogeneous firms – if microeconomic information is available based on panel-data – or by using simple algorithms analogous to mechanical-play in order to apply simple OLS-regression methods on a filtered/transformed input-output relation.

In the following chapter we want to give an overview of the implementation of hysteresis in economics, with an emphasis on two aspects: (1) the differentiation between micro- and macroeconomic hysteresis including an outline of an adequate aggregation procedure, and (2) different methods applied in econometrics in order to capture economic path-dependency empirically.

Keywords: band of inaction, labor market, market entry/exit, non-ideal relay, option (value), path-dependence, play-hysteresis, Preisach model, sunk (adjustment) costs, uncertainty, unit root dynamics, wait-and-see strategy

INTRODUCTION

Relations between economic variables are often characterized by a situation where initial conditions and the past realizations of economic variables matter. I.e. past (transient) exogenous disturbances and past states of the economic system do have an influence on the current economic relations. Typical examples are the dynamics of (un)employment in business cycles, i.e. the dynamics of the so called "natural" (equilibrium) rate of unemployment, and the dynamics of the nexus of exchange rate and exports. Since the standard characteristics of hysteresis apply – i.e. permanent effects of a temporary stimulus, resulting in path-dependent multiple equilibria – these economic phenomena are correctly titled as "hysteresis" [24]. In labor economics applying the concept of hysteresis was especially disseminated by *Phelps* [58], *Sachs* [64], *Blanchard/Summers* [20], and *Lindbeck/Snower* [46]. To foreign trade theory hysteresis was introduced by *Kemp/Wan* [43], *Baldwin/Krugmann* [7, 10], and *Dixit* [27].

Analogous to magnetism, the pattern of hysteresis depends on the scope: Based on sunk-adjustment costs (e.g. entry costs of starting export activity on foreign markets) *micro*economic behavior (e.g. of *single firms* on export markets) shows a discontinuous switching-pattern (being active on a foreign market or not) as described by a non-ideal relay, analogous to the magnetism of a single iron crystal. Correspondingly, *macro*economic dynamics of *aggregate* economic variables (e.g. the exports volume of a *whole country,* based on an aggregation over firms with heterogeneous cost structures) show a pattern similar to the well known hysteresis-loop of an entire piece of iron. The aggregate macroeconomic loop is characterized by a smooth/continuous transition between different "branches" of the loop, occurring with changes in the direction of the (e.g. exchange rate) movement.

Empirical Research in economics is using different methods in order to capture path-dependent effects. First econometric approaches tried to describe these effects by time-series processes with unit- or zero-root dynamics. However, since unit-root-dynamics are not related to genuine multiple equilibria but on the order of integration of the time series, these first attempts were expanded by more sophisticated time-series models integrating structural breaks, threshold-cointegration or non-linear autoregressive distributed lag-models. Another branch of empirical studies tries to keep closer to the original concept of the macro-loop, trying to apply an explicit *Mayergoyz-Preisach* aggregation procedure [49, 50, 63] for heterogeneous firms – if microeconomic information is available based on panel-data – or by using simple algorithms analogous to mechanical-play in order to apply simple OLS-regression methods on a filtered/transformed input-output relation.

In the following chapter neither a detailed description of the various techniques nor a comprising presentation of all manifold applications can be presented (see the references for this purpose). Rather, we want to illustrate the character of hysteresis based on standard examples including their economic intuition in order to give an overview of the distinct concepts and ways to implement hysteresis in economics. This overview emphasizes two aspects: First the differentiation between micro- and macroeconomic hysteresis including an outline of an adequate aggregation procedure, and second different methods applied in econometrics in order to capture economic path-dependence empirically.

THEORETICAL BACKGROUND FOR HYSTERESIS IN ECONOMICS

Microeconomic Hysteresis Based on Sunk Costs

A change of the relevant forcing variables typically leads to a change in the economic behavior of the observed unit(s). However, in the case of hysteresis a removal of a merely temporary change back to the initial value of the forcing variables does not induce a complete change back to the initial behavior. A typical economic mechanism for this path-dependent pattern is founded on *sunk adjustment costs* (see [7, 8]).

The typical microeconomic example is as follows: A firm which is previously not selling on a market has to bear market entry costs, e.g. for setting up distribution networks or for advertising, which are firm specific and cannot be regained after market entry. Thus, ex-post these market entry 'investments' are sunk costs. A market entry is only profitable, if the sunk entry costs are covered by revenues, thus, the market price must exceed the unit costs for triggering market entry. If a temporary high market price has led to a market entry of a firm, a subsequent removal of this price increase back to the initial level will – as long as the variable costs are covered – not induce a market exit. Summarizing, the same price level may result in different states of the firm's activity, dependent on the history of the price level.

Adjustment costs may occur on the supply side of the market (as outlined above in the simple example) or on the demand side of a market (see [34], p. 638, for a systematization in supply and demand side of factors generating hysteresis): Another important supply side reason is based on learning-by-doing, since former production and gaining experience leads to higher productivity and, thus, later on to lower costs. On the demand side, experience effects of consumers may lead to demand-carry over effects; i.e. a temporary price decrease, inducing additional sales and a higher willingness to pay may result in a permanently higher demand, if consumers accumulate positive experience with the firm's product. The defining aspect is that, though all these factors play only a transient role, the result is a difference between the ex-ante decision (before e.g. the sunk costs are paid) and the ex-post situation (after these 'investments' are carried out). As the relevant marginal costs or revenues were changed, the same exogenous environment path-dependently results in different reactions.

Thus, a transient disturbance can have permanent effects on economic equilibria – which is the constituting characteristic of economic hysteresis.

'Non-ideal Relay'-Hysteresis in Economics

In order to illustrate an economic situation which is characterized by the most elementary form of hysteresis – the so called '*non-ideal relay*' – we model the simple introductory microeconomic example [38]: A firm j decides in period t whether or not to supply one unit of a product ($x_{j,t} = 1$ or $x_{j,t} = 0$). For producing this unit "variable" costs c_j have to be paid, and additionally, if the firm has not produced and sold on the market in the preceding period (t–1), fixed starting costs k_j. This sunk 'investment' k_j is completely firm specific and decays immediately when the firm stops producing. Hence, the costs of an active firm is path-dependent: If inactive in the preceding period (production/sales $x_{j,t-1} = 0$) it has to pay both components, but if it was active before (supply $x_{j,t-1} = 1$) only the variable part is relevant. The firm's (unit) cost function is:

$$(1) \quad K_{j,t} = \begin{cases} 0 & \text{if } x_{j,t} = 0 \\ c_j + k_j & \text{if } x_{j,t} = 1 \wedge x_{j,t-1} = 0 \\ c_j & \text{if } x_{j,t} = x_{j,t-1} = 1 \end{cases} \qquad \text{with } c_j, k_j \geq 0 \tag{1}$$

If the (price-taking) firm is (for the time being) assumed to be myopic, i.e. only looking at the present periods costs and revenues and not considering the future advantage of being active in the current period, it compares the current price p_t with the current costs. The resulting supply behavior of the firm is illustrated in Figure 1: If the (forcing variable) price p_t starts to rise continuously from a zero level, the (previously inactive) firm starts producing/selling, if the price exceeds the sum of both costs $\alpha_j = (c_j + k_j)$ as an entry *trigger* in point B. Market entry leads to a jump from the (x_j=0)-inactivity-line to the (x_j=1)-activity-line (point C). If the firm in later periods faces a price decrease (e.g. in point A), it will continue to produce as long as the variable costs (as an exit trigger $\beta_j = c_j$) are covered – since ex-post the sunk entry cost k_j are not relevant any more. The variable costs are the exit trigger. The lowest-price limit of the firm depends on past market/production activity. Two different path-dependent equilibria are possible between points F and B (or E and C, respectively). A switch from one equilibrium-branch to the other takes place when the triggers are passed – otherwise the activity status remains the same. Therefore, the area FB (or CE) can be described as a *'band of inaction'* ([7], pp. 7, and [12], p. 11). This elementary form of hysteresis is called *'non-ideal relay'* in [45], since the switch between the two possible equilibria is triggered by *two different* values ([21], pp. 23).[1] Since for a non-ideal relay only a transition between two branches is possible, a major characteristic of hysteresis – the *'multibranch non-linearity'* –exists in a very simple way ([49], p. 603).

[1] On a micro level, i.e. on the level of a single iron-crystal, magnetic hysteresis shows exactly this pattern ([44], pp. 401).

Passing of microeconomic triggers usually results from "large shocks". Thus, papers implicitly relying on "band of inaction"-type hysteresis, point out the difference between large shocks triggering permanent effects and small ones that do not. (See e.g. [10, 12] and [31]).

The ex-ante/ex-post difference of the relevant costs and revenues leading to a band of inaction applies in an analogous way to different markets. The important example of labor demand is briefly outlined in the following. Potential mechanisms inducing labor market hysteresis are:

- Hiring and firing costs (as sunk employment adjustment cost): the employer has to pay for hiring and training new employees; and firing costs occur e.g. for severance pay or due to the loss of firm specific human capital of the fired employees. Both hiring and firing costs could not be regained after the employment adjustment is executed.
- Due to learning effects, the productivity of workers is rising in the course of a longer employment. In contrast, unemployment leads to a depreciation of the quality of labor due to a lack of training-on-the-job. These mechanisms are increasing the willingness-to-pay for active (already employed) workers relative to unemployed workers. Thus, the time-path of employment is determining individual wages and employment opportunities.

Thus situations triggering hiring are different from the level that triggers firing, i.e. a hysteresis band of inaction arises: the wage rate that induces hiring is lower than the wage rate that causes firing. Consequently, transient labor market disturbances can have permanent effects on the so called "natural" equilibrium rate of unemployment.

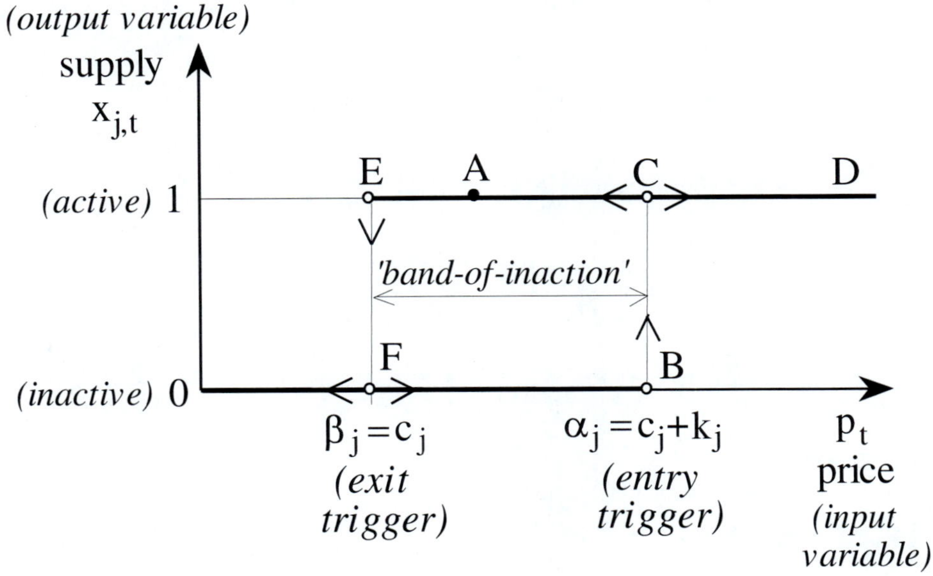

Figure 1. 'Non-ideal relay' of a firm with sunk entry costs.

"Mechanical Play" in Microeconomics

Another very simple type of hysteresis – which is in economic modelling not as common as the non-ideal relay pattern – is a dynamics similar to the phenomenon of mechanical play.[2] In microeconomics '*play-hysteresis*' results from *variable* sunk adjustment costs. The firm j is now able to produce and sell the variable amount $x_{j,t} \geq 0$, based on the cost function:

$$K_{j,t} = c_j \cdot x_{j,t}^2 + k_j \cdot D \cdot (x_{j,t} - x_{j,t-1}) \qquad \text{with: } D = \begin{cases} 0 & \text{if } x_{j,t} \leq x_{j,t-1} \\ 1 & \text{if } x_{j,t} > x_{j,t-1} \end{cases} \tag{2}$$

In addition to the standard average variable costs $(c_j \cdot x_{j,t})$, the firm has to pay *variable adjustment costs* for an increase $(x_{j,t} - x_{j,t-1}) > 0$ of production. Adjustment costs are linear in the size of the increase, i.e. k_j per unit of the expansion. To keep the model simple, again a "myopic" firm – only considering current payments – is assumed. Previous production $(x_{j,t-1})$ determines the "capacity", which is costly to be increased – and which has to be written of completely if not employed for one period. If the price level p_t is exogenously given, i.e. if the firm is a price taker and only adjusts its quantity according to profit maximization (based on the rule "marginal revenue = price = marginal costs"):

$$p_t = 2 \cdot c_j \cdot x_{j,t} + k_j \cdot D \quad \Rightarrow \quad \begin{cases} p_t = 2 \cdot c_j \cdot x_{j,t} & \text{if } x_{j,t} \leq x_{j,t-1} \\ p_t = 2 \cdot c_j \cdot x_{j,t} + k_j & \text{if } x_{j,t} > x_{j,t-1} \end{cases} \tag{3}$$

Marginal costs are different in situations with expanding production compared to situations with constant or decreasing production: the conditional optimum for increasing production $x_{t;j}^{up*}$ (including capacity adjustment costs k_j) and $x_{t;j}^{do*}$ as the optimum conditional on constant or decreasing production:

$$x_{j,t}^{up*} = \frac{p_t - k_j}{2 \cdot c_j} \quad \text{and} \quad x_{j,t}^{do*} = \frac{p_t}{2 \cdot c_j} \qquad \text{with: } x_{j,t}^{do*} > x_{j,t}^{up*} \Leftrightarrow k_j > 0 \tag{4}$$

The actual optimum $x_{j\,t}^{*}$ results from the comparison of the two conditional optima with the past production $x_{j,t-1}$:

$$x_{j,t}^{*} = \begin{cases} x_{j,t}^{up*} = \dfrac{p_t - k_j}{2 \cdot c_j} & \text{if } x_{j,t}^{up*} > x_{j,t-1} \\ x_{j,t-1} & \text{if } x_{j,t}^{up*} \leq x_{j,t-1} \leq x_{j,t}^{do*} \\ x_{j,t}^{do*} = \dfrac{p_t}{2 \cdot c_j} & \text{if } x_{j,t}^{do*} < x_{j,t-1} \end{cases} \tag{5}$$

[2] See [38] for a more comprehensive presentation of the following example. See [45], pp. 6 and [21], pp. 24 and pp. 42 for a general treatment of the play-operator. For an example of play-hysteresis in economics see [25], Fig. 2, p. 472 where price-stickiness as a result of menu-costs is analysed.

Figure 2 illustrates the resulting play-hysteresis loop of the optimal (output) $x_{j,t}^{*}$ for changes in the current price p_t (as input variable). Starting (in the origin O) with a zero price and a zero capacity ($x_{j,t-1} = 0$), a price increase will induce no production (increase) until in point A a threshold is exceeded (where enlargement costs k_j are covered). For larger price increases the production reacts according to line AFB. If the price rises up to point B, the production increases as well. However, for a subsequent price decrease, the production shows no reaction along the line BCD, until a threshold is passed in point D. As in the case of a non-ideal relay, no reaction of the output variable occurs for a particular input area. However, in the case of play the position of the inaction-band is itself path-dependent (and not invariant as for a non-ideal relay). Initially, its position is defined by the line OA and later by BD (or by EF).

Play is represented by the horizontal distance between the right line AFB (describing a production increase) and the left OED line (representing decreasing production generated by a price decline). As a consequence of *variable* sunk adjustment (capacity enlargement) costs, there is a difference between the two conditional optima $x_{t;j}^{up*}$ and $x_{t;j}^{do*}$, resulting in a constant supply of the firm if the direction of the price/quantity movement is changed – until the thresholds, determined by the past production $x_{j,t-1}$, are passed.

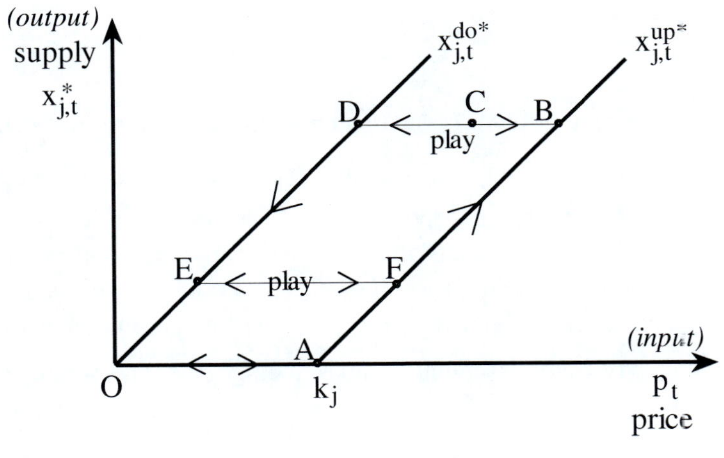

Figure 2. A microeconomic reaction similar to 'mechanical play'.

Both microeconomic hysteresis loops, non-ideal relay and play, show some common features: multibranch non-linearity is characterized by two main branches and the change between these branches requires passing of thresholds, while passing of the thresholds results in permanent effects. However, for play-loops the change is not discontinuously abrupt but smoothly. However, the non-ideal model case can be interpreted as a special case of the play model where only a discrete (0,1) supply reaction is feasible.

Uncertainty and Option Value Effects

A forward-looking firm considers future effects of a present sunk cost 'investment'. If the exogenous variable (price) is stochastic, a real option approach applies [59, 60, 27, 19] and

[16]. An inactive firm deciding on a present entry or to stay passively, will include the option to enter later as a potential alternative. A price which is presently covering costs, may in a stochastic situation, decrease in the future. By staying passive the firm can avoid future losses if this situation will realize. Moreover, an instantaneous entry kills the option to enter later and to '*wait-and-see*' if the future price movement will turn out to be (un)favorable. Thus, in a stochastic situation, the sunk costs and, additionally, an option value of waiting have to be covered in order to trigger an entry. Thus, uncertainty implies an upward shift of the entry trigger price.

Consider a firm with an infinite planning horizon. Again the previously inactive price taker firm j has to decide in period t whether or not to enter, paying sunk costs k_j, and to supply one unit, produced with costs c_j. Assume a non-recurring single stochastic change in the price level in the next period (t+1), which can be either positive (+u) or negative (–u) (with $u \geq 0$), with the same probability of ½ : $p_{t+1} = p_t \pm u$ and $E_t(p_{t+1}) = p_t$: After this shock, the price is assumed to remain constant for the rest of the infinite future. Under uncertainty, the three alternatives which have to be compared are (1) to enter now (in t), (2) to enter in the next period t+1, or (3) to enter never. A *present* entry in period t leads to an expected present value $EPV^{enter,j}_{;t}$, if discounting is based on the interest rate $i > 0$ (and using the formula of the present value of an annuity due):

$$EPV^{enter}_{j,t} = \frac{(1+i)\cdot(p_t - c_j)}{i} - k_j \qquad \Rightarrow \quad EPV^{enter}_{j,t} > 0 \quad \text{if} \quad p_t > \alpha^c_j = c_j + \frac{i}{1+i}\cdot k_j \tag{6}$$

Naively looking only at a positive present value of a present entry (neglecting the option to wait), would result in a certainty-equivalent entry trigger price $[\alpha^c_j = c_j + \frac{i}{1+i}\cdot k_j]$. However, additionally considering the option to wait, the firm differentiates between two potential u-realizations. If the firm waits in t, the firm can use its option to enter in t+1 conditional on a (+u)-realization (with probability ½). In case of (–u)-realization the firm will remain passive (and avoids losses). Consequently, the expected present value of the wait-and-see strategy $EPV^{wait,j}_{;t}$ is, based on discounted values of entry costs and annuity (since now the conditional entry is one period later):

$$EPV^{wait}_{j,t} = \frac{1}{2}\cdot\frac{1}{1+i}\cdot\left(\frac{(1+i)\cdot(p_t + u - c_j)}{i} - k_j\right) \tag{7}$$

With this present value of waiting in mind the firm will enter the market in period t, only if a current entry is more valuable than waiting:

$$EPV^{enter}_{j,t} > EPV^{wait}_{j,t} \qquad \Leftrightarrow \quad p_t > \alpha^u_j = c_j + \frac{i}{1+i}\cdot k_j + \frac{u}{1+2\cdot i} = \alpha^c_j + \frac{u}{1+2\cdot i} \tag{8}$$

An increase in uncertainty enlarges the option value to enter later, since with a larger u the potential payoff conditional on a positive price change increases, leaving the downside payoff unchanged, since the firm will not enter in t+1 if the price falls. As a result, the entry trigger price including option value effects is $\alpha^u_{;j}$ augmented by the term $[u/(1+2\cdot i)]$. Thus, uncertainty leads to a widening of the band of inaction, aggravating the hysteresis property of the firm's behavior. However, the qualitative property of micro hysteresis as a non-ideal relay has not changed.

Aggregation and Macroeconomic ("Strong") Hysteresis

In the following the application of the *Mayergoyz-Preisach*-procedure [49], [63] is outlined. Being close to the original concept of hysteresis in magnetism, this method is based on the explicit aggregation of non-ideal relay agents ($j = 1, ..., n; n \gg 0$) with heterogeneity in their cost structures, resulting in heterogeneous entry/exit triggers α_j and β_j (see *Amable et al.* [1, 2, 4], *Cross* [22, 23], *Göcke* [36], *Piscitelli et al.* [61], and *Mota / Vasconcelos* [52] for applications of the *Preisach* model in foreign trade and in labor market economics).

Every potentially active firm j is characterized by a α_j/β_j-set of entry/exit triggers. In an α_j/β_j-diagram (since $\alpha_j \geq \beta_j$) all firms are represented by points in a triangle area above the 45°-line (see Figure 3). Generally, the aggregation procedure can be performed without restrictions concerning the distribution of the firms over the triangle area Points on the 45°-line represent non-hysteretic firms ($k_j = 0 \Rightarrow \alpha_j = \beta_j = c_j$). Firms with a position above the 45°-($\alpha = \beta$)-line are characterized by a non-ideal relay supply– the distance from the ($\alpha = \beta$)-line determined by k_j.

To avoid a long description of the past development, a simple situation with an initial price $p = 0$ is assumed, implying no firm is initially active. Now, a rising price leads to an entry of the firms with the lowest entry trigger $\alpha_j = (c_j + k_j)$. Aggregate supply increases, as traced in Figure 3 (a), with a growing space of the hatched triangle $S_{t;}^{+}$ representing the (active) firms which entered (and $S_{t;}^{-}$ representing the inactive firms). For a rising price the triangle $S_{t;}^{+}$ grows via an upward shift of the horizontal borderline. The corresponding aggregate/macro supply reaction is delineated in Figure 4 by the path OA.

Figure 3 (b) outlines the effects of a subsequent price decrease (after a local price maximum p_1^M was reached). Firms leave the market, if their exit triggers β_j are underbid and the triangle $S_{t;}^{+}$ representing active firms now loses space as illustrated by a left vertical shift of the $S_{t;}^{-}$-$S_{t;}^{+}$-borderline. In Figure 4 the corresponding path is BC.

If, after reaching the local minimum p_1^m, there is again a rise in the price level, as depicted in Figure 3 (c) only the right-horizontal part of the $S_{t;}^{+}$-borderline is shifted upwards (as long as the old maximum p_1^M is not passed), resulting in a macro reaction as path CD in Figure 4. A subsequent price decrease is illustrated in Figure 3 (d): The lower vertical part of the borderline is shifted to the left (corresponding to DE in Figure 4). Several cycles of rising and decreasing prices will result in a "staircase"-borderline of the $S_{t;}^{+}$-area of active firms, where the coordinates of the staircase are determined by past extrema of the price-movement.

If later on the "old" local minima are underbid by even lower prices, or if local maxima are subsequently surpassed by higher prices, the corresponding staircase-corner (and the memory of the "old" extrema) is erased from the macro system ([49], p. 605). If, however, a new local maximum p_2^M is lower than the "old" maximum p_1^M (as in Figure 3 (c)), the memory of p_1^M remains.

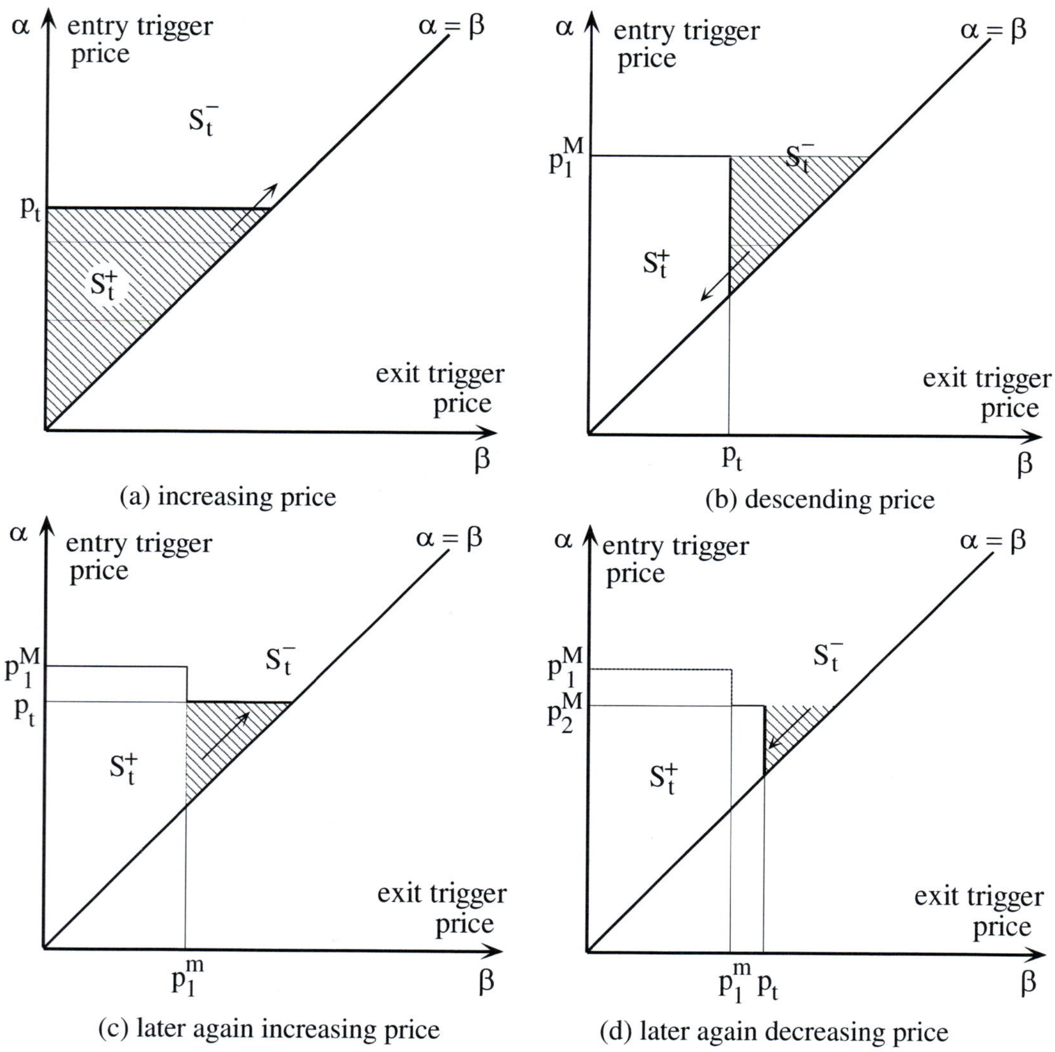

Figure 3. *Preisach* aggregation procedure.

Summarizing, the aggregate system shows a memory of non-erased ('non-dominated') past input extrema – graphically represented by the "staircases" in the borderline of the area $S_t;^+$ of active firms. Aggregation results in a reinforcement in the type of hysteresis: For the aggregate loop a branch-to-branch transition occurs with *every local extremum* of the path of the input variable, and this transition is *continuously*, while at the microeconomic firm level the passing of triggers is necessary, causing a discontinuous switch/jump in order to induce permanent remanence effects. Therefore, this type of macro-hysteresis is called "*strong*" hysteresis (e.g. [1, 4]; see [21], pp. 22, for typical characteristics of the macro hysteresis loop). The *distribution* of the heterogeneous firms in the α,β-space determines the *curvature* of the branches of the macro loop. The more clustered the firms are in a specific area (i.e. the less heterogeneous the firms are), the more "curved" are the branches. In the limit case of homogeneous firms, these similar firms are all represented by a single point in the α/β-diagram – and the macro-loop will degenerate to a non-ideal relay.

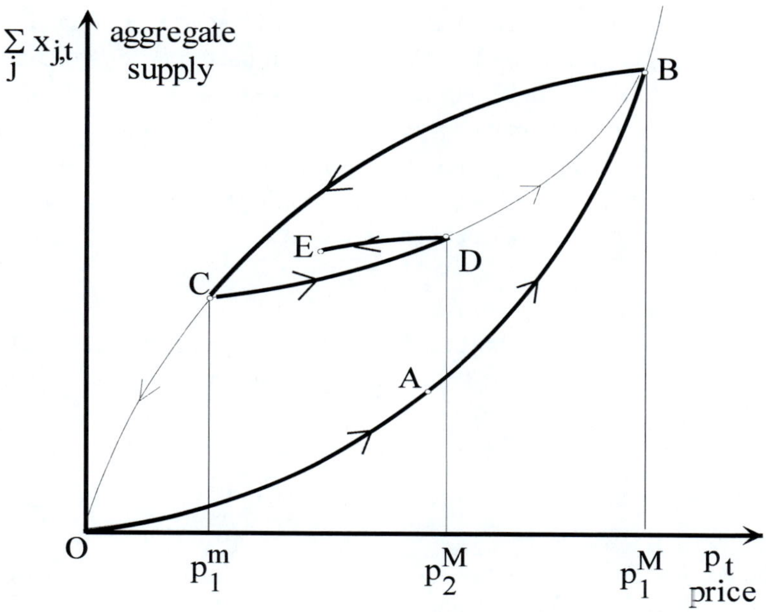

Figure 4. Aggregate macroeconomic hysteresis loop.

Approaches to Address Economic Hysteresis Empirically

The different types of hysteresis require specific modeling in empirical research. Thus, a variety of methods is used to describe permanent effects of temporary shocks empirically. In the next sections some econometric approaches that attempt to capture the hysteretic dynamics adequately are briefly outlined. A common method to describe persistence effects is by stochastic unit root difference equations (e.g. via cointegration models) which is adequate, if permanent effects are actually based on the degeneration of the adjustment dynamics. For non-ideal relay dynamics the switches between the branches may be addressed by (0,1)-dummy variables as additional variables in a regression equation. However, identifying the trigger values for the branch-to-branch transition remains a problem, since it has to be done as an endogenous part of the estimation [14]. Methods to integrate "strong"/aggregate hysteresis-loops in empirical models are e.g. presented in [16, 17] and [61]. These approaches – based on past extrema – attempt to calculate a transformed "hysteresis variable" in order to capture the dynamic pattern.

Time-series Models

Linear Difference Equations with Unit Roots

As a simple way to describe persistent effects economists often try to describe hysteretic dynamics with models containing linear difference (differential) equations exhibiting unit (zero) roots (see [53], as an example). Since zero root dynamics in continuous time can be considered analogously, here only unit root dynamics in discrete time are outlined explicitly. (For a general discussion of unit/zero roots and hysteresis see [2, 3, 4] and [40], pp. 43.) However, unit root dynamics show some differences compared with "genuine" hysteresis. According to *Amable et al.* ([2], p. 8) unit root dynamics should be named 'persistence', but in economics this term is already applied for processes "close to" unit root dynamics, with a slow adjustment towards a unique equilibrium. Real hysteretic dynamics are rather captured by non-linear difference/differential equations ([45], pp. 90; [21], pp. 122), and not by linear difference equations with unit roots as an oversimplification.

A Single Equation Model with a Unit Root

Unit root dynamics are now illustrated by an inhomogeneous first order difference equation with constant coefficients, a state variable Y_t and only one exogenous variable R_t [33], using the lag-operator L:

$$Y_t = b \cdot Y_{t-1} + R_t \Rightarrow Y_t - b \cdot L\, Y_t = R_t \Rightarrow (1 - b \cdot L) \cdot Y_t = R_t \tag{9}$$

The characteristic root of $[1 - b \cdot z = 0]$ is $z = (1/b)$. All roots have to be outside the complex unit circle, in order to have a stable steady state equilibrium ([48], pp. 11). Thus, for $0 < |b| < 1$ the steady state (with: $Y^* = Y_t = Y_{t-1}$ and $R^* = R_t = R_{t-1}$) results:

$$Y^* = \frac{1}{1-b} \cdot R^* \tag{10}$$

For $0 < b < 1$ a stable equilibrium and for $b > 1$ an explosive non-equilibrium path results. However, for $b = 1$, there is a unit root ($z = 1$), resulting in:

$$Y_t = Y_0 + \sum_{\tau=1}^{t} R_\tau \tag{11}$$

Thus, for a non-constant exogenous variable ($R^* =;/ 0$) a constant steady state Y^* results in the case of $0 < b < 1$. In a unit root situation ($b = 1$) the state variable only remains constant if no exogenous impulse occurs (i.e. $R = 0$). In econometrics, a time series with a unit-root is called to be "*integrated of order one*". As in the case of hysteresis, the unit-root system is path-dependent, as the current state under a unit root depends on the initial conditions and on the past realizations of the exogenous variable. However, there are some differences: In the case of "genuine" hysteresis there is a path-dependence of *locally stable* equilibria (under "slow control", after adjustment has taken place), while in the case of unit-root dynamics the permanent effects are based on an *indifference between equilibria*, i.e. a *degeneration of the adjustment dynamics*. A branch-to-branch transition in genuine hysteresis results from a *local instability* resulting in path-dependent *structural breaks* ([9], p. 245). For unit root dynamics a temporary shock results as well in a different equilibrium, but this is based on *global indifference* between different equilibrium states ([1], pp. 12). Due to this difference in the equilibrium situation, unit roots and genuine hysteresis are different in two aspects: (1) For unit root processes the memory is unselective, summing up every past shock while the memory of a (macro-) hysteretic system is selective, since only the non-dominated past extrema are resulting to a branch-to-branch transition ([23]). (2) Equal but opposite shocks will in the case of linear unit roots leave the equilibrium unchanged, whereas under hysteresis two successive shocks with opposite sign result in a new equilibrium and a remanence effect in a non-linear way ([4] and [61], p. 59).

Unit Root Dynamics in a Multivariate System

Of course, unit root dynamics can be generalized to multivariate systems of difference equations. We use a standard error correction model as an example (see [38]). As a first equation we use eq. (11) again. In our example Y could be national income which is exposed to an exogenous shock R. As a second equation, we assume an error correction mechanism for the aggregate consumption C: In the long run, consumption C is proportional to income Y (with a ratio 'c', called the "propensity to consume"). Short-run deviations from this equilibrium relation $C = c \cdot Y$ are adjusted in the following period by a ratio 'd'. Consumption is affected by another exogenous shock Z. The error correction system is:

$$Y_t = b \cdot Y_{t-1} + R_t \qquad \Rightarrow \quad (1 - b \cdot L) \cdot Y_t = R_t$$

$$\Delta C_t = d \cdot (C_{t-1} - c \cdot Y_{t-1}) + Z_t \qquad \Rightarrow \quad d \cdot c \cdot L \cdot Y_t + [1 - (1+d) \cdot L] \cdot C_t = Z_t \tag{12}$$

with: $0 < b \leq 1$, $-2 \leq d \leq 0$ and $0 \leq c \leq 1$

Using matrix and lag-operator representation, $A(L) \cdot y_t = r_t$, the system can be restated:

$$\begin{bmatrix} 1-b \cdot L & 0 \\ d \cdot c \cdot L & 1-(1+d) \cdot L \end{bmatrix} \cdot \begin{bmatrix} Y_t \\ C_t \end{bmatrix} = \begin{bmatrix} R_t \\ Z_t \end{bmatrix} \tag{13}$$

The characteristic polynomial of the lag-operator matrix A(L) and its roots z_1 and z_2 are:

$$\det(A(L)) = (1 - b \cdot L) \cdot [1 - (1+d) \cdot L] = 0 \quad \Rightarrow \quad z_1 = \frac{1}{b} \text{ and } z_2 = \frac{1}{1+d} \tag{14}$$

In the case of $b = 1$ and $d = 0$ unit roots ($|z_i| = 1$) result. If the first equation shows a unit root ($b = 1$) – while the second equation has a root outside the unit circle ($-2 < d < 0$), the whole system shows indifference and path-dependence concerning the levels of Y_t and C_t (i.e. these variables are integrated of order one), however, the relation between both ($C_t = c \cdot Y_t$) acts as a stable attractor (see [2], Figure 7 and 9, for an illustration). This situation is called "cointegration" of C_t and Y_t. In such a situation, a deviation of the relation $C_t = c \cdot Y_t$ is reduced by an adjustment process, and $C = c \cdot Y$ can be seen as a long-run equilibrium.

Therefore, the error correction concept is equivalent to cointegration between the variables of stochastic processes in time series analysis, as shown by *Engle / Granger* [29]. (For an introduction to cointegration see [30] and [48], pp. 346).

However, "unit root hysteresis" is a borderline case since the adjustment process degenerates to indifference only if the roots meet the complex unit circle *exactly* ([2], p. 14). For unit roots only near to the unit circle either a slow dynamic away from an unstable equilibrium or a gradual adjustment towards a definite stable equilibrium will result. Since economic time series are relatively short compared to the duration of economic adjustment processes, and since economic variables are consecutively exposed to external disturbances, an empirically significant differentiation between unit roots dynamics and situations close to a unit root is often impossible and statistical procedures testing on unit roots show very little power in differentiating between both cases, see [26].

(Typically in labor economics) cointegration models were extended by potential structural breaks of the (long-run) equilibrium relations in the sense of [56] and [6]. However, the introduction of structural breaks in the cointegration/equilibrium relation leads to a rejection of degenerated unit roots adjustments in a lot of cases where unit roots could not be significantly rejected before explicitly considering structural breaks. If a unit root is rejected after extending the model by structural breaks, this is (by these authors) interpreted as a rejection of "hysteresis", since path-dependence based on unit-roots is seen as the main characteristic of hysteresis. In contrast, the structural breaks are in a so called "structuralist" view à la *Phelps* [57] thought as reflecting changes in unobserved variables (e.g. of institutions, labor law etc.) resulting in a shift of the "non-accelerating rate of unemployment" (NAIRU) as a (for a while) stable equilibrium. Shifts in the equilibrium rate of unemployment are not seen as hysteresis, but only indifferent unit-root adjustment dynamics are (see e.g. [5], [47] or [41] as recent examples for this kind of (mis-)interpretation). From the point of view of "genuine" hysteresis (as presented in this chapter), this unit-root simplification falls short of the complex dynamics including branch-to-branch transitions between locally stable equilibria, i.e. path-dependent structural breaks. Thus, structural breaks are not counter evidence but rather a consequence of genuine hysteresis.

Non-linear Time Series Models

Nowadays, time series econometricians usually apply non-linear techniques in order to test for hysteretic effects in economic relationships. *O'Shaughnessy* [54], for example, provides an overview with regard to hysteresis in unemployment. However, *Hughes Hallet / Piscitelli* [39] state that it is a challenging task to find hysteretic behavior in economic data because no explicit test for it exists.

Generally, it is possible to distinguish between several forms of non-linearity in time series models. On the one hand, there are authors modeling asymmetric adjustment to a long-run, i.e. cointegrating relationship (see, for example [42]). This is generally referred to as threshold cointegration. On the other hand, *Verheyen* [68] applies the non-linear autoregressive distributed lags approach (NARDL) of *Shin et al.* [65] to export demand and tries to derive stylized facts of export demand which are in accordance with hysteresis.

One problem of conventional linear cointegration models is the fact that these assume a linear error correction mechanism. However, various reasons exist why this adjustment might be non-linear. Just think of menu costs which imply that adjustment occurs only if a certain threshold value is surpassed. Otherwise, within this certain band of menu or transaction costs no adjustment occurs. Furthermore, strategic behavior could lead to different adjustment patterns depending on whether the deviation from the long-run equilibrium is positive or negative. If these arguments hold, conventional residual-based cointegration tests as the well-known *Engle / Granger* [29] approach might fail to detect cointegration although a stable long-run equilibrium relationship exists.

In this vein, *Kannebley* [42] applies the threshold cointegration model, introduced by *Balke / Fomby* [13], to Brazilian exports. He finds evidence in favor of non-linear behavior in nine out of 16 models for which he is able to detect a cointegrating relationship. According to his results adjustment to equilibrium takes place faster in cases when the deviations were below a certain threshold value. Relating these results to the trade hysteresis model of *Belke / Göcke* [16], [17], *Kannebley* [42] concludes that market exit costs might be higher than market entry costs.

While threshold cointegration models incorporate non-linearities as a discrete regime shift, smooth transition (STR) models popularized by *Teräsvirta* [66], [67] do not assume an abrupt switch between different regimes but rather allow for a gradual move from one regime to another and thus, allow so to speak for a continuum of regimes. Such a pattern is probably more plausible on a macroeconomic level as different firms might face different market entry and exit costs. In such a case, aggregation of individual firms would lead to no certain threshold value that triggers strong reactions as outlined in the previous sections but to a path with moderate and strong reactions. In STR models, system dynamics change according to a transition variable which might be an observable economic quantity. In case of trade relationships this could be for example the exchange rate. Model dynamics could differ between cases of currency appreciations or depreciations. For example, *O'Shaughnessy* [54] applies such a framework to UK unemployment and capacity data.

Recently, *Verheyen* [68] applied the NARDL approach of *Shin et al.* [65] to export demand equations in order to derive some stylized facts of trade relationships. He especially focuses on possible non-linear exchange rate effects which should exist if hysteresis is present in export activity as shown in the previous sections. The NARDL approach proposes to split up a regressor into its partial sums and incorporate these series in the regression function.

Verheyen does so and incorporates partial sums of exchange rate series into export demand equations. Precisely he distinguishes two cases. Firstly, he differentiates between cases of currency appreciations and depreciations. His results indicate that exports react stronger to currency depreciations than to appreciations, which is an indication of strategic behavior of exporting firms in the sense that they might adjust their margins in order to cushion unfavorable price developments. Secondly, in order to test for hysteretic effects, he distinguishes between three cases when calculating the partial sums: large appreciations, small changes in exchange rates and large depreciations. As the sunk cost hysteresis model outlines large changes in exchange rates should lead to stronger reactions in exports than moderate changes. However, his results do not strongly correspond to such a pattern. One reason for this finding might be that the changes in exchange rates he considers might still be too small for hysteretic effects to occur.

Applying the NARDL method and using a 4-fold decomposition (i.e. small or large appreciations vs. small or large depreciations) *Fedoseeva* [32] shows for export prices of German sugar confectionary that the pricing-to-market reactions of German exporters are neither symmetric nor linear.

One obvious reason which makes it difficult to find hysteretic effects in macroeconomic data is of course that the sunk cost hysteresis model is a microeconomic phenomenon. After aggregation, then, on a macroeconomic level various phenomena of non-linearity could look like hysteresis.

Models Based on "Strong" Hysteresis

Now, two different methods to integrate "strong"/aggregate hysteresis-loops in empirical models are presented: (1) the *Piscitelli* [61] method as a direct application of the Preisach-dynamics and (2) the *Belke / Göcke* [15] play-algorithm. Both methods are based on calculating artificial "hysteresis variables" from the original forcing variables in order to integrate path-dependent multibranch-effects.

An Algorithm Describing Preisach-Dynamics

Piscitelli et al. [61, 62] developed an algorithm to transform the forcing variables of an economic relation in a way that the Preisach-aggregation method is reproduced. Based on an assumption of the distribution of the α_j/β_j-set of entry and exit triggers of the heterogeneous firms and on the weight of these firms, the weight of the $S_t;^+$-area of active firms (as outlined with Figure 3) can be calculated. In a first step, the series of non-dominated extrema has to be identified and then the changes of the $S_t;^+$-area with changes in the forcing variable are calculated. Usually, the most simple assumption of a uniform weight/distribution is applied, and thus only the acreage of triangles and quadrangles (i.e. quadratic functions) has to be computed in order to describe the aggregate loop (as depicted in Figure 4). The *Piscitelli*-algorithm is e.g. applied in [52] to estimate hysteresis in the Portuguese labor market. A graphical representation of this transformation would look like the aggregate macro-loop in Figure 4, if the original variable was depicted on the abscissa and its transformed hysteresis variable on the ordinate.

A disadvantage of the *Piscitelli*-algorithm is the distribution assumption of the α_j/β_j-sets representing firms, since this distribution determinates the curvature of the aggregate loop. However, as shown in [39] via simulations, the results are quite stable concerning this simplifying uniform weight distribution assumption. Nevertheless, a situation with increasing uncertainty, will induce a widening of the band of inaction of non-ideal relays, i.e. an increase in the distance of the α_j/β_j-points from the 45°-(α=β)-line. The consequences of these dynamics in the α_j/β_j-distribution could not be addressed with this method, based on a static distribution of the α_j/β_j-sets.

The Linearized Play-model

Belke / Göcke [16], [17], as an advancement of [35] and [37], develop an algorithm based on a kind of macroeconomic/generalized play-dynamics in order to approximate macroeconomic "strong" hysteresis-loops.[3] They assume linear partial segments with two

[3] For an application of the play-model see [51].

different slopes: a play-section with a weak reaction (not necessary with a zero slope) and so called "spurt" sections with a stronger reaction of the dependent variable on changes of the forcing variable, i.e. with a steeper slope.[4] Figure 5 gives an example: Start with point A (price p_0) located on the downward leading (left) spurt line. Now, the price changes the direction and increases, entering the play area and resulting in a weak play reaction results (point B) until the entire play area is passed (p_1, point C). A further increase to p_2 induces a strong response of y along the (right) upward leading spurt line (point D). A following price decrease first (up to p_3) takes place on a new play-line DE, which is shifted upwards and to the right compared to the "old" play-line AC. A further price decrease (to p_4, point F) then runs on the left spurt-down-line. Again, the play area is shifted, now downwards and to the left (to line FG) by a past spurt movement. Summarising, persistent effects result, if movements go beyond a play area and take place on a spurt-line: As in the standard case of play-hysteresis, there are no permanent effects from small variations taking place only inside the play-area.

Based on summing up the p-movements on the spurt-line the *Belke / Göcke*-algorithm calculates an artificial shift variable. This "*spurt variable*" s_t integrates all spurt movements which had led to shifts of the play area. The spurt variable s_t is just the series of the original forcing variable p_t where all small movements inside the play areas are filtered out. Since filtering is based on the play width γ, the resulting spurt variable s_t depends on the size of γ. As a result a standard linear equation of the following type has to be estimated (e.g. by OLS):

$$y_t = \text{constant} + \alpha \cdot p_t + \beta \cdot s_t(\gamma) + \text{function(further variables)} \qquad (15)$$

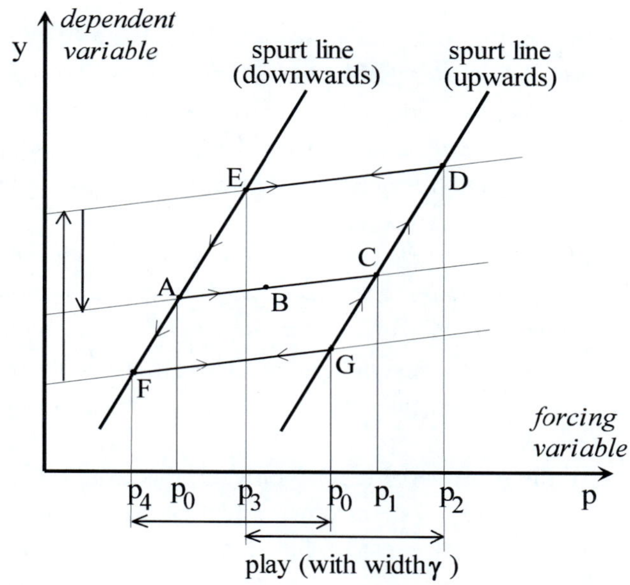

Figure 5. Linear play-hysteresis and spurt areas.

[4] See [59], pp. 980, [28], pp. 15, for "spurts" based on a microeconomic sunk cost mechanism.

The coefficient α is the low slope inside the play areas, and the coefficient β of the filtered series s_j is the *difference in slope* between the weak play and the strong spurt reaction. The complex path-dependent dynamics are captured in a simple linearized way, by adding only two new parameters: (1) the width of the play area γ, and (2) the slope difference β of spurt sections compared to play sections.

Since uncertainty results in a widening of the band of inaction on a micro-level, increased macroeconomic uncertainty will result in a shift of all α/β-points away from the $\alpha=\beta$-45°-line. This would result in a widening of the play area on the aggregate level. The algorithm was extended to allow for such effects (see [17] for a version with a variable play-width related to varying uncertainty).

An Example for Estimating Play-Hysteresis

Next, we present an example of estimating play-hysteresis in German exports to Japan, in analogy to [18]. The data are mainly from Eurostat: We use the quarterly export value in euros from 1991 to 2012. The regarding exchange rates are quarterly averages of monthly averages reported by Deutsche Bundesbank (before 2000 converted DM data). GDP-deflators were used for calculating real exports and the real exchange rate. As additional explanatory variables we include the real GDP of Japan as the destination country (one quarter lagged and seasonally adjusted), a linear trend, and 3 quarterly dummies (Q1, Q2, Q3). Eq. (15) is estimated with the FM-OLS method in the following form,

$$EXP_t = CONSTANT + \alpha \cdot ER_t + \beta \cdot SPURT(\gamma) + GDP_{t-1} + TREND + Q1 + Q2 + Q3 \quad (16)$$

where EXP_t is the real export value (from Germany to Japan) and ER_t is the real exchange rate in indirect quotation. The spurt variable was calculated based on filtering ER as the original variable (using play width γ) and is represented by $s(\gamma) = SPURT(\gamma)$. In a first step – for reasons of comparison - we estimated eq. (16) without the non-linear spurt component to get the standard/linear model and received the regression results in table 1 (using the software-package EViews).

In order to estimate the optimal play width, we run a grid search for different potential values of the play width γ – and choose the γ-value which results in the spurt-regression with the highest R^2. The grid search referring to the R^2-level for different γ-values and the estimated value of $\gamma = 17.25$ is represented in Figure 6. The dynamics of the exchange rate and of the spurt variable - where movements inside the play are filtered out - are depicted in Figure 7.

If the coefficient β of the spurt variable is significant, this is an indication of non-linear play-hysteresis. As the exchange rate is defined in indirect quotation, we expect (and received in both estimations) a negative coefficient of the exchange rate and of the spurt – as an appreciation of the exporter's currency usually leads to a reduction of exports. In case of the non-linear model the sum of the coefficients of the exchange rate and the spurt variable $(\alpha + \beta)$ should be negative and represent the stronger spurt reaction, while the coefficient α now represents a weak play reaction. Thus, compared to the linear standard regression, the coefficient α of the original variable (ER) should be smaller in size (or even insignificant) if the filtered spurt variable is added in the non-linear play-model. A comparison of both regressions (linear standard model versus non-linear play-model) shows, that the inclusion of non-linear play-dynamics improves the fit (indicated by an increase of the adjusted R^2 and by the t-value of the spurt variable). Furthermore, the theoretical assumptions about the direction and the size of play- versus spurt-reactions are corroborated if the estimation results for the α- and β-coefficients are compared.

Table 1. Linear standard regression

Dependent Variable: EXP				
Method: Fully Modified Least Squares (FMOLS)				
Sample: 1995Q1 2012Q4				
Included observations: 72				
Cointegrating equation deterministics: CONSTANT TREND Q1 Q2 Q3				
Variable	Coefficient	Std. Error	t-Statistic	Prob.
CONSTANT	-6372.275	1778.814	-3.582317	0.0007
ER	α = -18.17429	2.602108	-6.984448	0.0000
JAPAN_GDP(-1)	0.009061	0.001651	5.486520	0.0000
TREND	-0.569324	4.181815	-0.136143	0.8921
Q1	-25.05944	102.4723	-0.244548	0.8076
Q2	-225.0373	102.5192	-2.195075	0.0317
Q3	-17.68709	102.3520	-0.172806	0.8633
R-squared	0.741734	Mean dependent var		3066.994
Adjusted R-squared	0.717894	S.D. dependent var		494.6592
S.E. of regression	262.7313	Sum squared resid		4486804.
Durbin-Watson stat	0.897385	Long-run variance		94212.79

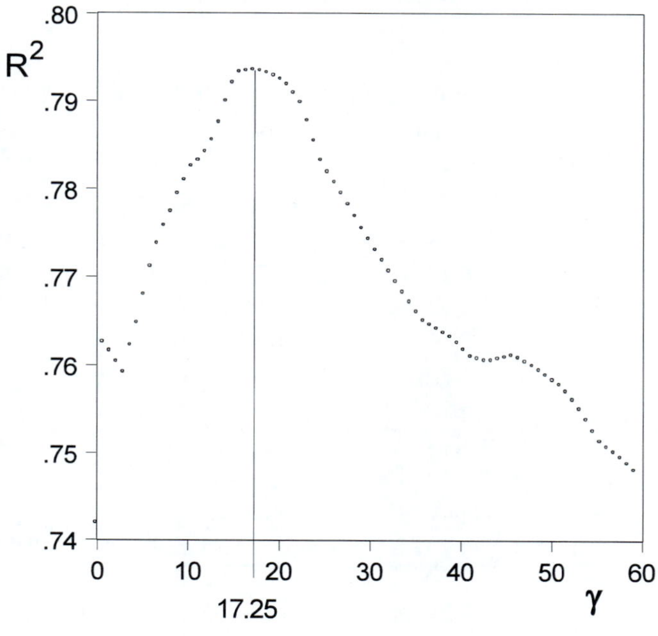

Figure 6. Grid-Search for different play-widths γ.

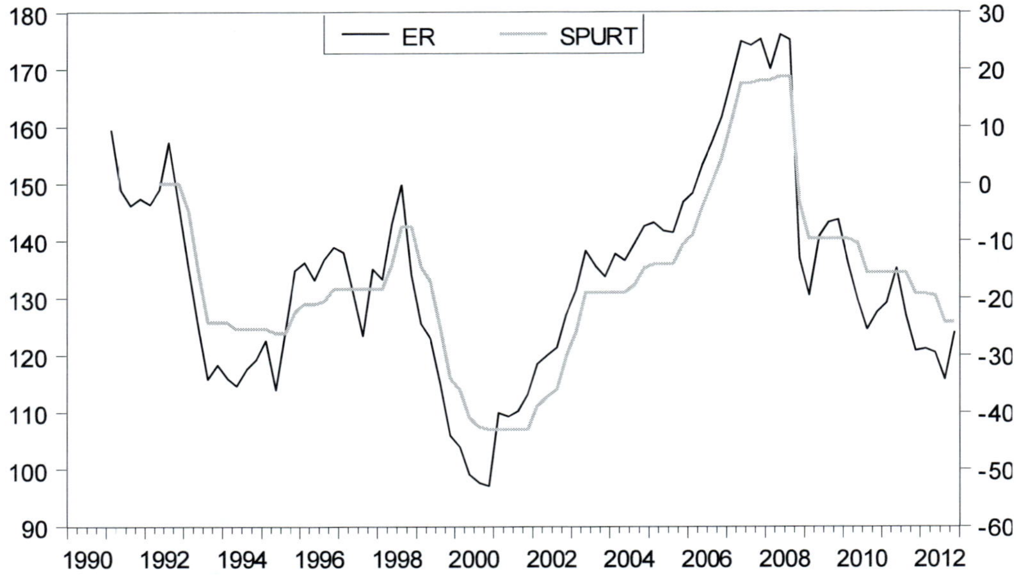

Figure 7. Exchange rate and resulting spurt variable (play width $\gamma = 17.25$).

The result of the non-linear play/spurt-regression using this optimal play width ($\gamma = 17.25$) is as follows:

Table 2. Non-linear play/spurt-regression

Dependent Variable: EXP
Method: Fully Modified Least Squares (FMOLS)
Sample: 1995Q1 2012Q4
Included observations: 72
Cointegrating equation deterministics: CONSTANT TREND Q1 Q2 Q3

Variable	Coefficient	Std. Error	t-Statistic	Prob.
CONSTANT	-8632.354	2054.556	-4.201567	0.0001
ER	α = -5.269178	5.644607	-0.933489	0.3541
SPURT	β = -15.83624	7.002543	-2.261499	0.0271
JP_GDP(-1)	0.009233	0.001669	5.533511	0.0000
TREND	0.107063	4.238531	0.025259	0.9799
Q1	-44.70860	103.1600	-0.433391	0.6662
Q2	-221.6490	103.2352	-2.147029	0.0356
Q3	-15.01675	103.0401	-0.145737	0.8846
R-squared	0.793414	Mean dependent var		3066.994
Adjusted R-squared	0.770818	S.D. dependent var		494.6592
S.E. of regression	236.8077	Sum squared resid		3588986.
Durbin-Watson stat	0.984079	Long-run variance		95446.87

CONCLUSION

This contribution has provided an overview of the phenomenon of hysteresis in economics. Especially unemployment dynamics and the export activity can show signs of hysteresis. Hysteresis generally refers to a situation in which there is a permanent change in a quantity that is triggered by some other variable which has changed only temporarily. Thus, the causal effect remains although the initial cause has vanished.

In economics, adjustment costs can result in hysteretic behavior. The microeconomic model presented above has shown that due to market entry and exit costs which are sunk ex post non-linear patterns of a firm's market activity could arise. Additionally, we have sketched that one has to distinguish between hysteresis on a micro, i.e. firm level and an aggregated macroeconomic scale.

Furthermore, various econometric approaches have been presented which try to identify hysteresis in economic relationships. While this is generally a challenging task – as there is no standard formal test of economic hysteresis – different empirical approaches have found patterns that are in accordance with this phenomenon. Nevertheless, progress in non-linear time series modeling and panel econometrics should provide better tools to model hysteretic behavior more adequately.

REFERENCES

[1] Amable, B., Henry, J., Lordon, F., Topol, R., (1991). Strong Hysteresis: An Application to Foreign Trade. *OFCE Working Paper* 9103.

[2] Amable, B., Henry, J., Lordon, F., Topol, R., (1992). Hysteresis: what it is and what it is not. *OFCE Working Paper* 9216.

[3] Amable, B., Henry, J., Lordon, F., Topol, R., (1993). Unit-Root-Hysteresis in the Wage-price Spiral Is Not Hysteresis in Unemployment. *Journal of Economic Studies* 20 1/2, 123-135.

[4] Amable, B., Henry, J., Lordon, F., Topol, R., (1994). Strong Hysteresis versus Zero-Root Dynamics. *Economic Letters* 44, 43-47.

[5] Ayala, A., Cuñado Eizaguirre, J., Gil-Alana, L.A., (2012). Unemployment Hysteresis – Empirical Evidence for Latin America. *Journal of Applied Economics* XV/2, 213-233.

[6] Bai, J., Perron, P., (1998). Estimating and Testing Linear Models with Multiple Structural Changes. *Econometrica*, 66/1, 47 – 78.

[7] Baldwin, R. E., (1989). Sunk-Cost Hysteresis. *NBER Working Paper*, 2911.

[8] Baldwin, R. E., (1990). Hysteresis in Trade. Franz, W. (Ed.) Hysteresis Effects in Economic Models, 19-34 (also published in: (1990). *Empirical Economics* 15,127-142).

[9] Baldwin, R. E., (1990a). Some Empirical Evidence on Hysteresis in Aggregate U.S. Import Prices. Gerlach, St., Petri, P.A. (Eds.) The Economics of the Dollar Cycle, 235-268.

[10] Baldwin, R. E., Krugman, P., (1989). Persistent Trade Effects of Large Exchange Rate Shocks. *Quarterly Journal of Economics* 104, 635-654.

[11] Baldwin, R. E., Lyons, R. K., (1994). Exchange Rate Hysteresis: The Real Effects of Large vs. Small Policy Misalignments. *European Economic Review* 38:1, 1-22.

[12] Baldwin, R. E., Lyons, R. K., (1989). Exchange Rate Hysteresis: The Real Effects of Large vs. Small Policy Misalignments. *NBER Working Paper* 2828.

[13] Balke, N. S., Fomby, T. B., (1997). Threshold cointegration. *International Economic Review* 38, 627–645.

[14] Belke, A., Göcke, M., (1997). Cointegration and Structural Breaks in German Employment – An Error Correction Interpretation. *Jahrbücher für Nationalökonomie und Statistik* 216, 129-152.

[15] Belke, A., Göcke, M., (1999). A Simple Model of Hysteresis in Employment Under Exchange Rate Uncertainty. *Scottish Journal of Political Economy* 46/3, 260-286.

[16] Belke, A., Göcke, M., (2001). Exchange Rate Uncertainty and Employment: An Algorithm Describing 'Play'. *Applied Stochastic Models in Business and Industry* 17, 181-204.

[17] Belke, A., Göcke, M., (2005). Real Options Effects on Employment: Does Exchange Rate Uncertainty Matter for Aggregation?. *German Economic Review* 62, 185-203.

[18] Belke, A., Göcke, M., Günther, M., (2013). Exchange Rate Bands of Inaction and Play-Hysteresis in German Exports – Sectoral Evidence for Some OECD Destinations. *Metroeconomica* 64, 152–179.

[19] Bentolila, S., Bertola, G., (1990). Firing Costs and Labour Demand: How Bad is Eurosclerosis?. *Review of Economic Studies* 57, 381-402.

[20] Blanchard, O. J., Summers, L. H., (1986). Hysteresis and the European Unemployment Problem. *NBER Macroeconomics Annual* 1, 15-78.

[21] Brokate, M., Sprekels, J., (1996). Hysteresis and Phase Transitions. *Applied Mathematical Sciences* 121, Springer, New York.

[22] Cross, R., (1993). On the Foundations of Hysteresis in Economic Systems. *Economics and Philosophy* 9, 53-74.

[23] Cross, R., (1994). The Macroeconomic Consequences of Discontinuous Adjustment: Selective Memory of Non-Dominated Extrema. *Scottish Journal of Political Economy* 41, 212-221.

[24] Cross, R., Allan, A., (1988). On the History of Hysteresis. Cross, R. (Ed.) *Unemployment, Hysteresis and the Natural Rate Hypothesis*, Oxford/New York, 26-38.

[25] Delgado, F. A., (1991). Hysteresis, Menu Costs, and Pricing with Random Exchange Rates. *Journal of Monetary Economics* 28, 461-484.

[26] Dickey, D. A., Fuller, W. A., (1979). Distribution of the Estimators for Autoregressive Time Series with a Unit Root. *Journal of the American Statistical Association* 74, 427-431.

[27] Dixit, A., (1989). Entry and Exit Decisions under Uncertainty. *Journal of Political Economy* 97, 620-638.

[28] Dixit, A., Pindyck, R. S., (1994). Investment under Uncertainty. Princeton, NY.

[29] Engle, R. F., Granger, C. W. J., (1987). Co-Integration and Error Correction: Representation, Estimation, and Testing. *Econometrica* 55, 251-276.

[30] Engle, R. F., Granger, C. W. J., (Ed.) (1991). Long-Run Economic Relationships – Readings in Cointegration, Oxford.

[31] Evans, G. W., Honkapohja, S., (1993). Adaptive Forecasts, Hysteresis, and Endogenous Fluctuations. *Economic Review* 1993, 3-13.

[32] Fedoseeva, S., (2013). (A)symmetry, (Non)linearity and Hysteresis of Pricing-to-Market: Evidence from German Sugar Confectionary Exports. *Journal of Agricultural & Food Industrial Organization* 11, 69-85.

[33] Franz, W., (1990). Hysteresis in Economic Relationships: An Overview. Franz, W. (Ed.) Hysteresis Effects in Economic Models, Heidelberg, 1-17. First published: (1990). *Empirical Economics* 15, 109-125.

[34] Froot, K. A., Klemperer, P. D., (1989). Exchange Rate Pass-Through When Market Share Matters. *American Economic Review* 79, 637-654.

[35] Göcke, M., (1994). An Approximation of the Hysteresis-Loop by Linear Partial Functions – Econometric Modelling and Estimation. *Jahrbücher für Nationalökonomie und Statistik* 213, 572-596.

[36] Göcke, M., (1994a). Micro- and Macro-Hysteresis in Foreign Trade. *Aussenwirtschaft – Schweizerische Zeitschrift für internationale Wirtschaftsbeziehungen*. 49, 555-578.

[37] Göcke, M., (2001). A Macroeconomic Model with Hysteresis in Foreign Trade. *Metroeconomica*, 52, 449-473.

[38] Göcke, M., (2002). Various Concepts of Hysteresis Applied in Economics. *Journal of Economic Surveys* 16, 167-88.

[39] Hughes Hallett, A. J., Piscitelli, L., (2002). Testing for Hysteresis against Nonlinear Alternatives. *Journal of Economic Dynamics & Control* 27, 303-327.

[40] Hule, R., (1996). Irreversibilität in der Ökonomik. PhD Diss., University of Innsbruck/Austria.

[41] Kanalıcı Akay, H., Nargeleçekenler, M., Yılmaz, F., (2011). Hysteresis in Unemployment – Evidence from 23 OECD Countries. *Ekonomický časopis* 59, 488-505.

[42] Kannebley, S. Jr., (2008). Tests for the Hysteresis Hypothesis in Brazilian Industrialized Exports: A Threshold Cointegration Analysis. *Economic Modelling* 25, 171–190.

[43] Kemp, M. C., Wan, H. Y. Jr., (1974). Hysteresis of Long-Run Equilibrium from Realistic Adjustment Costs. Horwich, G., Samuelson, P. A. (Eds.) Trade, Stability and Macroeconomics, Essays in Honour of Lloyd Metzler, New York, 221-242.

[44] Kneller, E., (1962). Ferromagnetismus, Springer, Berlin.

[45] Krasnosel'skii, M. A., Pokrovskii, A. V., (1989). Systems with Hysteresis, Springer, Berlin.

[46] Lindbeck, A., Snower, D. J., (1986). Wage Setting, Unemployment and Insider-Outsider Relations. *American Economic Review, Papers and Proceedings* 76, 235-239.

[47] Liu, D., Sun, C., Lin, P., (2012). Hysteresis Hypothesis in Unemployment and Labour Force Participation Rates – Evidence from Australian States and Territories. *Australian economic papers* 51, 71-84.

[48] Lütkepohl, H., (1991). Introduction to Multiple Time Series Analysis, Berlin.

[49] Mayergoyz, I. D., (1986). Mathematical Models of Hysteresis. *IEEE Transactions on Magnetics* 22, 603-608.

[50] Mayergoyz, I. D., (2003). Mathematical Models of Hysteresis and Their Applications. Elsevier, New York.

[51] Mota, P. R., Varejão, J., Vasconcelos, P. B., (2012). Hysteresis in the Dynamics of Employment. *Metroeconomica* 63, 661–692.

[52] Mota, P. R., Vasconcelos, P. B., (2012). Nonconvex Adjustment Costs, Hysteresis, and the Macrodynamics of Employment. *Journal of Post-Keynesian Economics*. 35, 93-112.
[53] O'Shaughnessy, T., (2000). Hysteresis in an Open Economy Model. *Scottish Journal of Political Economy* 47, 156-182.
[54] O'Shaughnessy, T., (2001). Unemployment hysteresis and capacity. *Working Paper*, St. Anne's College, Oxford.
[55] O'Shaughnessy, T., (2011). Hysteresis in Unemployment. *Oxford review of economic policy* 27, 312-337.
[56] Perron, P., (1989). The Great Crash, the Oil Price Shock and the Unit Root Hypothesis. *Econometrica* 57, 1361–1401.
[57] Phelps, E. S., (1994). Structural Slumps: The Modern Equilibrium Theory of Unemployment, Interests and Assets. Harvard University Press, Cambridge.
[58] Phelps, E. S., (1972). Inflation Policy and Unemployment Theory – The Cost-Benefit Approach to Monetary Planning, London.
[59] Pindyck, R. S., (1988). Irreversible Investment, Capacity Choice and Uncertainty. *American Economic Review* 78, 969-985.
[60] Pindyck, R. S., (1991). Irreversibility, Uncertainty, and Investment. *Journal of Economic Literature* 29, 1110-1148.
[61] Piscitelli, L., Cross, R., Grinfeld, M., Lambda, H., (2000). A Test for Strong Hysteresis. *Computational Economics* 15, 59-78.
[62] Piscitelli, L., Grinfeld, M., Lambda, H., Cross, R., (1999). On Entry and Exit in Response to Aggregate Shocks. *Applied Economics Letters* 6, 569–572.
[63] Preisach, F., (1935). Über die magnetische Nachwirkung. *Zeitschrift für Physik* 94, 277-302.
[64] Sachs, J. D., (1986). High Unemployment in Europe: Diagnosis and Policy Implications. *NBER Working Paper Series* 1830.
[65] Shin, Y., Yu, B., Greenwood-Nimmo, M. J., (2011). Modelling Asymmetric Cointegration and Dynamic Multipliers in a Nonlinear ARDL Framework. *Social Sciences Research Network*, SSRN-abstract-id=1807745.
[66] Teraesvirta, T., (1994). Specification, estimation and evaluation of smooth transition autoregressive models. *Journal of the American Statistical Association* 89, 208–218.
[67] Teraesvirta, T., (1998). Modeling economic relationships with smooth transition regressions. Giles, D. E. A. (Ed.) Handbook of Applied Economic Statistics. CRC press, New York, 507–552.
[68] Verheyen, F., (2013). Exchange rate nonlinearities in EMU exports to the US. *Economic Modelling*, 32, 66-76.

In: Hysteresis
Editor: José Carlos Dias

ISBN: 978-1-63321-336-4

Chapter 3

HYSTERETIC PROPERTIES OF MAGNETIC COMPOSITE MICROWIRES

A. Chizhik[1,*], A. Stupakiewicz[2], V. Zablotskii[3], A. Zhukov[1,4], A. Maziewski[2] and J. Gonzalez[1]

[1]Departamento de Física de Materiales, Facultad de Química, Universidad del Pais Vasco, UPV/EHU, San Sebastián, Spain
[2]Laboratory of Magnetism, Faculty of Physics, University of Bialystok, Bialystok, Poland
[3]Institute of Physics ASCR, Prague, Czech Republic
[4]IKERBASQUE, Basque Foundation for Science, Bilbao, Spain

ABSTRACT

This Chapter is devoted to the study of magnetic hysteresis in metallic microwires using the magneto-optical Kerr effect by means of both magnetometry and microscopy. It was found different mechanisms of the magnetization reversal process in dependence of the external magnetic field configuration. The analysis of obtained results has been performed using the theoretical model which proposes the existence of four magnetization states with different chirality. The visualization of the effect of a giant Barkhausen jump was reported. The nucleation of surface magnetic domains has been studied over a wide range of magnetic field frequencies and amplitudes.

Keywords: amorphous magnetic wire, magneto-optic Kerr effect, magnetic domains

INTRODUCTION

The mechanism of magnetization reversal in magnetic wires is the topic of much discussion in experimental and theoretical work where magnetization processes and domain

[*] Corresponding Author address: Departamento de Física de Materiales, Facultad de Química, Universidad del Pais Vasco, UPV/EHU, San Sebastián, Spain. Email: oleksandr.chyzhyk@ehu.es.

walls (DW) motion have been studied [1-4]. From the opportunity of high-frequency magnetic properties (such as the giant magnetoimpedance (GMI) effect), the study of the details of the magnetization reversal in the outer shell of the microwires is particularly significant [5, 6]. We have found the lack of information in this area during our systematic study of magnetic and magneto-optical properties of glass covered amorphous microwires. Earlier, we have studied a series of glass covered microwires of different compositions, metallic nucleus diameters (0.8–100 μm) and glass covering thickness (2–50 μm). The microwires of nominal compositions of $Co_{67}Fe_{3.85}Ni_{1.45}B_{11.5}Si_{14.5}Mo_{1.7}$ (metallic nucleus radius 11.2 μm, glass coating thickness 3 μm, ratio of metallic nucleus diameter to total microwire diameter $\rho = 0.79$) and $Fe_{5,71}Co_{64,04}B_{15,88}Si_{10,94}Cr_{3,40}Ni_{0,03}$ (metallic nucleus radius 50 μm, glass coating thickness 20 μm, ratio of metallic nucleus diameter to total microwire diameter $\rho = 0.7$) have a great variety of surface magnetic domain structures. These microwires were selected for the detailed study of the magnetic field induced modification of surface domain structure.

EXPERIMENTAL TECHNIQUES

The magneto-optical Kerr effect (MOKE) technique is recommended as a very valuable and informative method for the study of surface domains and magnetization reversal in micrometer and sub-micrometer ferromagnetic materials [7]. The originality of the idea of our experiments consists to use the MOKE without external magnetic field: the studies have been performed in the presence of circular magnetic field that was produced by electric current flowing along the microwire. To study the magnetization reversal of the wire surface we use the MOKE hysperesis loop tracer (magnetometer) in longitudinal (L-MOKE) (Figure 1(a)) and transversal (T-MOKE) (Figure 1(b)) configurations. For L-MOKE the rotation of the angle of the light polarization is proportional to the magnetization component, which was parallel to the plane of the light. For T-MOKE the rotation angle is proportional to the magnetization component, which was perpendicular to the plane of the light incidence.

Magnetic domain imaging has been performed with magneto-optical polarizing microscopy (Figure 2) in the L-MOKE (Figure 2(a)) and the polar (P-MOKE) (Figure 2(b)) configurations. The images of the domain structures obtained by a L-MOKE microscopy show the difference of in-plane magnetization components. For the polar configuration of MOKE, we can observe the magnetic areas with out-of-plane component magnetization and in this way identify domains with the magnetization of opposite directions. The relevance of MOKE method is determined by the fact that the MOKE magnetometer and MOKE microscope are the complementary tools that permit to determine the directions of the magnetization in the surface of studied microwires and study in details the surface magnetization reversal.

The special peculiarity of the microwires is the non-planar surface. The light reflected from the cylindrical surface of the microwire forms of the conic surface. To avoid a distortion of the magneto-optical signal related to the reflection from the non-planar surface of the wire, the part of light which corresponds to a small area of the wire surface was selected with the diaphragm. The part of the surface of the microwire from which the light hits the detector could be considered as almost a plane. The diaphragm cuts part of the light, which corresponds to the angle of about 1°. An electric current flowing along the wire produced a circular magnetic field. An axial magnetic field was produced by the pair of Helmgolz coils.

MAGNETIZATION SWITCHING VIA THE GIANT BARKHAUSEN JUMP

Here the wire of nominal compositions $Co_{67}Fe_{3.85}Ni_{1.45}B_{11.5}Si_{14.5}Mo_{1.7}$ mentioned in Introduction has been studied. We have used two configurations of the MOKE microscopy experiments. In the first, the transformation of domain structure was obtained from comparison of remanent domain structure images obtained in zero magnetic field and images obtained after application of the pulses of electric current. The pulses have a constant time duration and increasing amplitude. In the second case, the domain structure evolution was determined after applying pulses of electric current with constant amplitude and increasing time duration.

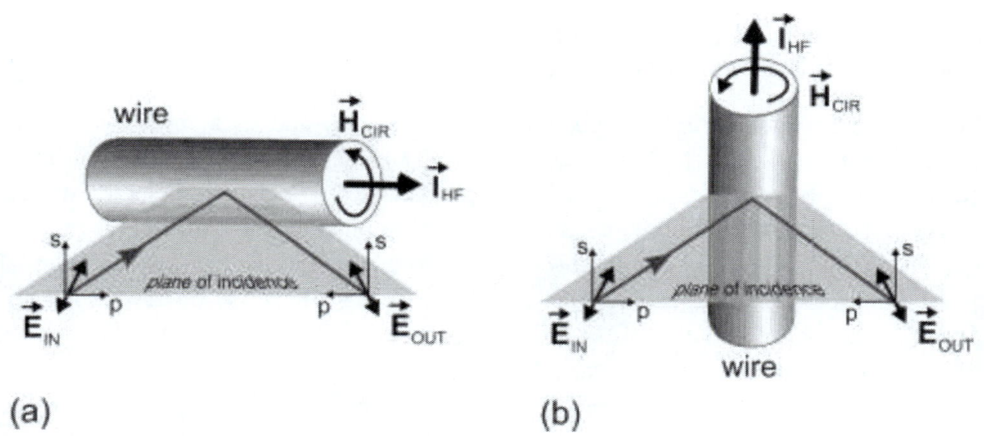

Figure 1. Experimental configurations MOKE magnetometry: L-MOKE (a) and T-MOKE (b).

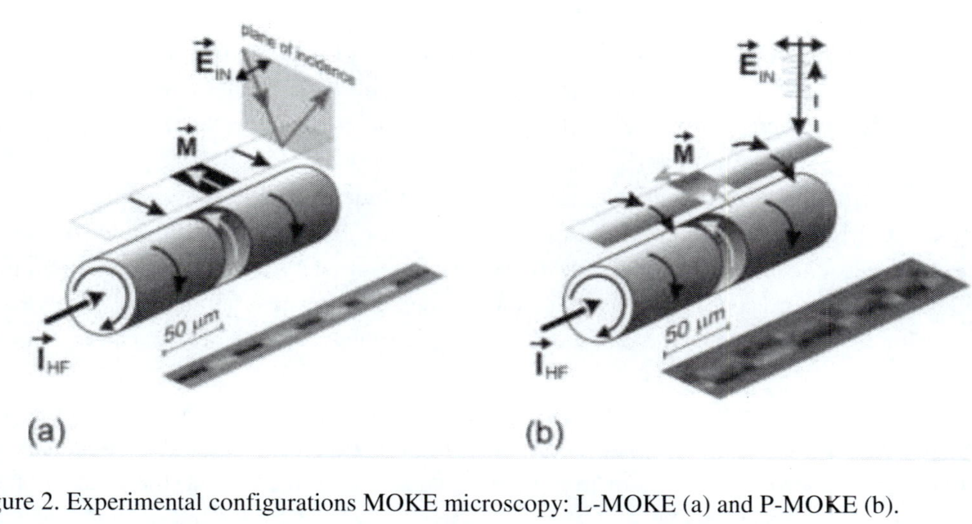

Figure 2. Experimental configurations MOKE microscopy: L-MOKE (a) and P-MOKE (b).

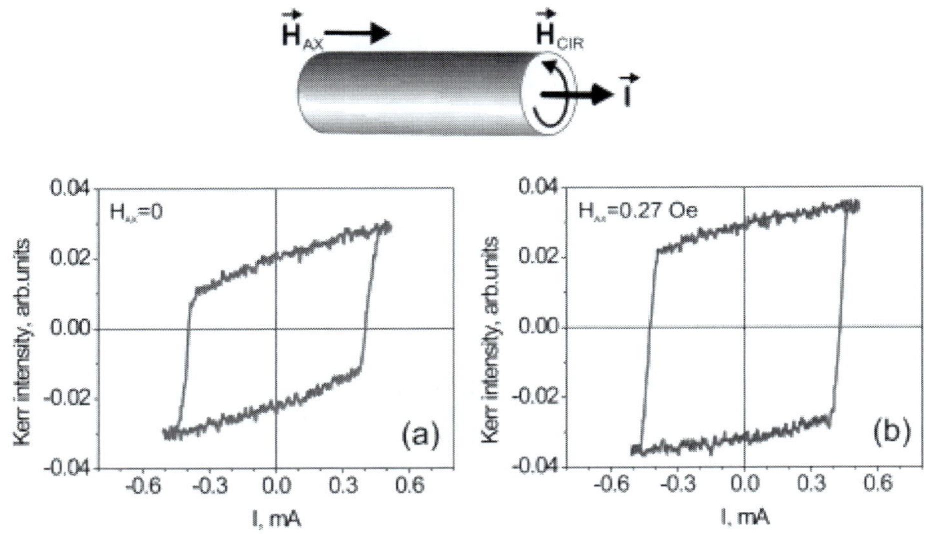

Figure 3. Hysteresis loops measured by MOKE as a function of the electric current (circular magnetic field) for different axial magnetic field: (a) H_{AX}=0; (b) H_{AX}= 0.27 Oe.

First, we performed MOKE magnetometer experiments and then investigation of the domain structure using the MOKE microscope. The dependencies of the Kerr intensity (proportional to Kerr rotation angle) on the circular magnetic field (using electric current I) are presented in Figure 3. These hysteresis loops demonstrate the surface magnetization reversal process [8]. The DC axial field (H_{AX}) was applied along the axis of microwire. The jumps of the circular magnetization related to Barkhausen jumps are observed (Figure 3). These jumps are related to the effect of circular magnetic bistability. It was found that applying an axial magnetic field (Figure 3b) changes the hysteresis curve: the jump of the circular magnetization becomes sharper. The observed change of the hysteresis loop has a relation with the transformation of the helical magnetic structure. In the presence of an axial field, the jump takes place between two circularly magnetized states. The amplitude of the electric current does not exceed 1 mA. For that reason, we performed the investigation outside the Joule heating.

We have two processes with different behaviour of magnetization reversal. The difference in the process of domain nucleation and domain walls propagation is reflected in this difference in velocity. To prove this assumption and to make clear the details of the Barkhausen jumps, we performed the MOKE microscope studies (Figures 4, 5).

We performed the experiments with increasing amplitude of the circular magnetic field as well as the experiments with increasing time duration of the circular magnetic field. A second configuration was chosen for $H_{AX} \neq 0$, taking into account the very sharp jump of the circular magnetization in the presence of the axial field. Time scanning was applied to observe the magnetization reversal because the circular field range of the existence of a giant Barkhausen jump is very short.

For the case of H_{AX}=0 the raise of the circular magnetic field causes the nucleation of the circular domains in the surface of the microwire (Figure 4), followed by the domain walls motion along the microwire. Three different stages were highlighted. First, the nucleation of two domains can be observed. Second, parts of other domains, which come into the field of vision, are observed along with the propagation of the nucleated domains. Finally, the joint motion of domain walls is finished by the disappearance of the disadvantageous circular domain (Figure 4(e)). Figure 5 shows the magnetization reversal process for the case of $H_{AX} \neq 0$. Increasing the time duration of the circular magnetic field we induce the nucleation of a circular domain followed by the motion of two domain walls moving into two opposite directions. This process we consider as a realization of the classic giant Barkhausen jump.

The axial magnetic field influences not only on the helical magnetic structure but also on the surface giant Barkhausen jump - the axial magnetic field changes the mechanism of formation of surface domain structures. When $H_{AX}=0$, we observe the process of multi-domain nucleation. It depends mainly on the local distribution of the density of nucleation centres on the surface of the microwire [9].

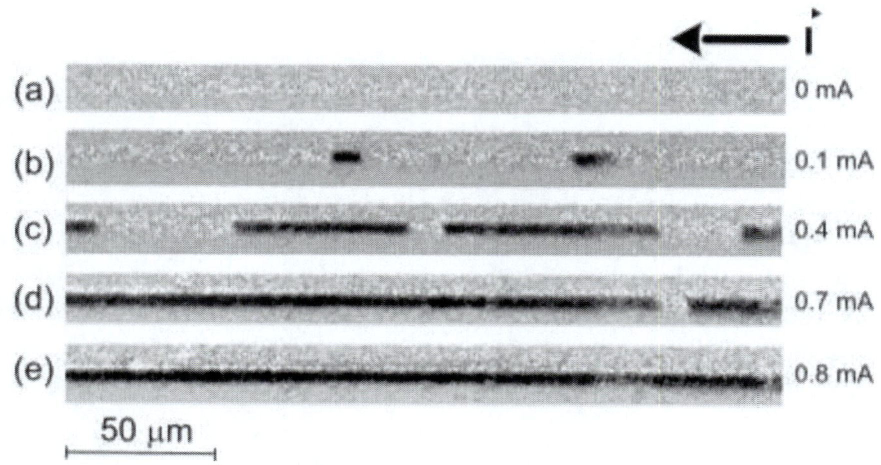

Figure 4. Images of circular magnetic domain structure transformation in the presence of electric current without axial magnetic field: (a) I=0; (b) I=0.1 mA; (c) I=0.4 mA; (d) I=0.7mA; (e) I=0.8 mA (H_{AX} =0).

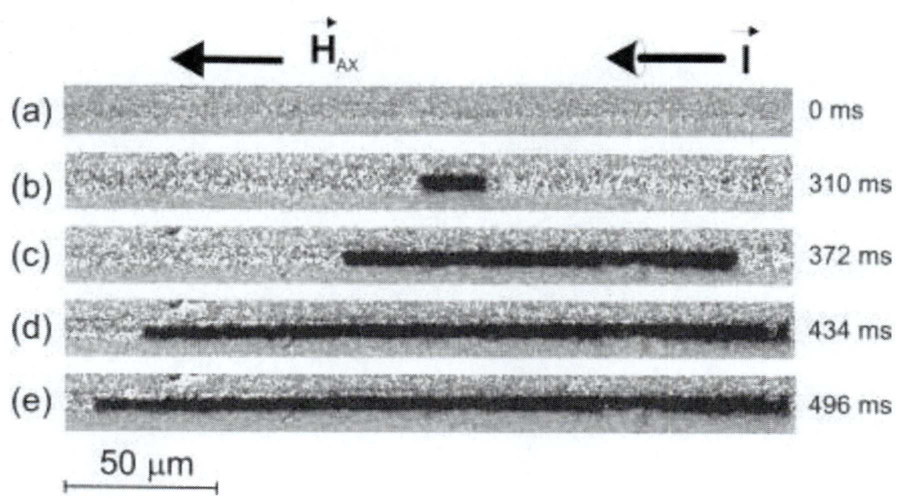

Figure 5. Images of circular magnetic domain, in the electric current I=0.38 mA with H_{AX}=0.27 Oe, recorded at different times: (a) t=0; (b) t=310 ms; (c) t=372 ms; (d) t=434 ms; (e) t=496 ms.

The axial magnetic field suppresses the nucleation of the circular domains. The quickest version of the giant Barkhausen jump is determined by propagation of two domain walls which follow the nucleation of one circular domain. The influence of the axial field on the helical structure and its influence on the surface nucleation process are in a strong correlation. The angle of the helicality (the direction of the magnetization in the surface area of the microwire) is controlled by the axial magnetic field. This angle determines the probability of the nucleation of the reversed domains and states the relative priority of the domain nucleation and the domain walls propagation.

CIRCULAR DOMAINS NUCLEATION

In this part we present the results of the experiments and theoretical approach performed in the wire of nominal compositions $Co_{67}Fe_{3.85}Ni_{1.45}B_{11.5}Si_{14.5}Mo_{1.7}$. Here we put our attention on a study of the nucleation of magnetic domains induced by a pulsed, oscillating magnetic field in ferromagnetic, glass-covered microwires. We have directed our efforts toward acquiring original experimental data and developing a model for the formation and transformation of the structure of surface magnetic domains. A specific feature of this study is the application of a pulsed circular magnetic field instead of the harmonic oscillation that is normally used in experiments of such type. This configuration is very useful because the meander-type waves contain a wide range of harmonic frequencies.

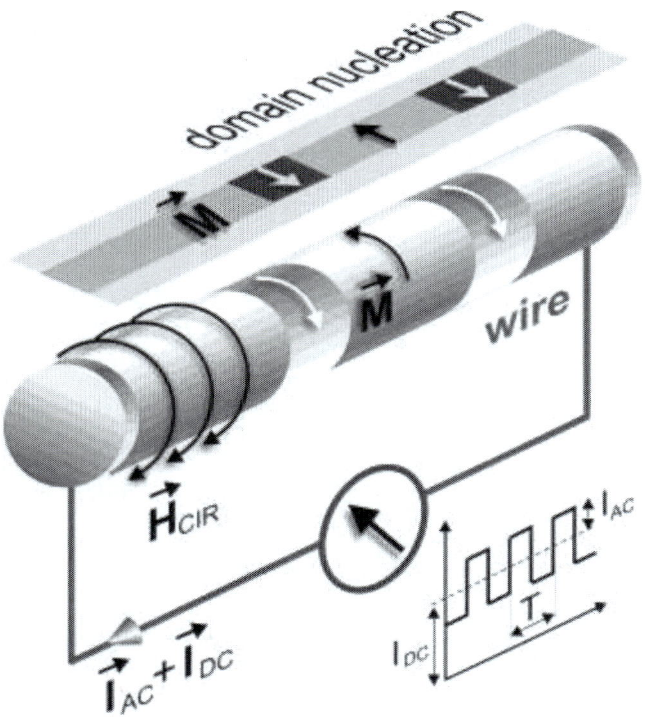

Figure 6. Schematic of the experimental configuration used to observe the nucleation of circular magnetic domains with L-MOKE geometry in microwires with pulses of electrical current. The inset shows rectangular pulses of the electric current which induce the circular magnetic field.

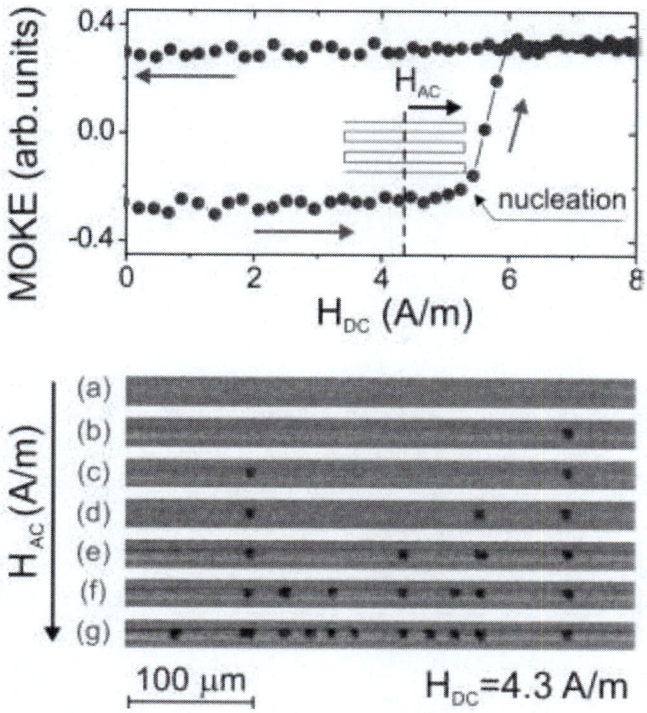

Figure 7. Top panel: reversal of the local magnetization in the wire through the application of a DC circular magnetic field (arrows show the direction of the magnetization process). Bottom panel: images of the domains nucleated at H_{DC}=4.3 A/m for (a) the initial state (H_{AC}=0) and different HAC amplitudes: (b) 2.48 A/m, (c) 2.62 A/m, (d) 2.64 A/m, (e) 2.70 A/m, (f) 2.89 A/m, and (g) 3.14 A/m.

In this Chapter we propose a description for domain nucleation, the main idea of which is the following. Upon an increase in the alternating circular magnetic field, the critical nucleus size changes periodically and approaches the mean defect size. The free energy to create a nucleus grows up to a maximum value at the critical nucleus radius. The height of this free energy peak corresponds to the energy barrier for the nuclei growth. On increasing magnetic field strength both the critical nucleus size and the energy barrier decrease obeying power-law functions. The energy barrier reaches a smallest value and the magnetic nuclei have the largest probability of growth, when the critical size closely approaches but does not yet reach the mean defect size.

A periodically pulsed electric current I_{AC} flowed through the wire to produce the alternating, circular magnetic field pulses H_{AC} (see Figure 6). The pulses of the electric current had a rectangular shape. The magnetic field H_{AC} was sweeping at the range from -4.2 A/m to 4.2 A/m. Taking into account that H_{CIR}=H_{AC}+H_{DC} and that H_{DC}=4.3 A/m the resulting value of H_{CIR} was positive during the performed experiments.

Figure 7 presents the MOKE magnetization curve measured with a DC circular magnetic field H_{DC}. Also a series of images that shows the transformation of the surface domain structure is presented in this figure. This transformation was induced by the magnetic fields H_{DC} and H_{AC} with a pulse frequency of 1 kHz. Figure 7(a) is the image of the initial state of the mono-domain structure with H_{DC}=4.3 A/m and H_{AC}=0. The number of magnetic domains increases as the amplitude of H_{AC} increases (Figures 7(b)-7(g)). The periods of these domains do not correspond to the total energy minima; therefore, such domains should be treated as metastable [3]. For the outer shell of a wire with only circular magnetic anisotropy, there is no magneto-static contribution to force the system to divide itself in circularly magnetized domains.

As a sketch of the obtained experimental results, figure 8 presents the dependence of the number of nucleated domains on the amplitude and frequency of H_{AC} from 1 to 300 kHz with H_{DC}=4.3 A/m.

A domain nucleus has magnetization that is opposite to that of the outer shell. In other words, the magnetization is locally parallel to the applied circular field within the nucleus. We assume that domain nucleation can occur on areas with the lowest value for the magnetic anisotropy constant. Such an area with modified magnetic parameters and/or geometry is called a defect. Thus, we consider a wire to be placed in an inverse magnetic field that is slightly lower than the coercivity field. An energy barrier must be crossed to reverse a nucleus. The height of the barrier (E), corresponds to the critical nuclei size (r_{cr}). The subcritical nuclei (with $r<r_{cr}(H_{CIR})$) become critical as the applied field is increased. In such a way, nuclei start to grow if $r \geq r_{cr}(H_{CIR})$. The average crossing time, τ for a barrier of height E is given by the Arrhenius law, $\tau=\tau_0 Exp(E/k_BT)$. It is commonly assumed that $\tau_0 \approx 10^{-9}$ s. The applied circular magnetic field changed as $H_{CIR}(t)=H_{DC}\pm H_m$, where H_m is the amplitude, and f is the frequency of the pulses of the H_{AC} field. The critical nucleus size change periodically and approach the mean defect size, r_0. The barrier value reaches its minimum, when the critical radius closely approaches but does not reach r_0 and the magnetic nuclei have a chance to grow. The nuclei grow through a finite barrier; i.e., $r_{cr}(H(t))$ is still larger than r_0 at all times.

For meander-like pulses, the time interval $\Delta t=T/2$ (where $T=1/f$) can be considered an attempt time for nuclei to cross the energy barrier. When $H_{CIR}(t)= H_{DC}+H_m$, the nuclei must overcome the lowest energy barrier during this time interval. If the barrier height is zero, $\tau=\tau_0$, which corresponds a GHz frequency range. There is no opportunity to observe domain nucleation if $\Delta t<\tau_0$.

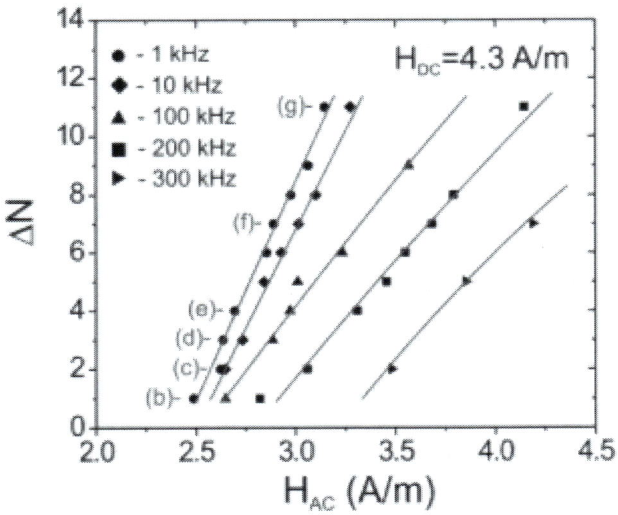

Figure 8. Number of nucleated domains as a function of the field amplitude at different frequencies with H_{DC}=4.3 A/m. The letters ((b)-(g)) correspond to the domain images shown in Figure 7. The solid lines represent the fit to Eq. (3) of ΔN as a function of H_{AC}.

For domain nucleation through a finite barrier, a nucleus will escape from the defect potential well if the sum of the field amplitude and H_{DC} increases. The height of the energy barrier depends on the magnetic interactions, the activation volume, *V*, and temperature, *T*. If the barrier height is minimal, according to the Arrhenius law, the mean crossing time for the barrier is given by

$$\tau = \tau_0 Exp\left(\frac{E - 2M_S V(H_{DC} + H_m)}{k_B T}\right) \tag{1}$$

where M_S is the saturation magnetization and k_B is the Boltzmann constant. The probability of domain nucleation (i.e., the probability for a nucleus to escape from the potential well) can be written as $p_{nuc}=1\text{-}2f\tau$, which gives the proportion, $\Delta N/N$, of domains nucleated during time Δt in a wire with only one type of barrier of height *E*. For a wire that contains defects with a distribution of energy barriers, *f(E)*, the relative number of nucleated domains is given by

$$\frac{\Delta N}{N} = \int_0^\infty f(E) p_{nuc} dE \tag{2}$$

where the distribution function satisfies the condition $\int_0^\infty f(E)dE = 1$. The distribution of energy barriers for the wire defects can be assumed to be Gaussian, $f(E)=c\,Exp(-(E-E_a)^2/2\sigma^2)$, where E_a and σ are the mean value and dispersion of the barrier height, respectively. Integrating Eq. 2 gives

$$\frac{\Delta N}{N} = 1 - 2cf\tau_0 Exp\left(\frac{E_a - 2(H_{DC} + H_m)M_S V + \sigma^2/(2k_B T)}{k_B T}\right) \tag{3}$$

where *c* is a numerical factor. The relative number of defects serving as centers for nucleation of domains depends on the amplitude and frequency of the ac-magnetic field and obeys Eq. 3. We fit the experimental data for ΔN (shown in Figure 8) to Eq. 3. We have assumed the following set of parameters for the wire and defects: $E_a/k_BT=10$; $\sigma=0.2E_a$; the mean defect size, $r_0\sim$ 20-50 nm; and $4\pi M_S$ = 5400 G. The results shown in Figure 8 demonstrate good agreement between the measured and the fitted data. Equation 3 allows us to calculate the amplitude and frequency bands in which domain nucleation can be either practically suppressed or effectively activated. On the (f, H_m)-plane of the phase diagram (Figure 9), a flat area exists in which the probability of domain nucleation is nearly zero, and domain nucleation is prevented. The wire defects become active centers for domain nucleation at threshold fields and frequencies (shown by the dashed curve in Figure 9). The number of nucleated domains grows progressively if the magnetic field amplitude increases and the frequency decreases. This number saturates at the lowest frequencies and highest amplitudes.

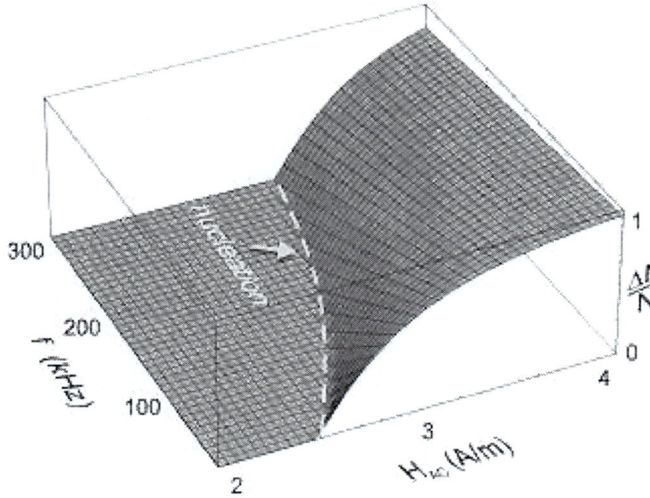

Figure 9. Relative number of nucleated domains as a function of the amplitude and frequency of the circular magnetic field pulse.

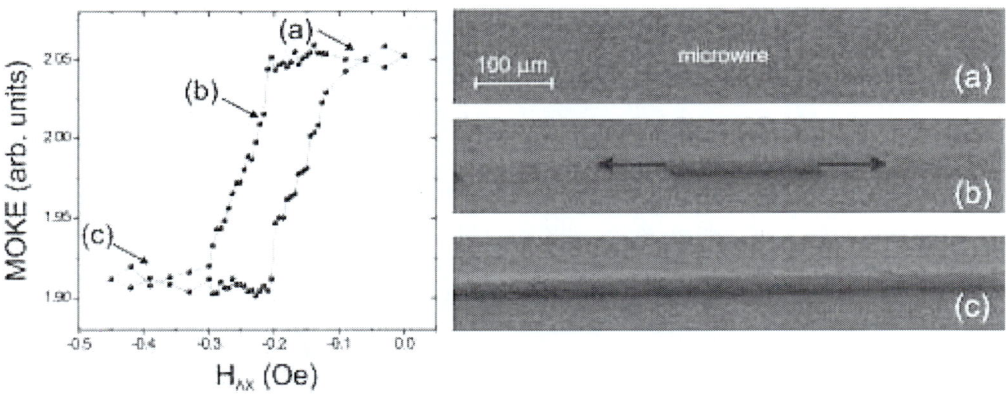

Figure 10. Images of surface magnetic domain structure transformation ((a)-(c)) and hysteresis loop measured by MOKE for: $H_{CIRC\ DC}=0$: (a) H_{AX} = -0.15 Oe;(b) H_{AX}=-0.22 Oe, c) H_{AX}=-0.39 Oe. The hysteresis loop obtained from images of domain structures.

Surface Domain Structure Transformation with Helical Magnetization State

In this part, we have focused on surface domain structure transformation in microwire of nominal composition $Fe_{5,71}Co_{64,04}B_{15,88}Si_{10,94}Cr_{3,40}Ni_{0,03}$ and metallic nucleus radius 50 µm. Experiments were performed using a magnetic field (H_Σ) which was the vector sum of alternating axial (H_{AX}) and DC circular magnetic ($H_{CIRC\ DC}$) fields. Vector H_Σ rotates in a cylindrical shape of the wire and follows the non-plane inclined trajectory.

Images of the domain structures obtained for the magnetic field applied parallel to the wire axis ($H_{CIRC\ DC}$ = 0) are presented in Figure 10. The observed magneto-optical contrast is proportional to the component of magnetization in the direction perpendicular to the wire axis. Figures 10(a)–10(c) show the domain structures obtained by decreasing axial field. The magnetization loop obtained from the MOKE images is also shown in Figure 10. Figure 10(a) corresponds to the mono-domain magnetic state (bright image, H_{AX} =-0.15 Oe). The opposite dark magnetic domain appears when the amplitude is equal -0.22 Oe (Figure10(c)). When the magnetic field sweep continues, the dark domains increase with a simultaneous decrease in bright domains (Figure 10(b)). Results from this experiment agree well with previous results on the long distance rapid motion of solitary circular DW [9].

The main difference between present results and previous results [10] is the inclined shape of the DWs. This case is more general, taking into account that the DWs which divide the circular domains are a special case of inclined DWs which divide the helical domains.

The results of the experiments conducted in the presence of a DC circular magnetic field, are presented in Figures 11-13. Successive changes in the mechanism of the magnetization reversal have been caused by increase in circular magnetic field. Here we present three different types of this mechanism as a dependence on $H_{CIRC\ DC}$ amplitude.

Figure 11 presents the transformation of the surface domain structure induced by the axial magnetic field for $H_{CIRC\ DC}$ = 0.48 Oe. Here we also begin from the mono-domain state. In the initial stage, magnetization reversal occurs in the same way as in the previous case: domain nucleation is followed by DW motion along the wire (Figure 11(a)), but the size of the nucleated domains is smaller. In the next step, the mechanism of magnetization reversal changes. In this case, the effect of domain suppression was observed (Figure 11(b)). The advantageous dark domain placed in the centre of the frame, disappears under the pressure of two moved DWs (marked by arrows in Figure 11). A key feature for this configuration is the periodical difference in angle of inclination of the DWs: the left DW of the bright domains is more inclined toward the axial direction than the right DW. The hysteresis loop in Figure 11 is related to the domain structure transformation. The sharp initial part of the hysteresis curve reflects nucleation of multiple domains. We also highlight the small kink observed near -1.43 Oe. The small plateau at the loop is related directly to the disappearance of the small dark domain in Figure 11(b) and results from the domain suppression effect.

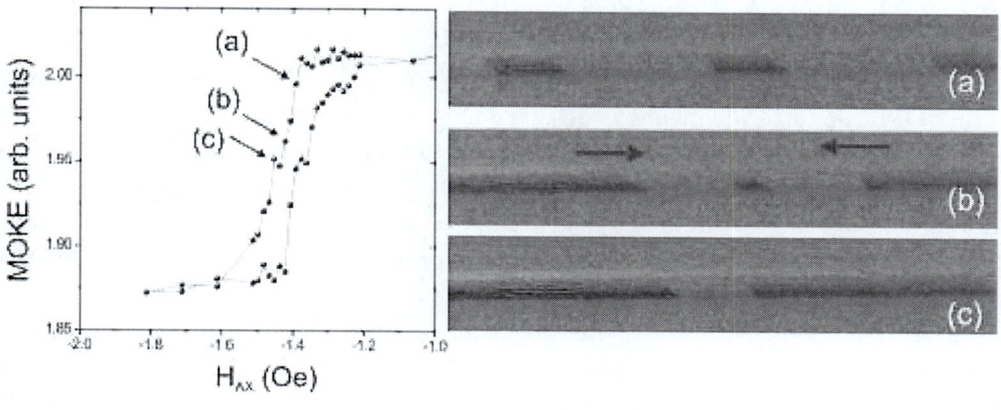

Figure 11. Images of surface magnetic domain structure transformation ((a)–(c)) and hysteresis loop measured by MOKE for $H_{CIRC\ DC}$ = 0.48 Oe and: (a) H_{AX} = -1.05 Oe;(b) H_{AX}= -1.41 Oe, c) H_{AX}= -1.5 Oe.

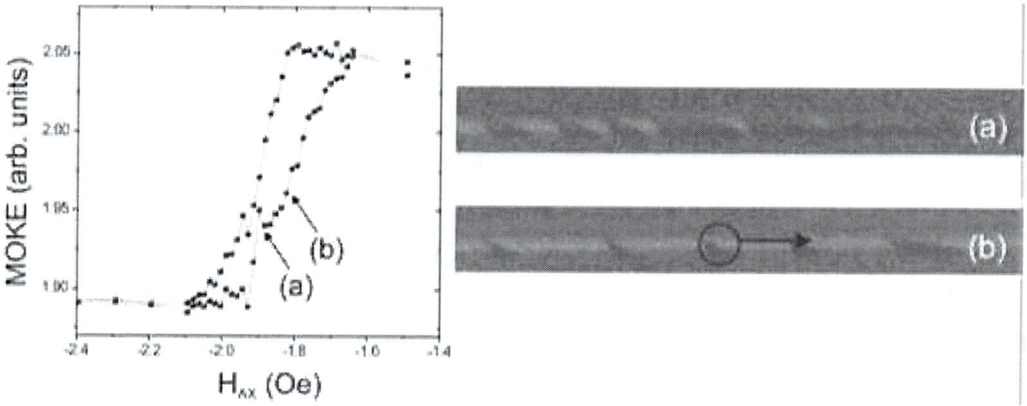

Figure 12. Images of surface magnetic domain structure transformation ((a)–(b)) and hysteresis loop measured by MOKE for $H_{CIRC\ DC}$ = 0.8 Oe and: (a) H_{AX} = -1.9 Oe; (b) H_{AX}= -1.86 Oe. Circle demonstrates the effect of domain drift.

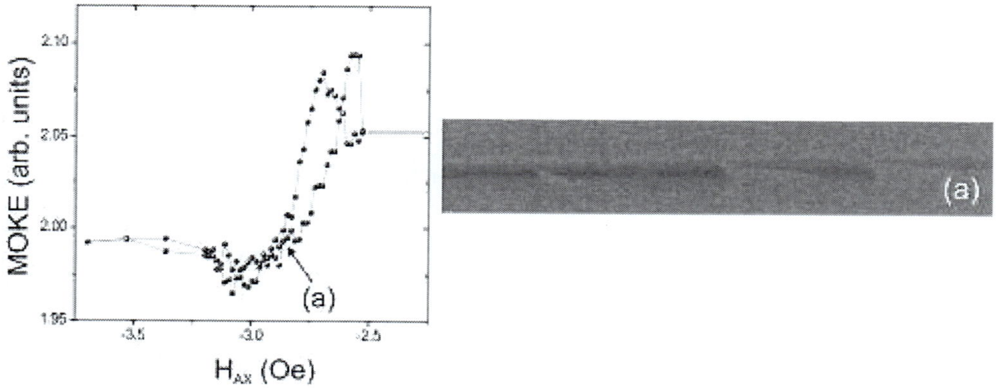

Figure 13. Image of surface magnetic domain structure (a) (H_{AX}= -2.85 Oe) and hysteresis loop measured by MOKE for $H_{CIRC\ DC}$ = 1.12 Oe.

An increase in circular magnetic field leads to the next modification in magnetization reversal: a drift of surface domains is observed (Figure 12). Under a DC circular magnetic field of 0.8 Oe, magnetization reversal occurs as the nucleation of multiple domains followed by the directed motion of the wedged domains as whole objects along the microwire. During the drift process the size of the shifted domains decreases up to the complete disappearance of wedged domains. Magnetization reversal is characterized by the greater number of moved DWs.

The last type of mechanisms of magnetization reversal observed is the formation and transformation of the vortex-like domain structure (Figure 13). The formation of a compact structure which can be qualified as a non-planar vortex-type structure was induced by the DC circular magnetic field increase.

Earlier, vortex-type surface magnetic structures were observed only in conventional microwires [11]. A key point of this experimental configuration is the high enough surface helical anisotropy confirmed by the inclined shape of the DW for $H_{CIRC\ DC}$=0. Another important factor is the value of the axial magnetic field at which magnetization reversal takes place. This value is one order of magnitude larger than the characteristic field for $H_{CIRC\ DC}=0$. The change in magnetic field configuration (Figures 10-13) that results in an increase in values of axial and circular fields, causes the formation of wedged domains. The angle of the wedge increases with increase in H_{Σ}. For the case in Figure 13, the growth in wedge angle causes the formation of the vortex structure. This structure is characterized by the appearance long DWs with fluent change in sign of curvature from "+" to "-". Under an axial magnetic field, a fluent DW motion and stepwise rearrangement of whole domain structure are observed. As a result of this rearrangement, the vortex-type domain transforms to a quasi periodical multi-domain structure with wedge-shaped domains.

The hysteresis loops in Figures10-13 show the minor loops. The part of the hysteresis curve related to the transformation in domain structure is shifted along the H_{AX} axis. The shift increases with increase of the circular field. The reason for this shift is the surface helical anisotropy. The circular field rotates the magnetization from the axial direction thereby making the axially induced magnetization reversal more difficult. In this situation, surface magnetization reversal occurs in a higher axial field. The hysteresis loops presented in the Figures 10-12 have a common behavior. The loops contain two main parts: a jump related to domain nucleation and a fluent part associated with DW motion and transformation of the domain structure. The hysteresis loop related to vortex formation has a specific feature of peaks in the field area at approximately -2.6 Oe. This is related to the overcoming of the energy barrier related to helical anisotropy. These peaks become significant when magnetization reversal occurs in the presence of a relatively high circular magnetic field.

To analyze the obtained MOKE experimental results we use the theoretical model presented in [3]. We consider that the crossed axial-circular magnetic field induces different types of helical magnetic structures. Based on previous calculations, we created a phase diagram (Figure 14) which shows helical magnetic phases (phases marked (1), (2), (3) and (4)) depending on the values of axial and circular magnetic fields. The parameter α is the angle between the anisotropy axis and the tangential direction. The angle between the magnetization and circular field is denoted φ [3].

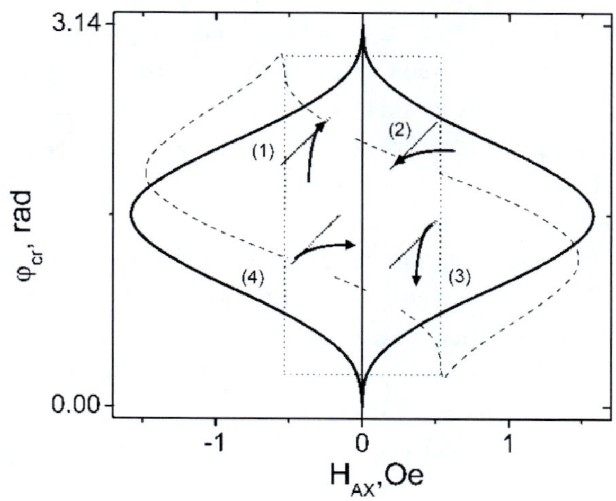

Figure 14. Phase diagram. The critical angle as functions of the axial applied magnetic field.

The phase diagram demonstrates the dependencies of angle φ_{cr} on axial magnetic field. φ_{cr} is the critical angle which corresponds to the moment of the magnetization switching. It was obtained by minimization of the total energy equation. There are two curves for two different angles of helical anisotropy, α [3]. The dashed line has been constructed for $\alpha \neq 0$. It is realized in our experiments, taking into account that the DWs have an inclined shape when $H_{CIRC\ DC}=0$. This confirms the helical (non-circular) character of the surface domain structure.

For $\alpha \neq 0$, a specific area exists where four helical magnetic states co-exist in stable or meta-stable conditions (shaded area in the phase diagram). The different angles of inclination of the DWs in Figures 11-13 confirm the co-existence of these helical states. We have found that the helical states are very sensitive to circular magnetic field. The increase in $H_{CIRC\ DC}$ results in competition between the different helical structures and also between the DW propagation and the domains nucleation processes. The main reason for the effects of domains drift and domain suppression is related to the co-existence of helical states of different types. Co-existence is realized in the specific multi-domain structure when the angle of the DW alternates along the wire. The DWs which separate different helical states have different angles of inclination and therefore, different velocity. When $H_{CIRC\ DC}$ is high enough, the helical structure of the different types is captured into the magnetization reversal initiating the simultaneous coexistence of four magnetic structures in the form of un-plane vortex.

CONCLUSION

Using an axial magnetic field as a unique external parameter we have achieved that the surface magnetization reversal is realized in the form of a giant Barkhausen jump. It was demonstrated that the giant Barkhausen jump is characterized by the quick motion of the solitary domain wall which overpasses long distances of hundreds of micrometers. We have shown that the extensive motion of the solitary domain wall really exists in magnetic wires and that this is the determinative property of the surface giant Barkhausen jump.

Also we demonstrate that a pulsed magnetic field induces domain nucleation in magnetic microwires. We observed domain nucleation in the MOKE images, which show stable, circularly magnetized domains. The number of active centres of domain nucleation depends on the frequency and amplitude of the pulsed magnetic field. Therefore, by choosing a specific mode for local magnetization reversal, we can suppress the motion of DW and simultaneously support the nucleation of domains.

Four mechanisms of magnetization reversal were found to exist depending on the configuration of external axial and circular magnetic fields. We can select the mechanism by varying the DC circular magnetic field value. This possibility is important taking into account the role of DW motion and domain structure transformation in the functioning of sensor elements using magnetic microwires.

REFERENCES

[1] Vazquez M., Chiriac H., Zhukov A., Panina L., Uchiyama T., (2011). On the state-of-the-art in magnetic microwires and expected trends for scientific and technological studies. *Phys. Stat. Sol. (a)* 208, 493-501.

[2] Parkin S. S. P., M. Hayashi M., Thomas L., (2008). Magnetic domain-wall racetrack memory. *Science* 320, 190-194.

[3] Chizhik, A., Zablotskii V., Stupakiewicz A., Gomez-Polo C., Maziewski A., Zhukov A., Gonzalez J., Blanco J.M., (2010). Magnetization switching in ferromagnetic microwires. *Phys. Rev.* B 82, 212401-212401-4.

[4] Chizhik A., Stupakiewicz A., Maziewski A., Zhukov A., Gonzalez J., (2012). High-frequency electric current influence on magnetization reversal and domain structure in Co-rich amorphous microwires. *IEEE Trans. Magn.* 48, 3800-3802.

[5] Jiles D. C., (2003). Recent advances and future directions in magnetic materials. *Acta Mater.* 51, 5907-5939.

[6] Mohri K., Humphrey F. B., Kawashima K., Kimura K., Mizutani M., (1990). Large Barkhausen and Matteucci effects in FeCoSiB, FeCrSiB, and FeNiSiB amorphous wires *IEEE Trans. Magn.* 26, 1789-1791.

[7] Chizhik A., Gonzalez J., (2013). Magnetic Microwires: A Magneto-Optical Study. *Pan Stanford Publishing Pte. Ltd., Singapore.*

[8] Chizhik A., Gonzalez J., Zhukov A., Blanco J. M., (2003). Circular magnetic bistability induced by tensile stress in glass-covered amorphous microwires. *Appl. Phys. Lett.* 82, 610-612.

[9] Ipatov M., Usov N. A., Zhukov A., Gonzalez J., (2008). Local nucleation fields of Fe-rich microwires and their dependence on applied stresses. *Physica B* 403, 379-381.

[10] Chizhik A., Stupakiewicz A., Maziewski A., Zhukov A., Gonzalez J., Blanco J. M., (2010). Direct observation of giant Barkhausen jumps in magnetic microwires. *Appl. Phys. Lett.* 97, 012502-012502-3.

[11] Chizhik A., Gonzalez J., Yamasaki J., Zhukov A., Blanco J. M., (2004). Vortex-type domain structure in Co-rich amorphous wires. *Journal of Applied Physics* 95, 2933-2935.

In: Hysteresis
Editor: José Carlos Dias

ISBN: 978-1-63321-336-4

Chapter 4

THEORETICAL AND EXPERIMENTAL STUDY OF MAGNETIC HYSTERESIS CYCLE

***R. Masrour*[1,2*], *O. Mounkachi*[3], *H. El Moussaoui*[3], *M. Hamedoun*[3], *A. Benyoussef*[2,3,4] *and E. K. Hlil*[5]**

[1]Laboratory of Materials, Processes, Environment and Quality, Cady Ayyed University, National School of Applied Sciences, Safi, Morocco.
[2]LMPHE, Faculté des Sciences, Université Mohammed V-Agdal, Rabat, Morocco
[3]Institute for Nanomaterials and Nanotechnologies, MAScIR, Rabat, Morocco
[4]Academie Hassan II des Sciences et Techniques, Rabat, Morocco
[5]Institut Néel, CNRS et Université Joseph Fourier, BP 166, F-38042 Grenoble cedex 9, France

ABSTRACT

The magnetic properties of materials and nanomaterials with the different structures are investigated. The magnetic hysteresis cycle measurement of hexagonal structure, GdSb alloys, MnAu superlattices, Ni/Au core/shell, and nanoferrites MFe_2O_4 (M=Fe, Nd, Co, Mn, Sn) are established by Monte Carlo simulations and experimental technical, Measurement System (MPMS-7XL, Quantum Design, Inc.). The saturation magnetization, remanent magnetization and coercive field are obtained for different materials and nanomaterials.

Keywords: magnetic hysteresis cycle; Monte Carlo simulations; experimental technical, measurement system; saturation magnetization, remanent magnetization; coercive field

* Corresponding author: Email: rachidmasrour@hotmail.com.

INTRODUCTION

Both semiconducting and ferromagnetic properties have been established in some of the rare earth mononitrides which thus attract interest for the potential to exploit the spin of charge carriers in semiconductor technologies for both fundamental and applied science by Natali et al. [1]. Narita studied the magnetic susceptibilities and spin structure of GdX [2,3]. The exchange parameters of GdX (X=P, As, Bi) are obtained by Larson et al. [4] and are in good agreement with experiment [1]. The metallic nanoparticles (NPs) or clusters, aggregates of several to millions atoms, have been emerging as a new type of important functional material [5]. They exhibit distinct properties (optical, electronic, magnetic, and so on) from those of individual atoms or their bulk counterparts due to the quantum-size and surface effects. More importantly, the properties are usually size- and structure-dependent, which have attracted tremendous research attention from either basic science or application viewpoints during the last two decades [6-17]. Rosengaard and Johansson [18] calculated the Curie temperatures of ferromagnetic body-centered-cubic (bcc) Fe, face centered cubic (fcc) Co and Ni using Monte Carlo (MC) simulations with the exchange parameters deduced from ab initio results of total energies for different configurations. Zhou et al. [19] used a similar procedure to study the magnetic phase transitions in face-centered-cubic (fcc) Fe and Mn antiferromagnets. More recently, it was shown that such an ab initio-Monte Carlo method also works well for low-dimensional magnetic systems [20]. The magnetic and electrical properties of practical antiferromagnetic Mn Alloys are studied by [21]. The magnetic and structural properties of MFe_2O_4 (M = Co, Mg, Mn, Ni) nanoferrite are reported in [22,23]. The magnetic and hyperfine states of iron doped tin dioxide milled powders were analyzed at low temperature [24]. The saturated magnetization of the superparamagnetic Fe_3O_4 nanoparticles prepared by the co-precipitation and combustion methods were found to be, respectively, ~63.5 emug^{-1} and 44.2 emug^{-1} at room temperature [25]. The magnetite nanoparticles display super-paramagnetic behavior which arises from the negligible energy barrier in the hysteresis of the magnetization loop of the particles as predicted by Bloch and Neel [26]. Diluted magnetic oxides (DMOs) received a major interest in the spintronics field for their important physical properties (spin polarization at the Fermi level, easy integration into semiconductor devices and other properties). Oxides doped with transition material are the most studied diluted magnetic semiconductors (DMSs) [27-36]. They have very interesting magneto-optical properties, major progress in this direction have been made, but few results on ferromagnetic properties, ferromagnetic, spin-glass or superparamagnetic at high temperatures in heterostructures based doped oxides have been made. To meet this need of making DMSs with controllable magnetic properties at high temperature, magnetic semiconductor nanostructures and bulks (ferromagnetic, superparamagnetic and spin-glass) are provided. The magnetic semiconductor is a tin oxide doped co-and tri-doped with several types of magnetic transition metals.

In this chapter, the thermal magnetizations and magnetic susceptibilities are computed in: GdSb alloys, Au/Ni and MnAu superlattices. The Néel temperature is deduced for each material. The magnetization versus the exchange interactions and crystal field are studied for a fixed size, a fixed temperature, a zero magnetic field and a zero crystal. The variation of the magnetization versus the crystal field is given for GdSb alloys and for Au/Ni core shell with different temperatures values. The magnetic hysteresis cycle is established. The thermal magnetizations of MnAu superlattices are given by using the Monte Carlo simulations for a fixed size and in the absence of external magnetic field. The MFe_2O_4 (M=Fe, Nd, Co, Mn, Sn) spinel ferrites are synthesized by Co-precipitation method. We see that the curves form a hysteresis loop is the dependence for studied nanoparticle ferrites. From this variation we determined the saturation magnetization, remanent magnetization and coercive field for all samples studied in this part. Finally, the magnetic characterization of $Sn_{0.95}Mn_{0.025}Co_{0.025}O_2$ was done by Magnetic Properties Measurement System (MPMS-7XL, Quantum Design, Inc). The magnetic properties of $Sn_{0.95}Co_{0.05}O_2$ nanocrystals have been studied by measuring the magnetization as a function of temperature in applied field of 500 Oe. Magnetic hysteresis loops of $Sn_{0.9}Cu_{0.05}Fe_{0.05}O_2$ nanocrystals were made at 5K and 300K, are shown.

THEORIES AND MODELS

Mean Field Theories

The Hamiltonian of the system is given by:

$$H = -\sum_{<i,j>} J_{ij}\vec{S}_i\vec{S}_j - H\sum_i S_i - \Delta\sum_i S_i^2 \tag{1}$$

where, H and Δ are the external magnetic field and the crystal filed, respectively. J_{ij} and S_i are the exchange interaction between the in i and j. The relationship between the exchange interactions and the Néel temperature, $T_N(K)$ is deduced by using the mean-field theory [37,38]. The expression obtained is written as follows:

$$\mathrm{J}(K) = \frac{T_N(K)}{4S(S+1)} \tag{2}$$

MONTE CARLO SIMULATIONS

We apply a standard sampling method to simulate the Hamiltonian given by Eq. (1). Cyclic boundary conditions on the MnAu superlattices were imposed and the configurations were generated by sequentially traversing the lattice and making single-spin flip attempts. The flips are accepted or rejected according to a heat-bath algorithm under the Metropolis approximation (see Figure1).

Our program calculates the following parameters, namely:

The thermal magnetization is defined as,

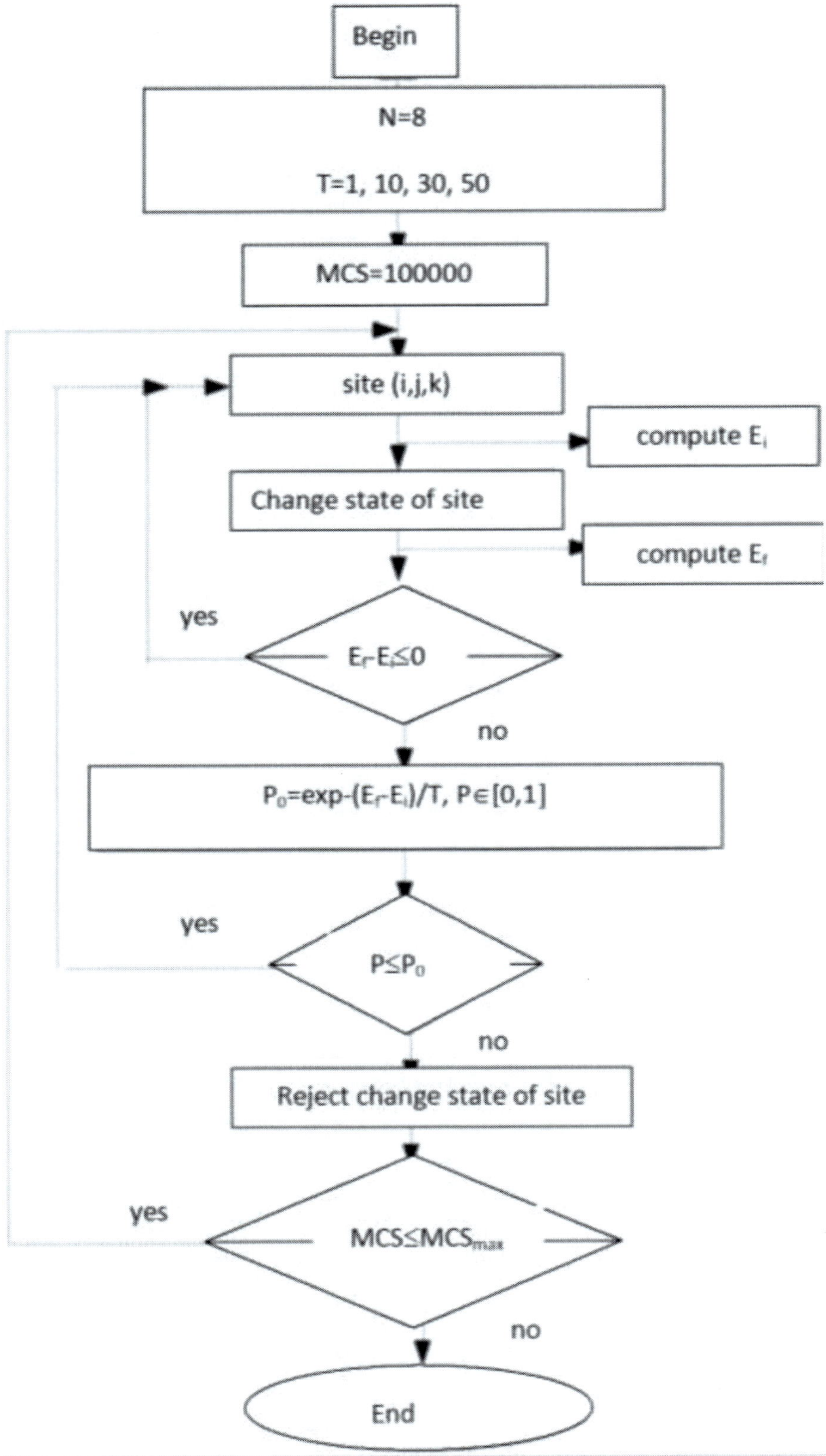

Figure 1. Flow chart of Monte Carlo simulations.

$$M=\left\langle \frac{1}{NNN}\sum_i S_i \right\rangle = \frac{1}{NNN}\sum_i \langle S_i \rangle = \frac{1}{NNN}\frac{\sum_i S_i e^{-\beta E}}{\sum_i e^{-\beta E}} \quad (3)$$

with E is the internal energy per site,

$$E=\frac{1}{NNN}\langle H \rangle \quad (4)$$

with

$$E=\frac{1}{NNN}\left\langle -J\sum_{\langle i,j\rangle} S_i S_j - H\sum_i S_i - \Delta\sum_i S_i^2 \right\rangle \quad (5)$$

and the corresponding magnetic susceptibility is given by:

$$\chi = \beta\left(\langle M^2 \rangle - \langle M \rangle^2\right) \quad (6)$$

where $\beta = \frac{1}{k_B T}$, T denotes the absolute temperature and $k_B = 1$ is the Boltzmann's constant.

Applications

Magnetic Hysteresis Cycle Study of GdSb Alloys

We apply a standard sampling method to simulate the Hamiltonian given by Eq. (1). Cyclic boundary conditions on the layers lattice were imposed and the configurations were generated by sequentially traversing the lattice and making single-spin flip attempts. We have used the Eq. (2) to calculate the exchange interaction in GdSb. The obtained value is J≈0.3 K. The variation of the magnetizations and magnetic susceptibilities versus the temperature of GdSb in absence and presence with N=8 size and for the exchange interactions J≈0.3 K, are given in Figure 2. The Néel temperature obtained is T_N=30.8 K. This value is between those reported in [39] (T_N=31 K) and [40] (T_N=28 K). Figure 3 shows the dependency of the magnetization on the exchange interaction *J*, for a fixed size (N=8 sites), and for a fixed temperature (T=10 and 30 K) and for absence of external magnetic field and crystal field. For a low value of *J*, the magnetization increases rapidly, to reach saturation (M_S=3.5) and reach saturation faster, when the temperature increases. The variation of magnetization versus the crystal field (Δ) of GdSb for H=0 with N=8 sites and the temperatures (T=10, 30 and 50 K) is shown in Figure 4. For a Δ negative, the magnetization increases up to a certain value of the crystal field and decreases rapidly. It reaches rapidly saturation value for a high temperature and a high value of a crystal field. The magnetization decreases with decreasing crystal field.

The magnetic hysteresis cycle of GdSb is obtained for N=8 sites (see Figure 5), for a fixed values of temperature (T=1, 30 and 50 K), and for a fixed value of the exchange interaction J≈0.3 K. In our FORTRAN program, we increase the external magnetic field and compute the magnetization until reaching the magnetization saturation corresponding to a magnetic field value (H_{max}). After that, we decrease the external magnetic field from H_{max} until reaching the inverted magnetization saturation ($-M_S$) which corresponds to ($-H_{max}$) field value. When decreasing the value of the external magnetic field, the magnetization undergoes zero value for the magnetic coercive field (H_C). For h=0, we found a remanent magnetization noted M_r. The saturation magnetization, coercive field and remanent magnetization are, respectively: $M_S=3.5$, $H_C=5$ and $M_r \approx 0$ and not influenced by the temperature.

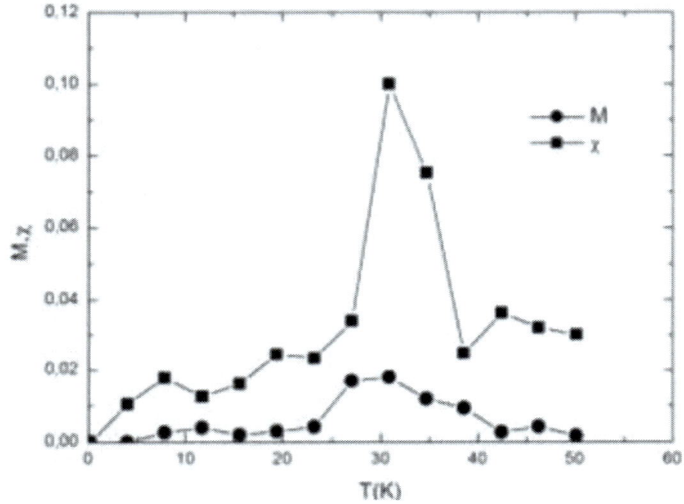

Figure 2. Magnetization and magnetic susceptibility versus the temperatures, of GdSb for h=0 and Δ=0 with N=8 sites.

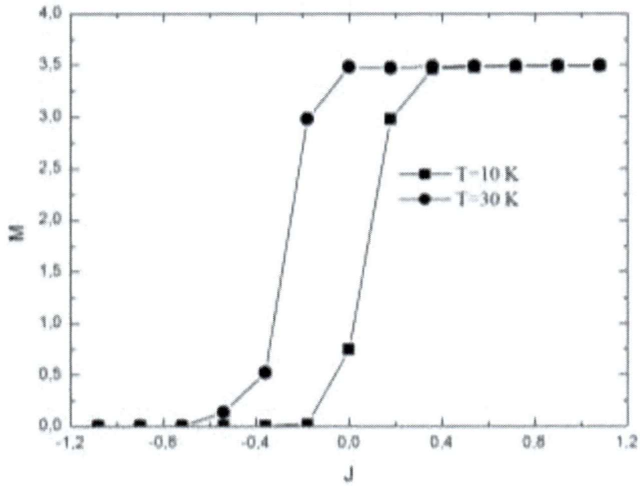

Figure 3. Magnetization versus the exchange interaction of GdSb for H=0 and Δ=0 with N=8 sites and different temperatures (T=10 and 30 K).

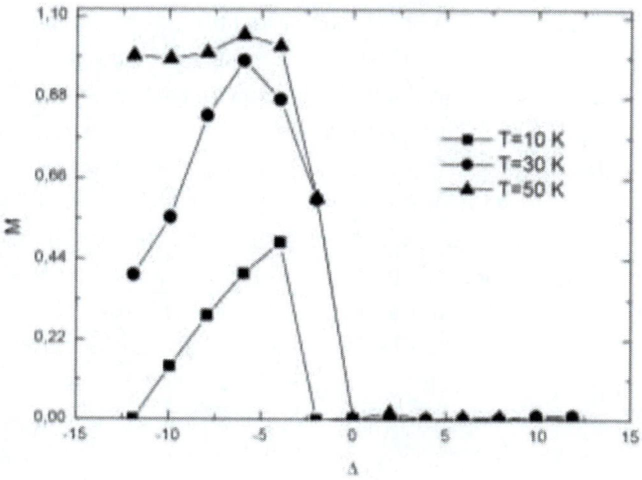

Figure 4. Magnetization versus the crystal field of GdSb for H=0 with N=8 sites and different temperatures (T=10, 20 and 50 K).

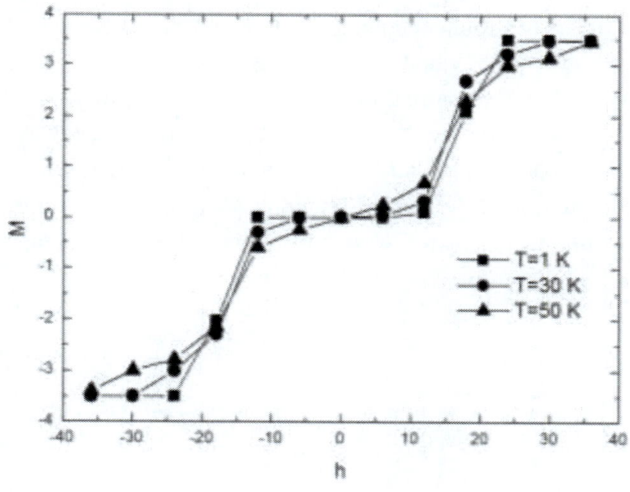

Figure 5. Magnetic hysteresis cycle of GdSb for different temperatures (T=1, 30 and 50 K) with N=8 sites.

Magnetic Hysteresis Cycle Study of Ni/Au Core/Shell

The cyclic conditions on the Au/Ni core shell are used for a fixed core/shell k, so that: 1≤ k≤ N, 1≤ i≤ N, and 1≤ j≤ N. Our data were generated with 10^5 Monte Carlo steps per spin, discarding the first 10^4 Monte Carlo simulations. Starting from different initial conditions, we performed the average of each parameter and estimate the Monte Carlo simulations, averaging over many initial conditions. The Eq (1) in the Au/Ni core shell case is given by:

$$H=-J_{Ni-Ni}\sum_{\langle i,j\rangle} S_iS_j-J_{Au-Au}\sum_{\langle i,j\rangle}\sigma_i\sigma_j-J_{Ni-Au}\sum_{\langle i,j\rangle nn'''} S_i\sigma_j-H\left(\sum_i\sigma_i+\sum_i S_i\right)-\Delta\left(\sum_i\sigma_i^2+\sum_i S_i^2\right) \quad (7)$$

To calculate the magnetization and magnetic susceptibility, we have used the Eqs. (3) and (6), respectively. We presented in Figure 6 the thermal total magnetizations and magnetic susceptibilities for a zero crystal field $\Delta=0$ and a zero external magnetic field H=0 of Au/Ni core shell. The sharp maximum in the total magnetic susceptibilities curves indicated a clear blocking behavior of the samples. The critical temperature increases from 6.28 to 15.88 K. The last value is comparable with those given by [41-43] (T_C=16K [42)]. This increasing indicating the coating enhanced the magnetic properties of the Au/Ni core shell. The total magnetization go from 1 at low temperature, to zero around 10 K and then become negative (~ -0.3) around 20 K. This change can be explained by the difference between the number of sites of Au and those of Ni. This reduction in moment might stem from the small size effect. Total magnetizations versus the external magnetic field for $\Delta=0$ and different values of the temperature of Au/Ni core shell for different temperatures are given in Figures. 7a and 7b. The total saturation magnetization of Ni/Au core/shell is equal 1 for a weak temperature and increases when the temperature increases. The same behavior is observed in the case of the coercive field decreases when increases the temperature (see Figures. 7a and 7b). We see from these figures, when the temperature in Ni/Au core shell becomes the superparamagnetic behavior at room temperature. This behavior is observed in [43,44]. In Figures. 8a and 8b, we have given the variation of the total magnetization versus the exchange interactions (J_{Ni-Au}). The total magnetization decreases when the exchange interaction J_{Ni-Au} increasing. The total magnetization decreases when the temperature and exchange interaction J_{Ni-Au} increasing. The dependence of total saturation magnetization and coercive field of the exchange interaction are observed also in last work [45]. The total magnetization versus the crystal filed (Δ) for different values of the temperature in Au/Ni core shell is given in Figure 9. The total magnetization in absolute value increases when a crystal filed increasing for $\Delta \geq -16$ and decreases when the temperature decreasing for $\Delta \leq -16$.

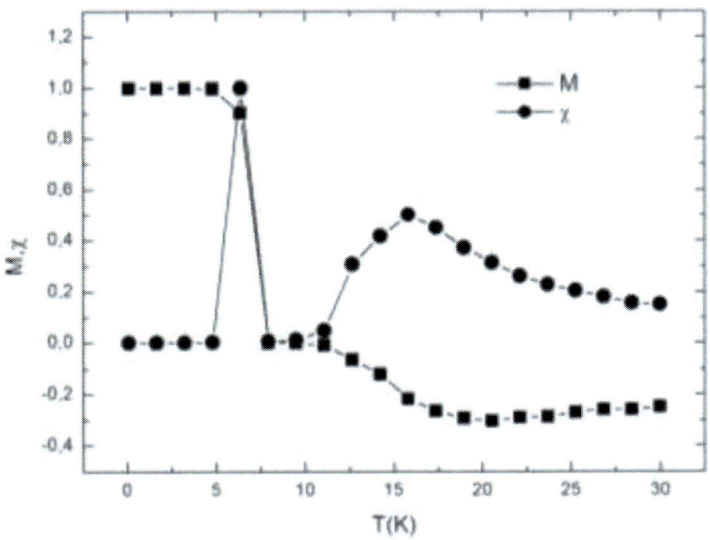

Figure 6. Total magnetizations and magnetic susceptibilities versus the temperature of Au/Ni core shell for $\Delta=0$, H=0, J_{Au-Au}=5 K, J_{Ni-Ni}=7.07 K and J_{Ni-Au}=-3 K.

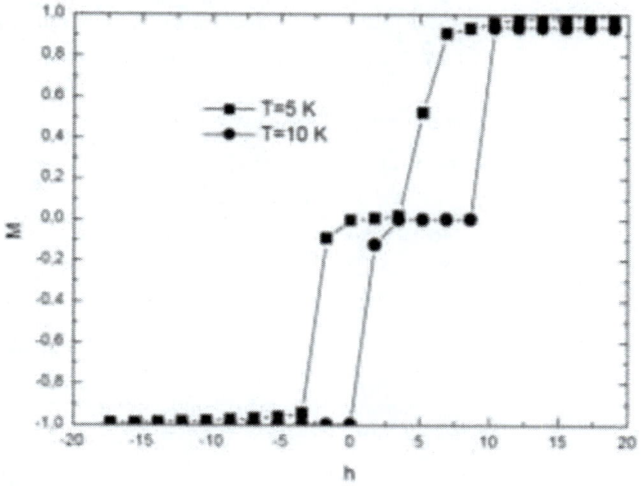

Figure 7a. Total magnetizations versus the external magnetic field for Δ=0 and for different values of the temperature of Au/Ni core shell with J_{Au-Au}=5 K, J_{Ni-Ni}=7.07 K and J_{Ni-Au}=-3 K.

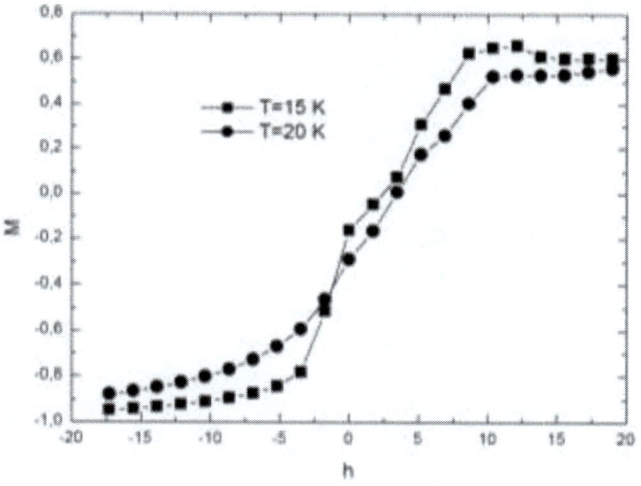

Figure 7b. Total magnetizations versus the external magnetic field for Δ=0 and for different values of the temperature of Au/Ni core shell with J_{Au-Au}=5 K, J_{Ni-Ni}=7.07 K and J_{Ni-Au}=-3 K.

Magnetic Hysteresis Cycle Study of MnAu Superlattice

The cyclic boundary conditions on the MnAu superlattices were imposed and the configurations were generated by sequentially traversing the lattice and making single-spin flip attempts. The flips are accepted or rejected according to a heat-bath algorithm under the Metropolis approximation (see flow chart of Monte Carlo simulations in annex). Cyclic conditions on the MnAu superlattice are used for a fixed layer k, so that: $1 \leq k \leq L$, $1 \leq i \leq N$ and $1 \leq j \leq N$. Our data are generated with 10^5 Monte Carlo steps per spin, discarding the first 10^4 Monte Carlo simulations. Starting from different initial conditions, we performed the average of each parameter and estimate the Monte Carlo simulations, averaging over many initial conditions. The Eq (1) in the MnAu superlattices case is given by:

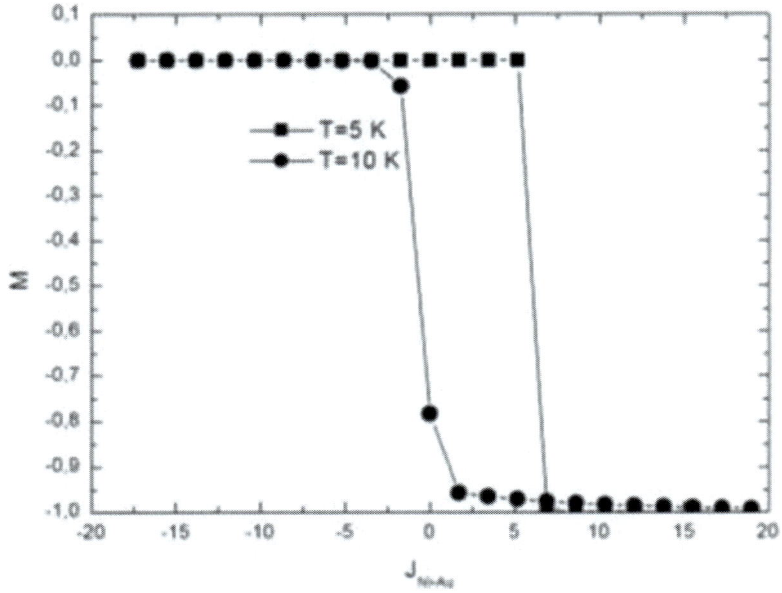

Figure 8a. Total magnetizations versus the exchange interaction ($J_{Ni\text{-}Au}$) of Au/Ni core shell for Δ=0, T=5 K, T=10 K, $J_{Au\text{-}Au}$=5 K and $J_{Ni\text{-}Ni}$=7.07 K.

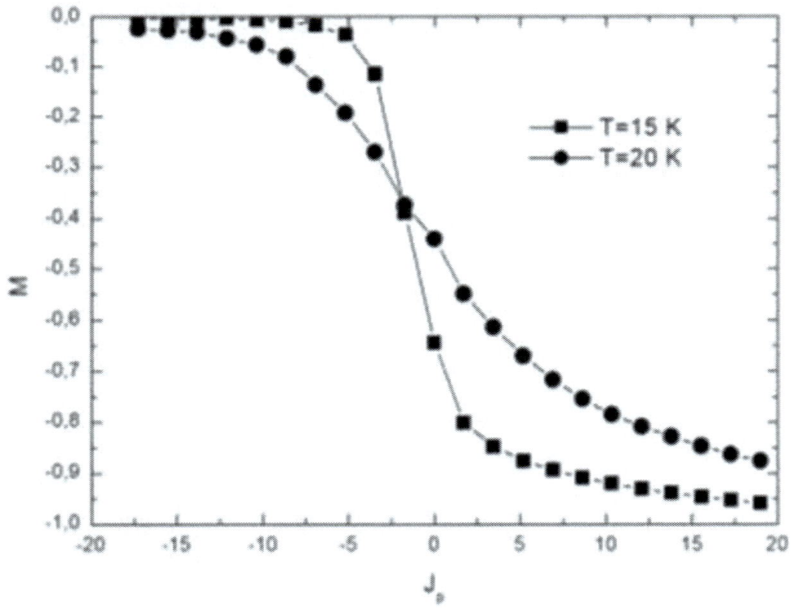

Figure 8b. Total magnetizations versus the exchange interaction ($J_{Ni\text{-}Au}$) of Au/Ni core shell for Δ=0, T=15 K, T=20 K, $J_{Au\text{-}Au}$=5 K and $J_{Ni\text{-}Ni}$=7.07 K.

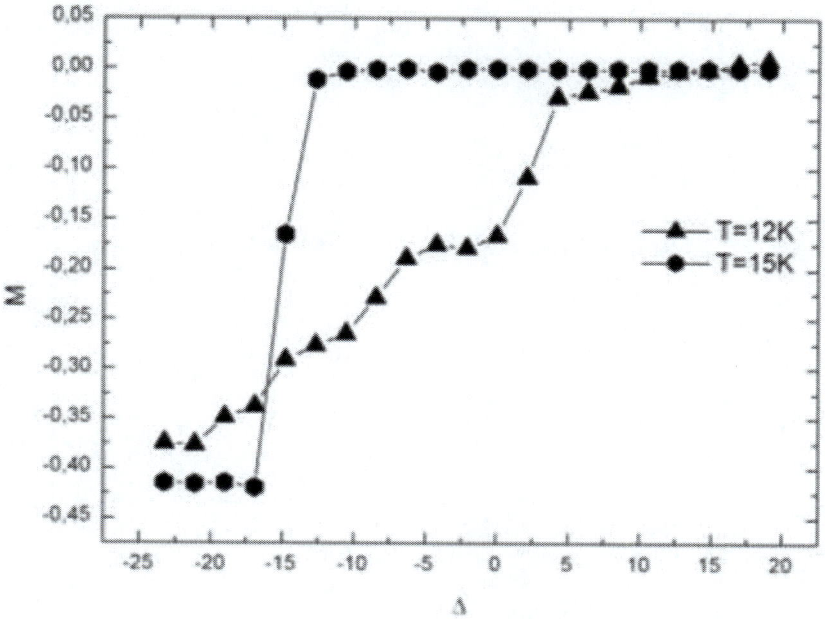

Figure 9. Total magnetizations versus the crystal filed (Δ) of Au/Ni core shell for T=12 K, T=15 K, J_{Au-Au}=5 K, J_{Ni-Ni}=7.07 K and J_{Ni-Au}=-3 K.

$$H = -J_a \sum_{<i,j>i,j\in k(plane)} S_i^k S_j^k - J_c \sum_{<i,j>i\in k,j\in(k+1)plane} S_i^k S_j^{k+1} - H \sum_{i\in\varepsilon(plane)} S_i^k \quad (8)$$

where $\langle i, j\rangle$ stand for the first nearest neighbor exchange interaction between the sites i and j with i∈k(plane) and j∈k+1(plane). Ja (or Jb), and Jc, are the exchange between the sites the site Si-Sj. The external magnetic field H is applied over all the spins of the lattice structure in the z-direction. The spin moment of $Mn^{3+}$is S=5/2. Figure 10 shows the variation of the thermal magnetization for antiferromagnetic MnAu super-lattices, for a zero external magnetic field and the exchange interaction (J_a=J_b=-23.85 K and J_c=-3.72 K) with different sizes (N_x=N_y=N). The Néel temperature is obtained (T_c=120 K). The peak of magnetic susceptibility is situated at the Néel temperature. At very low temperature, the magnetization M_S=5/2. This is in good agreement with the stand state phase diagrams. In Figure 11, we presented the magnetic hysteresis cycle of MnAu superlattice with a temperature T=120 K. From this figure, we deduced the magnetic coercive is H_C=200 and the saturation magnetization M_S=5/2. Figure 12 shows the variation of the magnetization versus the interface exchange interaction (J_c) between the MnAu superlattices for N=10, L=4 layers. For a J_c negative the magnetization it remains constant. For a J_c positive the magnetization increases up to its saturation value with absence of external magnetic field. The magnetization decreases up to its saturation value (-5/2) with a presence of external magnetic field.

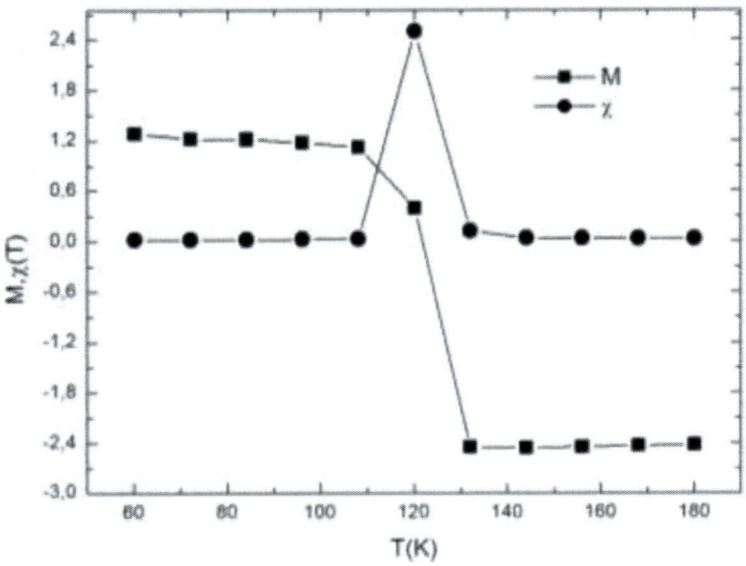

Figure 10. Magnetization and magnetic susceptibility versus the temperature of MnAu superlattices for H=0, N=10, L=4 layers.

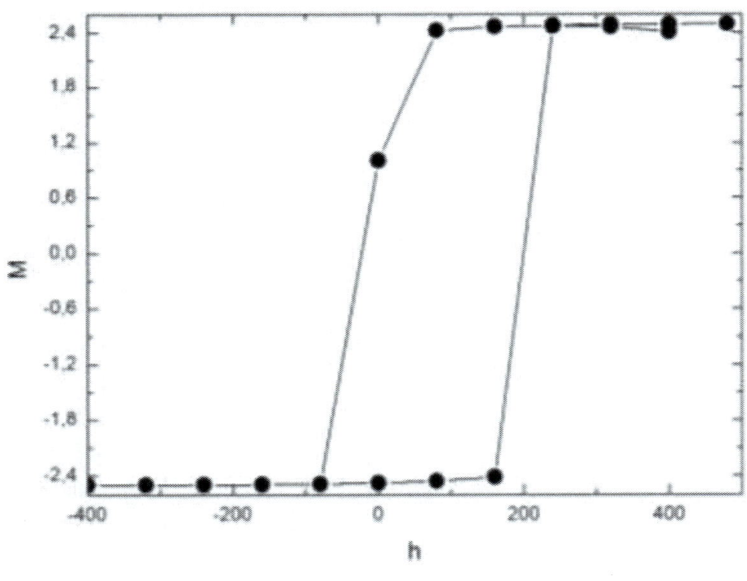

Figure 11. Magnetic hysteresis of MnAu superlattice with N=10, L=4 layers and for T=120 K.

Experimental Study of Magnetic Hysteresis Cycle

Nanoferrites

Magnetic properties of $CoFe_2O_4$ and $MnFe_2O_4$ nanoferrites

The magnetization as a function of temperature in applied field of 500 Oe (0.05T) for $CoFe_2O_4$ and $MnFe_2O_4$ ferrite nanoparticules are shown in Figure 13. It can be clearly seen that all the systems show ferromagnetic behavior. The corresponding Curie temperatures for $CoFe_2O_4$ and $MnFe_2O_4$ ferrites are 810 K and 620 K, respectively. It is clear that the Curie temperature for Co-doped ferrite is greater than that observed for Mn-doped ferrite nanoparticles. This is because the super-exchange interactions through oxygen atoms as J_{AB} (A–O–B) for $CoFe_2O_4$ more interesting than that for $MnFe_2O_4$.

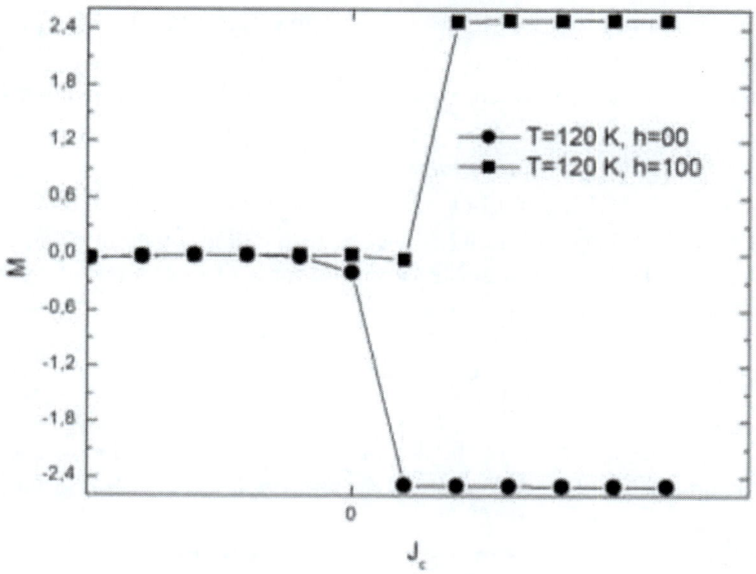

Figure 12. Magnetization versus the exchange interaction J_c for MnAu superlattice with unit- N=10, L=4 layers.

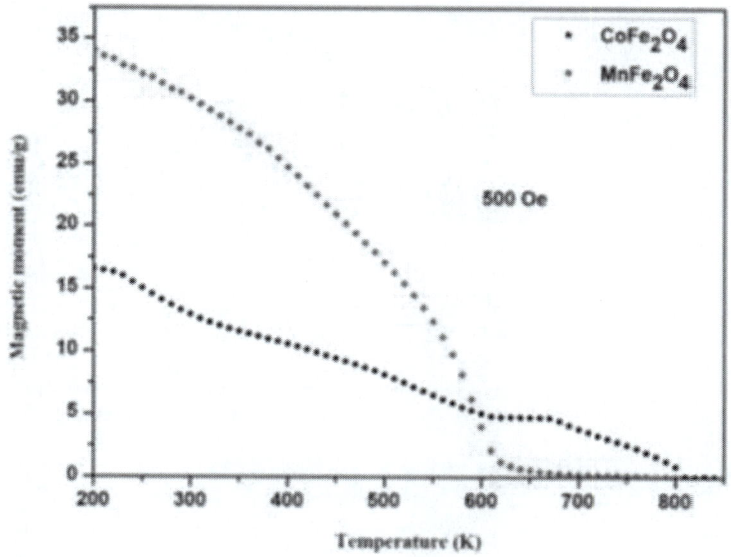

Figure 13. Temperature dependence of magnetic moment for $CoFe_2O_4$ and $MnFe_2O_4$ nanoparticles systems under an applied magnetic field of 500 Oe.

Figure 14 shows the magnetic hysteresis curves for studied nanoparticle ferrites at 300 K. It is known that magnetization is related with the magnitude of electron magnetic dipole moment. In this context, keeping in mind that Mn has 5 μ_B and Co has 3 μ_B as their magnetic dipole moments. From Figure 14, it is clear that the observed magnetization for Mn-doped ferrite is greater than that observed for Co-doped ferrite nanoparticles. In contrast, a different situation is observed in the coercive field. However, the measured coercive field for $MnFe_2O_4$ is lower than that measured for $CoFe_2O_4$; this is possible since that the Mn is a soft magnetic material in regard to Co. The coercive field value, the saturation magnetization and the remanent magnetization were extracted from the M-H curves and given in table 1.

Magnetic properties of Fe_3O_4 nanoferrites

The zero field cooled (ZFC) and field cooled (FC) curves are shown in Figure 15, which have low magnetic anisotropy of Fe_3O_4. The corresponding blocking temperature of the nanoparticles is 150K. The magnetic anisotropy constant K of the samples as-prepared can be calculated by the followed formula [46]:

$$K = 25k_BT_BV^{-1} \tag{9}$$

T_B is the blocking temperature of the samples, and V is the volume of a single particle. The calculated magnetic anisotropy constant K of our sample is 0.87 10^6 ergs/cm^3.

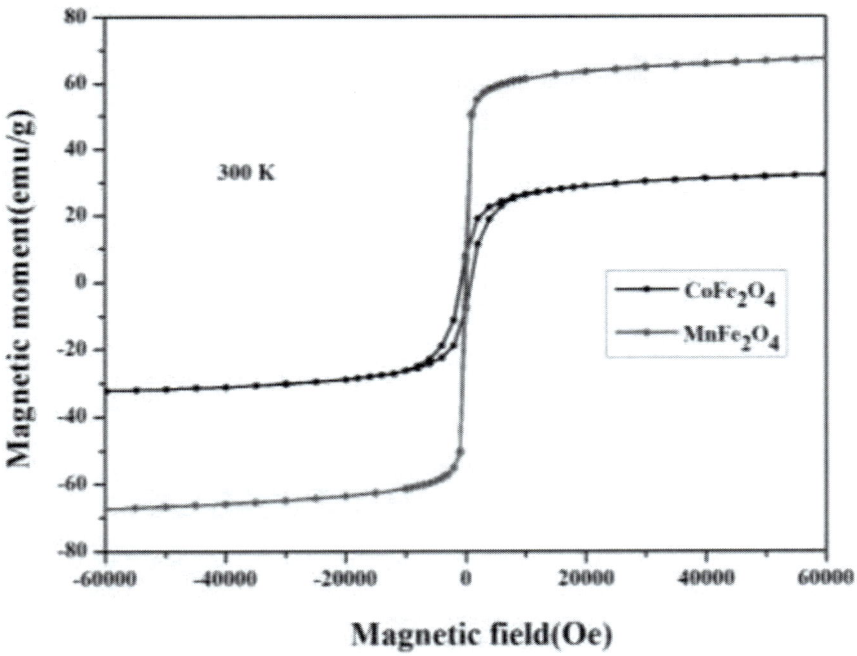

Figure 14. The magnetic hysteretic curves for $CoFe_2O_4$ and $MnFe_2O_4$ ferrites systems at 300K.

Figure 16 shows the variation of magnetization M (emu /g) with the applied magnetic field H (Oe) at 300 K for the Fe_3O_4 sample. The values of magnetization, remanent and coercive field measured at room temperature (300 K) are M_s=65.04 emu/g, M_r=2.53 emu/g and H_c=41.02 Oe, respectively.

Table 1. The value of the saturation magnetization, the remanent magnetization and the coercive field value for $CoFe_2O_4$ and $MnFe_2O_4$ nanoparticles systems

samples	M_s(emu/gr)	M_r(emu/gr)	M_r/M_s	H_c(Oe)
$CoFe_2O_4$	32	10	0.31	800
$MnFe_2O_4$	67	12	0.17	60

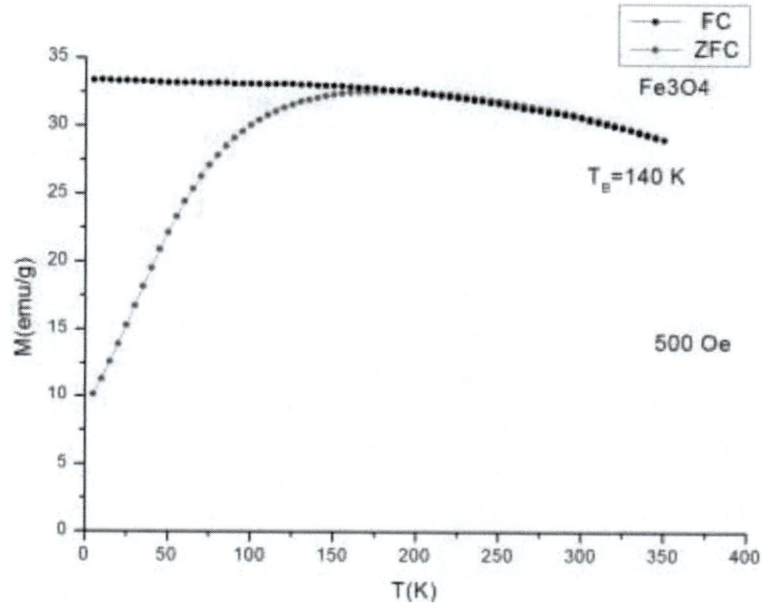

Figure 15. Zero-field-cooled (ZFC) and field-cooled (FC) curves for the as-synthesized Fe_3O_4 nanoparticles under an applied magnetic field 500 Oe.

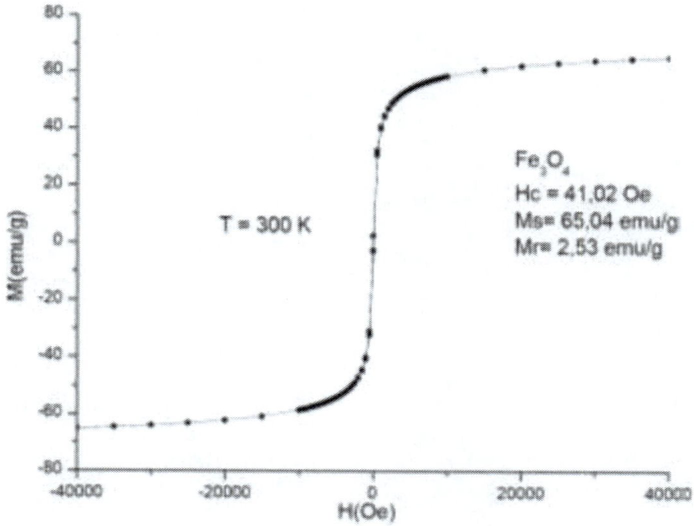

Figure 16. The magnetization M (emu/g) versus the applied magnetic field (Oe) at 300K for Fe_3O_4 nanoparticles.

Magnetic hysteresis cycle measurement of nanoferrites $NdFe_2O_4$

Figures 17 and 18 show the magnetic hysteresis curves for $NdFe_2O_4$ ferrite nanorods with different calcinations temperatures ($NdFe_2O_4$_400°C and $NdFe_2O_4$_600°C) at 5K and 300K. It is clearly observed that the magnetization and coercivity are both increased with diameter of nanorods. This may be elucidated by Bloch walls of the nanorods have increased with the growth of grains. It is known that the energy of a magnetic particle in external field is proportional to its particle sizes via the number of magnetic molecules in a single magnetic domain. With the increasing heat treatment temperature, the particle sizes (diameter of nanorods) will be increased and the crystallization will be more complete. The $NdFe_2O_4$ nanorods were formed through the agglomeration of magnetic nanoparticles. At 300 K, the saturation magnetization values of $NdFe_2O_4$ nanorods annealed at 400 °C are 26 emu/g, which are smaller than $NdFe_2O_4$ rods annealed at 600 °C values of 34 emu/g. The agglomeration of magnetic nanoparticles increased interaction between them. The reduction of Ms relatively to $NdFe_2O_4$ nanorods annealed is mainly due to the decreased particle size and spin canting at the surface of ferrite $NdFe_2O_4$. From the reported results, it can be seen that magnetic properties are also influenced by the synthesis process. Obtained M_s values for neodymium spinel ferrites nanorods are varying between 26 and 34 emu/g. Ms values for $NdFe_2O_4$ calcined at 400°C and 600°C are found to be 26 emu/g and 34 emu/g, respectively. Our results are in agreement with the literature if the reduced net magnetic moment and/or reduced Ms for Fe_3O_4 and $NiFe_2O_4$ ferrites upon rare-earth ion substitution are taken into account [47, 48]. The coercive field value, the saturation magnetization and the remanent magnetization are given in table 2.

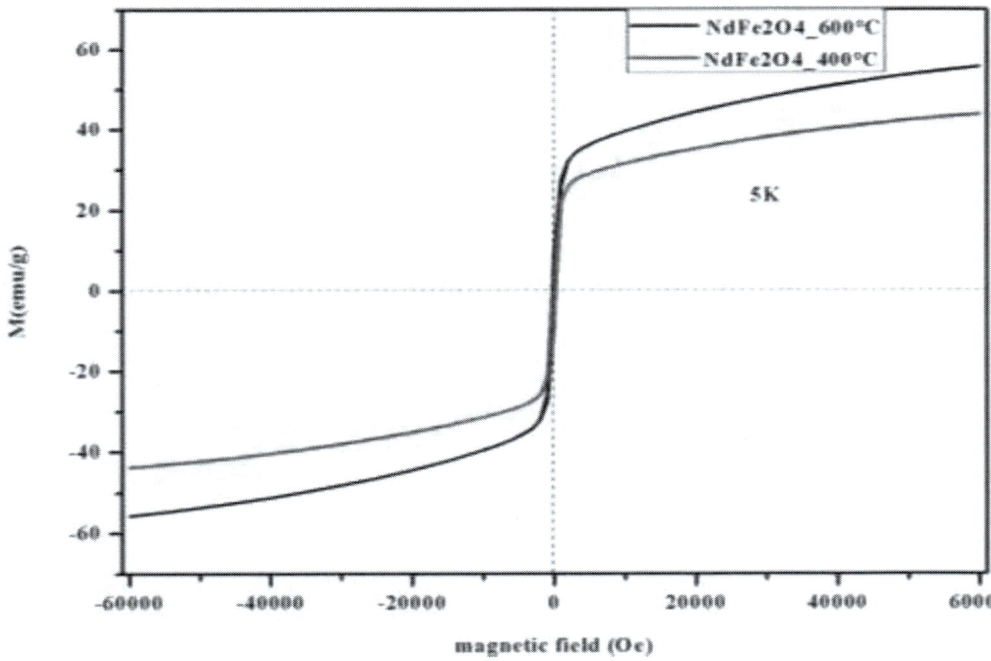

Figure 17. The magnetic hysteretic curves for $NdFe_2O_4$ nanorods at 5K.

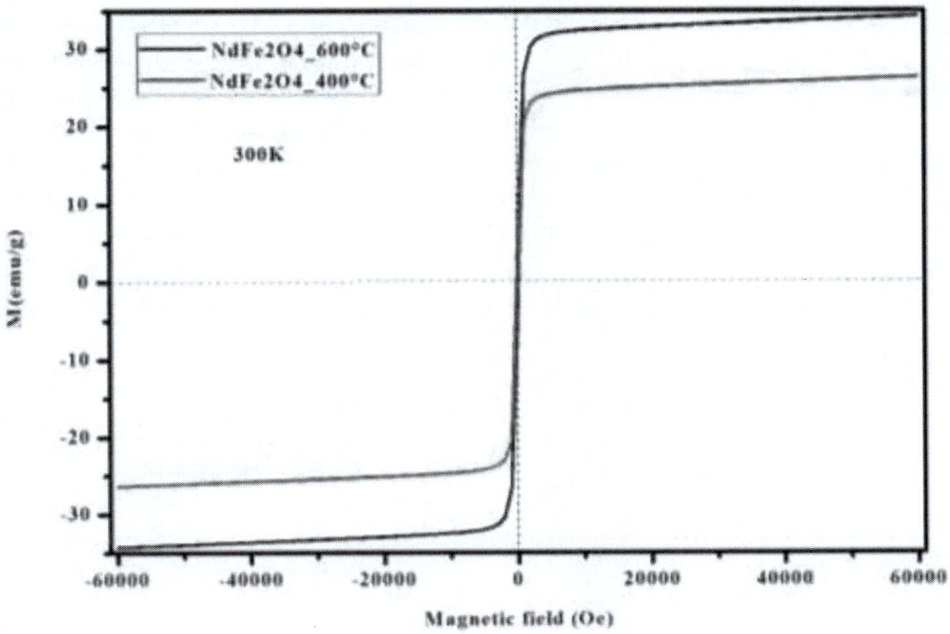

Figure 18. The magnetic hysteretic curves for $NdFe_2O_4$ nanorods at 300K.

Table 2. Magnetic parameters of $NdFe_2O_4$ nanorods

Samples	H_c (Oe)	M_s(emu/g)	M_r(emu/g)	M_r/M_s
$NdFe_2O_4$_ 400 °C at 5K	200	44	6	0.14
$NdFe_2O_4$ _600 °C at 5K	270	56	10	0.18
$NdFe_2O_4$ _ 400 °C at 300 K	151	26	4.22	0.16
$NdFe_2O_4$ _ 600 °C at 300K	171	34	6.55	0.19

The magnetic properties of $NdFe_2O_4$ nanorods ($NdFe_2O_4$_400°C and $NdFe_2O_4$_600°C) have been studied by measuring the magnetization as a function of temperature in applied field of 500 Oe. Typical M–T for both samples curve are shown in Figure 19. where it can be seen that the magnetization in the temperature range of 200–500K is stable and decreases sharply with the increase of the temperature in range 500K to 900 K. The corresponding Curie temperatures are shown in table 3. Indeed, Figure 20 shows a plot of inverse dc magnetic susceptibility ($1/\chi$=H/M) as a function of temperature for $NdFe_2O_4$. As can be seen from this figure, the plot follows a linear Curie–Weiss type behaviour, χ=C/ [T-θ]. The corresponding Curie–Weiss temperatures for both samples are presented in table 3. magnetic field of 500Oe.

Figure 21 shows the zero field cooled (ZFC) and field cooled (FC) curves of nanorods NdFe2O4 ($NdFe_2O_4$_400°C and $NdFe_2O_4$-600°C) under an applied magnetic field of 500 Oe. The blocking temperature (T_B) of both samples decreases with increases diameter of nanorods. The blocking temperature of our samples is given in table 4. Our result is more impotent than the earlier reports on nanocrystalline spinel ferrites [49]. Magnetic anisotropy has been a key issue in understanding the relaxation behaviors of nanosized particle systems. Magnetization vector in a particle prefers a direction along a crystallographic axis, called the easy axis, which corresponds to the minimum of magnetic anisotropy energy. This anisotropy energy acts as an energy barrier that limits the rotation of magnetic moments away from the easy axis. At room temperature, it is possible for the moments of noninteracting, single-domain particles to escape over the energy barrier by means of thermal agitation with relaxation time, given by the Néel–Arrhenius equation: $\tau = \tau_0 \mathrm{esp}(\Delta E_a / k_B T)$. where $\tau_0$is the attempt time of the order of 10^{-9}– 10^{-13} s. In the simplest case, $\Delta E_a = KV$ is the anisotropy energy barrier for small applied field. Table 4 shows the value of blocking temperature (T_B), magnetic anisotropy constant, lattice volume (V), relaxation time and the anisotropy energy barrier applied magnetic field curves for the as-synthesized $NdFe_2O_4$ nanorods under an applied magnetic field of 500 Oe.

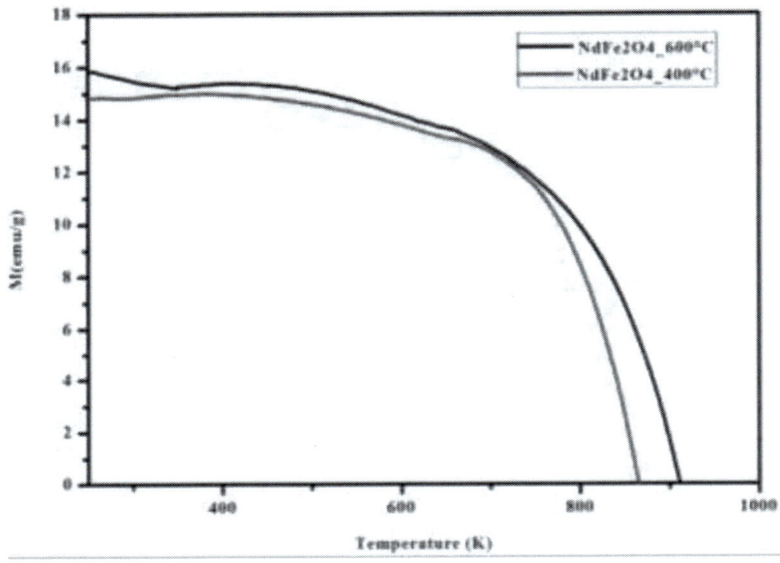

Figure 19. Temperature dependence of magnetization M (emu/g) for $NdFe_2O_4$nanorods.

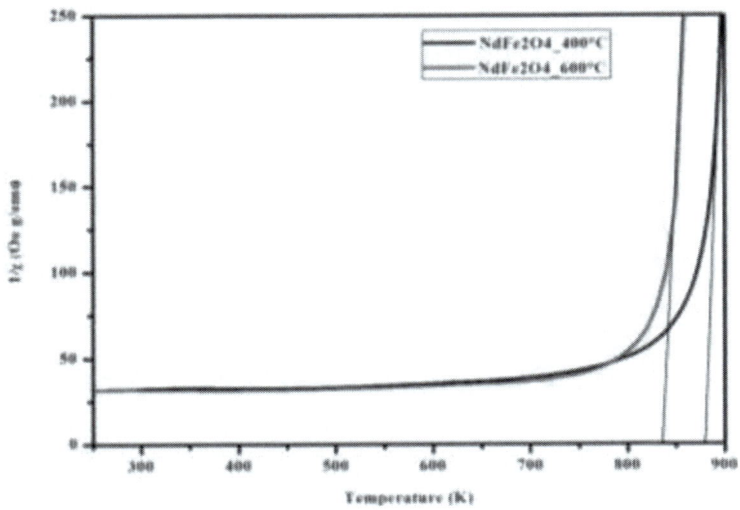

Figure 20. Temperature dependence of inverse AC magnetic susceptibility ($1/\chi$) for $NdFe_2O_4$ nanoparticles. Transition temperature is indicated by the straight lines.

Table 3. The value of Curie temperatures (T_c) and Curie–Weiss temperatures (θ_c) of $NdFe_2O_4$ nanorods

Samples	T_C(k)	θ_c(k)
$NdFe_2O_4$ _400°C	870	835
$NdFe_2O_4$_600°C	911	880

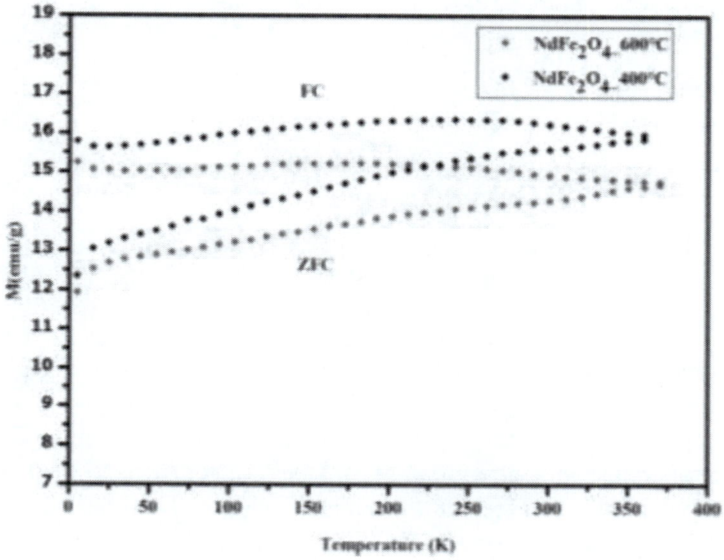

Figure 21. Zero-field-cooled (ZFC) and field-cooled (FC) curves for the as-synthesized $NdFe_2O_4$ nanonods under an applied magnetic field of 500 Oe.

Table 4. The value of blocking temperature (T_B), magnetic anisotropy constant, lattice volume (V), relaxation time and the anisotropy energy barrier applied magnetic field curves for the as-synthesized $NdFe_2O_4$ nanorods under an applied magnetic field of 500 Oe

samples	T_B(k)	K* 10^{+5} (ergs/cm^3)	ΔE_a 10^{-5} (ergs)	τ (ns)
$NdFe_2O_4$_400°C	370	2.823	1277.42	24.58
$NdFe_2O_4$_600°C	362	0.821	1249.80	12.62

Magnetic hysteresis cycle measurement of $SnFe_2O_4$ nanoferrites

Figure 22 shows the variation of magnetization M (emu /g) with the static applied magnetic field H recorded at 3K, 25K, 125K and 325K for the $SnFe_2O_4$ sample. As a normal behavior, the magnetization increases with increasing the applied magnetic field and attains its saturation value for fields higher than 1 KOe. The dependence of saturation magnetization M_s, determined by extrapolation of the magnetization curve to H=0. Figure 21 showed the saturation magnetization increase remarkably with decreasing temperature and Temperature dependence of remanent magnetization for the $SnFe_2O_4$.

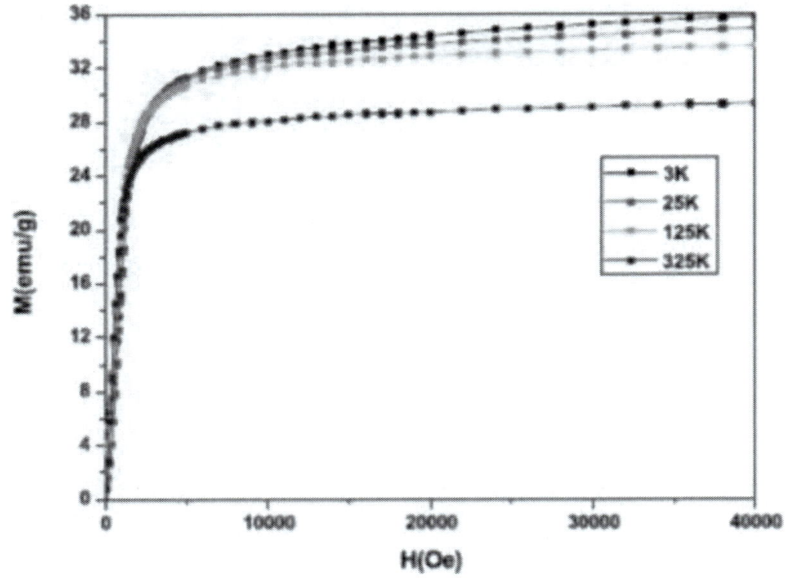

Figure 22. The magnetic hysteretic curves for $SnFe_2O_4$ ferrites systems at different temperatures.

The magnetic properties of $SnFe_2O_4$ have been studied by measuring the magnetization as a function of temperature in applied field of 500 Oe (0.05T). Typical M–T curve is shown in Figure 23, from which one can see that the magnetization in the temperature range of 200–700K increases sharply with the decrease of the temperature.

The value of the saturation magnetization, the remanent magnetization and the coercive field value for $SnFe_2O_4$ nanoparticles are given in table 5.

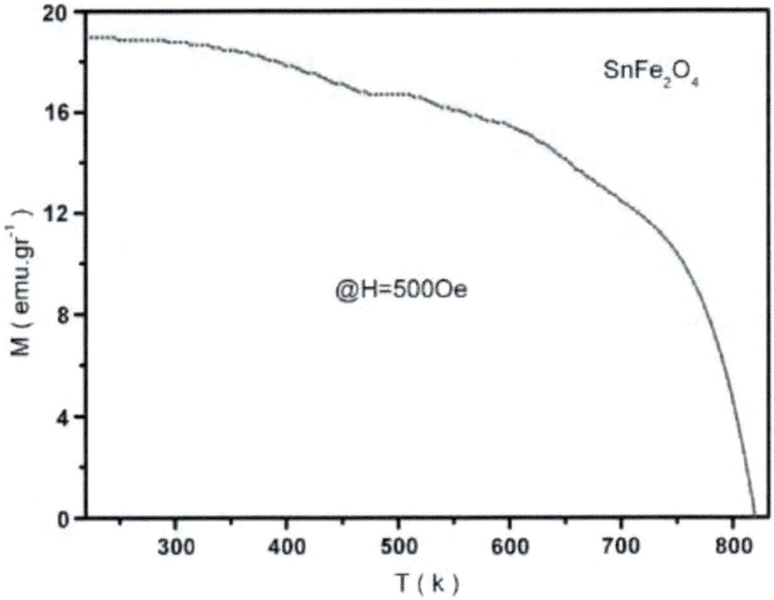

Figure 23. Temperature dependence of magnetic moment for $SnFe_2O_4$ nanoparticles systems under an applied magnetic field of 500 Oe.

Table 5. The value of the saturation magnetization, and the remanent value for $SnFe_2O_4$ nanoparticles systems

Samples	M_s(emu/g)	M_r(emu/g)	M_r/M_s
$SnFe_2O_4$	29	0.5	0.017

Diluted Magnetic Oxides (DMOs)

Hysteresis of Spin glass SnO_2 based diluted magnetic oxides

The magnetic characterization of $Sn_{0.95}Mn_{0.025}Co_{0.025}O_2$ was done by Magnetic Properties Measurement System (MPMS-7XL, Quantum Design, Inc). Figure 24 shows the magnetic field dependence of the magnetization recorded at 5K, and 300K for the $Sn_{0.95}Mn_{0.025}Co_{0.025}O_2$ bulk sample. In side of Figure 24 a plot of inverse dc magnetic susceptibility ($1/\chi$=H/M) as a function of temperature for $Sn_{0.95}Co_{0.025}Mn_{0.025}O_2$ in a field of 10000 Oe was shown. The inverse magnetic susceptibility follows a linear dependence with T at high temperatures; the resulting values of the Curie-Weiss temperature were negative, this means that the magnetic impurities (Mn and Co) are coupled antiferromagnetically. This interaction yields frustration of magnetic moments, which is believed to be a possible reason for a spin-glass (SG) phase. In addition the saturation magnetization increase remarkably with decreasing temperature, the thorough magnetic saturations of $Sn_{0.95}Mn_{0.025}Co_{0.025}O_2$ samples are still not achieved at the applied field of 60 kOe. The lack of saturation of moment reflects the fact that the studied alloy is not a ferromagnetic but a spin glass. Therefore, the value of coercivity (H_c) decreases with increasing temperature. This interaction yields frustration of magnetic moments, which is believed to be a possible reason for a spin-glass (SG) phase.

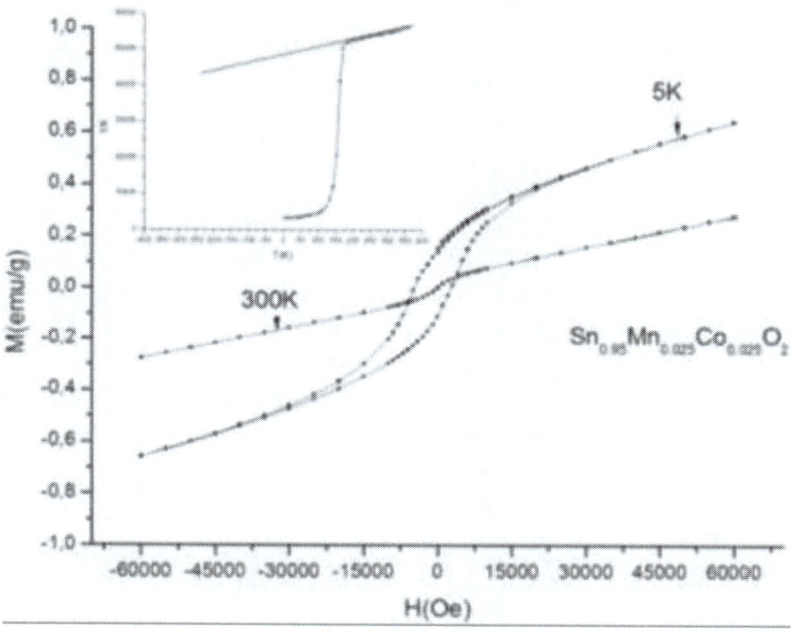

Figure 24. Magnetic hysteresis cycle of $Sn_{0.95}Mn_{0.025}Co_{0.025}O_2$.

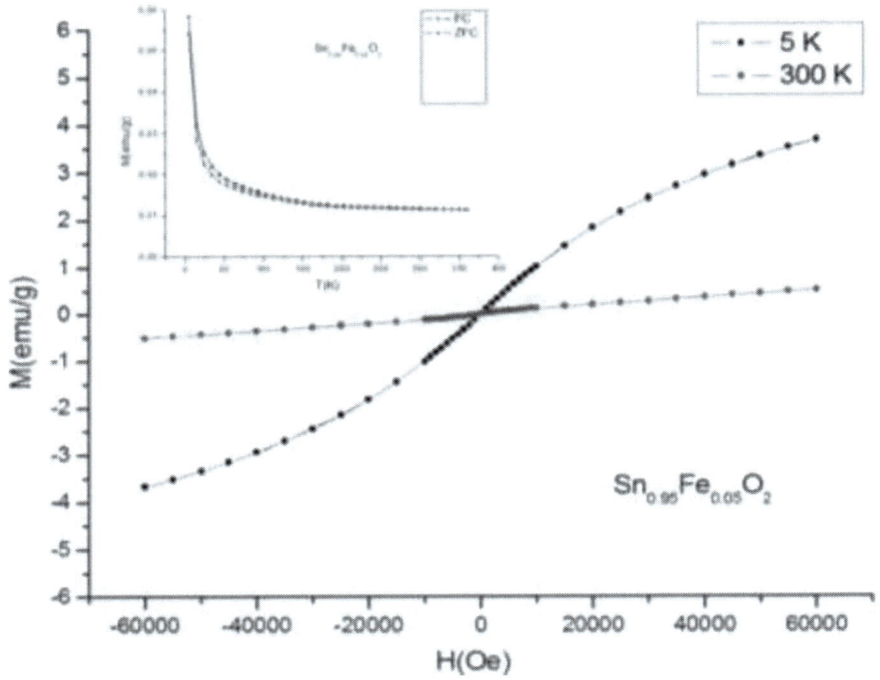

Figure 25. Magnetic hysteresis cycle of $Sn_{0.95}Fe_{0.05}O_2$.

Inside of Figure 25 shows magnetization versus temperature of $Sn_{0.9}Fe_{0.05}O_2$ nanocrystals in a field of 500Oe, the magnetization (M) increases rapidly with the decrease of temperature (T) and shows a small distinct splitting between ZFC and FC measurement at low temperature, implying the presence of soft ferromagnetic character. The concave nature of magnetization versus temperature, M(T) shown in Figure 25, and the behaviour with steep rise observed at low temperature was explained in terms of polaron-percolation-theory. It is worth noting that at low temperature, the steep rise in magnetization seems to be intrinsic to diluted magnetic semiconductors. This behaviour is due to ferromagnetism in clusters, a percolation type of this clusters immured a back ground of paramagnetic or spin glass moment, grow in size as the temperature is lowered until long-range order extends through the whole system. As suggested in the case of other magnetic oxides, the absence of ferromagnetism at high temperature can be due to the absence of free carriers which mediate the interaction between magnetic ions. Figure 25 shows the magnetic field dependence of the magnetization (M, H) recorded at 5 K for $Sn_{0.9}Fe_{0.05}O_2$. The $Sn_{0.9}Fe_{0.05}O_2$ nano-powders have a magnetization loop typical for paramagnetic systems with no indication of long range ferromagnetism, the coercive field and the remanent magnetization being very close to zero. The paramagnetic or spin glass behaviour, being linear, has been fitted to a straight line with an intercept. Since the intercepts are identical for positive-and negative-field regions, the high-field exhibits the paramagnetic behaviour.

Hysteresis of ferromagnetic SnO2 based diluted magnetic oxides

The magnetic properties of $Sn_{0.95}Co_{0.05}O_2$ nanocrystals have been studied by measuring the magnetization as a function of temperature in applied field of 500 Oe. (M,T) curves are shown inside of Figure 26, which have ferromagnetic character. The corresponding Curie temperature is Superior at room temperature (T_c=330K). Figure 26 shows magnetic field dependence of magnetic moment at 5K and room temperature (300K) of $Sn_{0.95}Co_{0.05}O_2$. This sample was found to be ferromagnetic at room temperature. This is due to the magnetic impurities (Cobalt) are coupled ferromagnetically. The saturation moment is (Ms=1.5 emu/g) and the coercive field is *(Hc=7Oe)* measured at room temperature (300 K). The appearance of ferromagnetism in the pristine Co doped SnO2 nanocrystalline may be attributed to effective exchange interactions between magnetic impurities (Cobalt) or by unpaired electron spins originating from the surface defects such as oxygen vacancy clusters instead of single neutral oxygen vacancies or by secondary phases.

Magnetic hysteresis loops of bulk $Sn_{0.94}Mn_{0.025}Co_{0.025}Fe_{0.01}O_2$ were made at 5K and 300K, are shown in Figure 27, the values of saturation magnetization (M_s), residual magnetization (M_r) and coercivity (H_c) decrease with increasing temperature. In comparison to $Sn_{0.95}Mn_{0.025}Co_{0.025}O_2$ doping with iron increases the saturation magnetization and decreases the magnetic field needed to reach it. Curie temperatures (Tc) were deduced from magnetization versus temperature measured using a SQUID magnetometer. Inside of Figure 27 shows magnetization versus temperature of $Sn_{0.94}Mn_{0.025}Co_{0.025}Fe_{0.01}O_2$ in a field of 500 Oe. The transition temperature is around 330K, its behaviour at low temperature is characteristic of ferromagnetic phase, which suggests the transition temperature from paramagnetic to ferromagnetic phases lower than room temperature. This means that the magnetic impurities (Mn and Co) are coupled ferromagnetically.

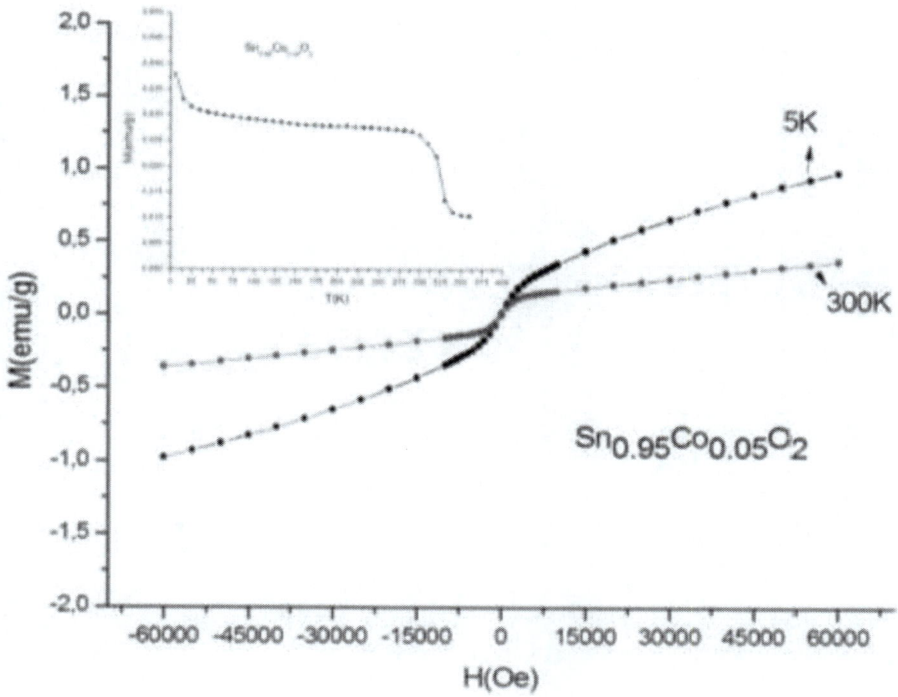

Figure 26. Magnetic hysteresis cycle of $Sn_{0.95}Co_{0.05}O_2$.

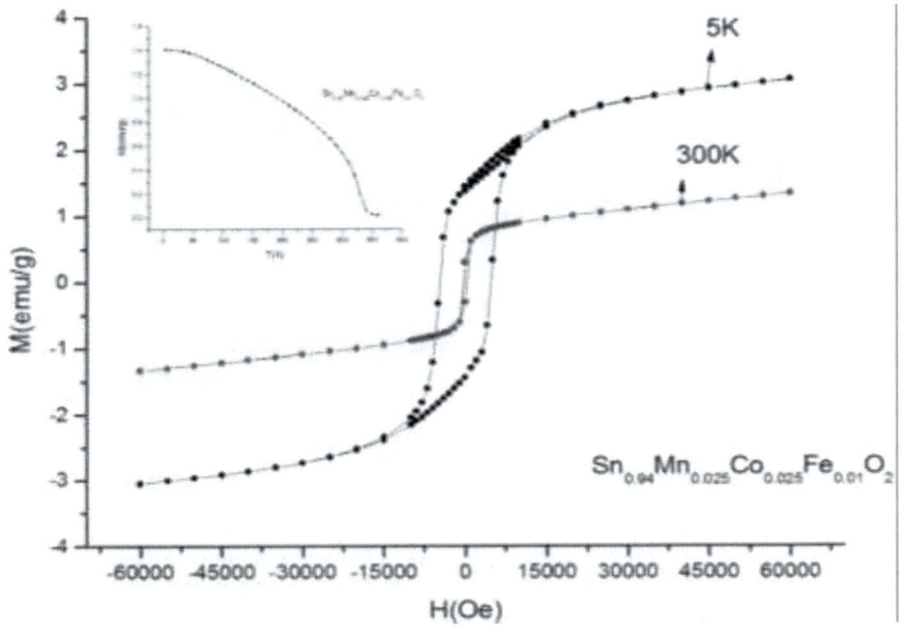

Figure 27. Magnetic hysteresis cycle of $Sn_{0.94}Mn_{0.025}Co_{0.025}Fe_{0.01}O_2$.

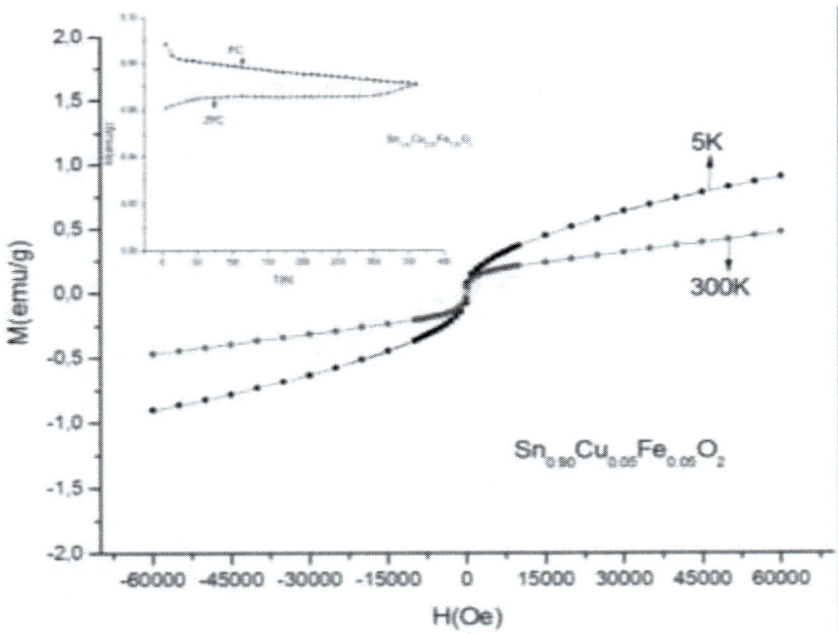

Figure 28. Magnetic hysteresis cycle of $Sn_{0.9}Cu_{0.05}Fe_{0.05}O_2$.

Hysteresis of superparamagnetic SnO_2 based diluted magnetic oxides

Magnetic hysteresis loops of $Sn_{0.9}Cu_{0.05}Fe_{0.05}O_2$ nanocrystals were made at 5K and 300K, they are shown in Figure 28, and the magnetization seems to approach saturation at low field.

The remanent and saturation magnetization decrease with increasing temperature, which indicate that the ferromagnetic coupling is enhanced remarkably from T=300K to 5K. Moreover, doping with Cu increases the saturation magnetization and decreases the magnetic field needed to reach it in comparison to $Sn_{0.9}Fe_{0.05}O_2$. Curie temperatures (Tc) were deduced from magnetization versus temperature measured using a MPMS. The bifurcation between ZFC and FC curves of $Sn_{0.9}Cu_{0.05}Fe_{0.05}O_2$ indicates a large thermal irreversibility with the onset near 360 K, which is characterized as the existence of superparamagnetism above room temperature, due to the nano-size effect. However, the coercivity H_C and the remanent M_r are very small, which suggest also that it will become superparamagnetic at higher temperatures.

Magnetic hysteresis loops of $Sn_{0.9}Mn_{0.05}Fe_{0.05}O_2$ nanocrystals were made at 5K and 300K, they are showing in Figure 29. The hysteresis loops close at low field and the magnetization seems to approach saturation at low field in the case of $Sn_{0.9}Mn_{0.05}Fe_{0.05}O_2$ and $Sn_{0.9}Cu_{0.05}Fe_{0.05}O_2$ nanocrystals. The remanence and saturation magnetization decrease with increasing temperature, which indicate that the magnetic coupling is enhanced remarkably from T=300K to 5K. Doping with Cu and Mn atoms increases the saturation magnetization and decreases the magnetic field needed to reach it in comparison to $Sn_{0.9}Fe_{0.05}O_2$. However, the coercivity H_c and the remanence M_r are very small, which suggest also that it will become superparamagnetic at higher temperatures.

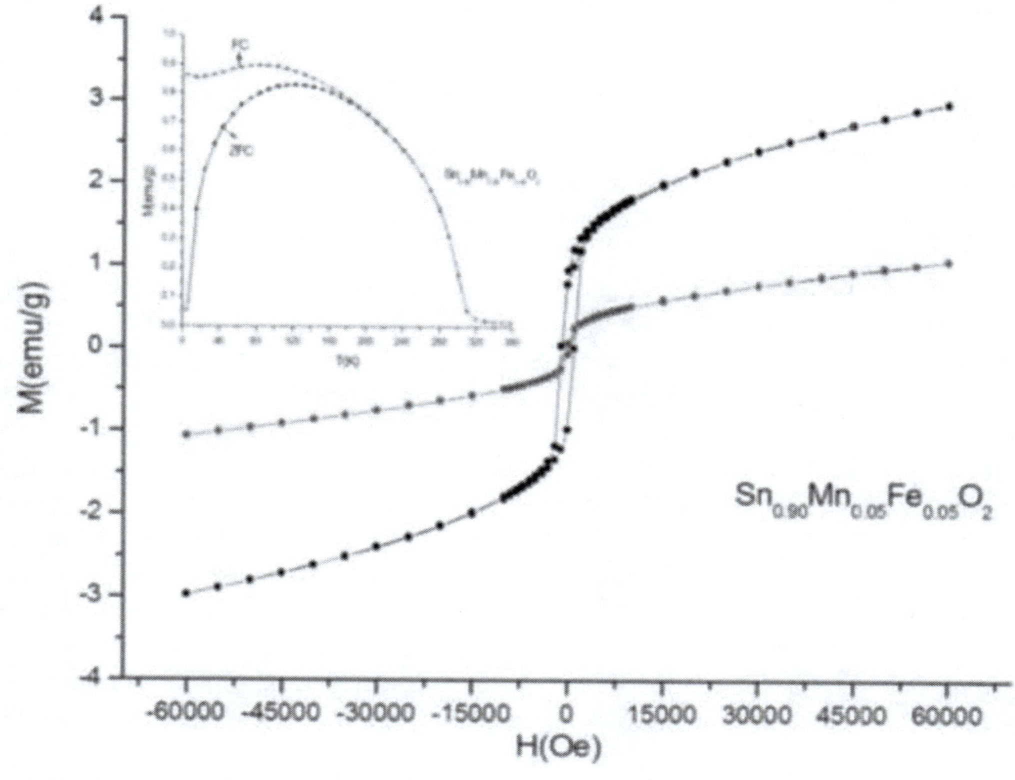

Figure 29. Magnetic hysteresis cycle of $Sn_{0.90}Mn_{0.05}Fe_{0.05}O_2$.

Conclusion

The magnetic properties of GdSb alloys, Au/Ni core shell and MnAu superlattice have been investigated using Monte Carlo simulations. The Néel or critical temperature of these systems is given and comparable with experiment results [40]. The magnetization versus exchange interactions is deduced for different values of the temperatures, zero external magnetic and crystal fields and a fixed sizes are obtained. The magnetic hysteresis cycle is deduced. The remanent, saturation magnetization and magnetic field coercive are deduced. The total magnetization increases in absolute value when the crystal field increases. In other hand the ferrite spinel MFe_2O_4 (M=Co, Mn, Fe, Nd and Sn) nanoparticles were prepared by the co-precipitation method. The effect of divalent atoms nature on the magnetic properties in spinel ferrites MFe_2O_4 (M=Co and Mn) was studied. Also, the effect of annealing temperature on the magnetic properties of $NdFe_2O_4$ has been investigated. The Magnetic parameters of all samples nanoparticles studied in this part were determinate. The ferromagnetic superparamagnetic and spin glass semiconductors (nanostructures and massive) are provided. Magnetic semiconductor is a tin dioxide co-doped and tri-doped. This compound is in the intrinsic ferromagnetic superparamagnetic and spin glass semiconductors phases. The magnetic phases are optimized by a suitable choice of the concentration and type of magnetic ion doping, doping and co-tri-doping.

References

[1] Natali, F., Ruck, B. J., Plank, N. O. V., Trodahl, H. J., Granville, S., Meyer, C. & Lambrecht, W. R. L. (2013). Rare-earth mononitrides. *Progress in Materials Science*, *58*, 1316-1360.

[2] Narita, A. (1988). Magnetic Susceptibilities and Spin Structures in Gadorinium Pnictides. I. Semiconductor Case. *J. Phys. Soc. Japan*, *57*, 251-260.

[3] Narita, A. (1988). Magnetic Susceptibilities and Spin Structures in Gadolinium Pnictides. II. Carrier Concentration Effect in the Semimetallic Case. *J. Phys. Soc. Japan*, *57*, 3961-3972.

[4] Larson, P. & Lambrecht, W. R. L. (2006). Electronic structure of Gd pnictides calculated within the LSDA+U approach. *Phys. Rev. B.*, *74*, 085108-085115.

[5] Johnston, R. L. (2002). *Atomic and Molecular Clusters*, Taylor and Francis, London.

[6] Spasova, M., Salgueirino-Maceira, V., Schlachter, A., Hilgendorff, M., Giersig, M., Liz-Marzan L. M. & Farle M. (2005). Magnetic and optical tunable microspheres with a magnetite/gold nanoparticle shell. *J. Mater. Chem.*, *15*, 2095-2098.

[7] Harigaya, K., Yamashiro, A., Shimoi, Y. & Wakabayashi, K. (2005). Magnetic vs. charge ordered states, and electric capacitance in zigzag nanographite ribbons. *Synth. Met.*, *152*, 317-320.

[8] Mornet S., Vasseur S., Grasset F. & Duguet E. (2004). Magnetic nanoparticle design for medical diagnosis and therapy. *J. Mater. Chem.*, *14*, 2161-2175.

[9] Ye, X. R., Lin, Y. H., Wang, C. M., Engelhard, M. H., Wang, Y. & Wai, C. M. (2004). Supercritical fluid synthesis and characterization of catalytic metal nanoparticles oncarbon nanotubes. *J. Mater. Chem.*, *14*, 908-913.

[10] Trewyn, B. G., Giri, S., Slowing, I. I. & Lin Victor, S.-Y. (2007). Mesoporous silica nanoparticle based controlled release, drug delivery, and biosensor systems. *Chem. Commun*, 3236-3245.

[11] Alivisatos, A. P. (1996). Semiconductor Clusters, Nanocrystals, and Quantum Dots. *Science*, *271*, 933-937.

[12] Murray, C. B., Kagan, C. R., Bawendi, M. G. & Annu. (2000). Synthesis and characterization of Monodisperse nanocrstals and close packed nanocrystal assemblies. *Rev. Mater. Sci.*, *30*, 545-610.

[13] Xia, Y., Yang P., Sun, Y., Wu Y., Mayers, B., Gates, B., Yin, Y., Kim, F. & Yan, H. (2003). One-Dimensional Nanostructures: Synthesis, Characterization, and Applications. *Adv. Mater.*, *15*, 353–389.

[14] El-Sayed, M. A. (2004). Small Is Different: Shape-, Size-, and Composition-Dependent Properties of Some Colloidal Semiconductor Nanocrystals. *Acc. Chem. Res.*, *37*, 326-333.

[15] Wang, X., Zhuang, J., Peng, Q. & Li, Y. D. (2005). A general strategy for nanocrystal synthesis. *Nature*, *437*, 121-4.

[16] Xia, Y., Xiong, Y., Lim, B., Skrabalak, S. E. & Angew. (2009). Shape-controlled synthesis of metal nanocrystals: simple chemistry meets complex physics. *Chem.*, *Int. Ed.*, *48*, 60-103.

[17] Chen, H. M., Chen, C. K., Chang, Y.-C., Tsai, C.-W., Liu, R.-S., Hu, S.-F., Chang W.-S. & Chen, K.-H. (2010). Quantum Dot Monolayer Sensitized ZnO Nanowire-Array Photoelectrodes: True Efficiency for Water Splitting . *Angew. Chem.*, *Int. Ed.*, *49*, 5966-5969.

[18] Rosengaard, N. M. & Johansson, B. (1997). Finite-temperature study of itinerant ferromagnetism in Fe, Co, and Ni. *Phys. Rev. B.*, *55*, 14975-14986.

[19] Zhou, Y.-M., Wang, D.-S. & Kawazoe, Y. (1999). Effective ab initio exchange integrals and magnetic phase transition in fcc Fe and Mn antiferromagnet. *Phys. Rev. B.*, *59*, 8387-8390.

[20] Wang, J.-T., Zhou, L., Wang, D.-S. & Kawazoe, Y. (2000). Exchange interaction and magnetic phase transition in layered Fe/Au(001) superlattices. *Phys. Rev. B.*, *62*, 3354-3360.

[21] Fukamichi, K., Umetsu, R. Y., Sakuma, A. & Mitsumata, C. (2006). *Handbook of Magnetic Materials*, *16*, 209-401.

[22] Costa, A. C. F. M., Silva, V. J., Cornejo, D. R., Morelli, M. R., Kiminami, R. H. G. A. & Gama, L. (2008). Magnetic and structural properties of NiFe2O4 ferrite nanopowder doped with Zn^{2+}. *J. Magn. Magn. Mater.*, *320*, e370–e372.

[23] Gherca, D., Cornei, N., Mentré, O., Kabbour, H., Minaud, S. D. & Pui, A. (2013). In situ surface treatment of nanocrystalline MFe2O4 (M = Co, Mg, Mn, Ni) spinel ferrites using linseed oil. *Applied Surface Science*, *287*, 490-497.

[24] Bilovol, V., Cabrera, A. F., Rodríguez-Torres, C. E. & Mudarra-Navarro, A. M. (2013). Study of magnetic state of $Sn_{0.9}Fe_{0.1}O_2$ powders at low temperature. *J. Magn. Magn. Mater.*, *344*, 188-1992.

[25] Kulkarni, S. A., Sawadh, P. S., Palei, P. K. & Kokate, K. K. (2014). Effect of synthesis route on the structural, optical and magnetic properties of Fe3O4 nanoparticles. *Ceramics International*, *40*, 1945-1949.

[26] Sun, J., Zhou, S., Hou, P., Yang, Y., Weng, J., Li, X. & Li, M. (2007). Synthesis and characterization of biocompatible Fe3O4 nanoparticles, *Journal of Biomedical Materials Research Part A*, *80*, 333-341.

[27] Ohno, H. (1998). Making Nonmagnetic Semiconductors Ferromagnetic. *Science*, *281*, 951-956.

[28] Žutić, I., Fabian, J. & Das-Sarma, S. (2004). Spintronics: Fundamentals and applications. *Rev. Mod. Phys.*, *76*, 323-410.

[29] De-Boeck, J., Van-Roy, W., Das, J., Motsnyi, V., Liu, Z., Lagae, L., Boeve, H., Dessein, K. & Borghs, G. (2002). Technology and materials issues in semiconductor-based magnetoelectronics. *Semicond. Sci. Technol.*, *17*, 342-354.

[30] De-Groot, R. A., Mueller, F. M., Van-Engen, P. G. & Buschow, K. H. J. (1983). New class of materials: half-metallic ferromagnets. *Phys. Rev. Lett.*, *50*, 2024–2027.

[31] Park, J.-H., Vescovo, E., Kim, H.-J., Kwon, C., Ramesh, R. & Venkatesan, T. (1998). Direct evidence for a half-metallic ferromagnet. *Nature*, *392*, 794–796.

[32] Soulen-Jr, R. J., Byers, J. M., Osofsky, M. S., Nadgorny, B., Ambrose, T., Cheng, S. F., Broussard, P. R., Tanaka, C. T., Nowak J., Moodera, J. S., Barry, A. & Coey, J. M. D. (1998). Measuring the Spin Polarization of a Metal with a Superconducting Point Contact. *Science*, *282*, 85-88.

[33] Stroppa, A., Picozzi, S., Continenza, A. & Freeman, A. J. (2003). Electronic structure and ferromagnetism of Mn-doped group-IV semiconductors. *Phys. Rev. B.*, *68*, 155203-155212.

[34] Akai, H. (1998). Ferromagnetism and Its Stability in the Diluted Magnetic Semiconductor (In, Mn)As. *Phys. Rev. Lett.*, *81*, 3002-3005.

[35] Akinaga, H., Manago, T. & Shirai, M. (2000). Material Design of Half-Metallic Zinc-Blende CrAs and the Synthesis by Molecular-Beam Epitaxy. *Jpn. J. Appl. Phys.*, Part 2, *39*, L1118-L1120.

[36] Wolf, S. A., Awschalom, D. D., Buhrman, R. A., Daughton, J. M., Von-Molnár, S., Roukes M. L., Chtchelkanova A. Y. & Treger, D. M. (2001). Spintronics: A spin-based electronics vision for the future. *Science*, *294*, 1488–1495.

[37] Holland, W. E. & Brown, H. A. (1972). Application of the Weiss molecular field theory to the B-site spinel. *Phys-Stat. Sol (a)*, *10*, 249-253.

[38] Van-Vleck, J. H. (1941). On the Theory of Antiferromagnetism. *J. Chem. Phys.*, *9*, 85-90.

[39] Larson, P. & Lambrecht, W. R. L. (2006). Electronic structure of Gd pnictides calculated within the LSDA+U approach. *Phys. Rev. B.*, *74*, 085108-085115.

[40] Shenoy, G. K., Friedt, J. M., Dunlap, B. D. & Vogt, O. (1974). Étude des structures magnétiques de CeSb et GdSb par spectropie Mossbauer.^{112}Sb. *J. Phys.*, *35*, Cl-43.

[41] Chen, D., Liu, S., Li, J., Zhao, N., Shi, C., Du, X. & J. Sheng. (2009). Structural stability and magnetism of γ'-Fe4N and CoFe3N compounds. *J. Alloys. Compounds*, *475*, 494-500.

[42] Chen, D., Li, J., Shi, C., Du, X., Zhao, N., Sheng, J. & Liu, S. (2007). Properties of Core-Shell Ni-Au Nanoparticles Synthesized through a Redox-Transmetalation Method in Reverse Microemulsion . *Chem. Mater.*, *19*, 3399-3405.

[43] Wang, J. & Zeng, X. C. (2009). *Nanoscale Magnetic Materials and Applications*, 35-65. ISBN: 978-0-387-85598-1.

[44] Caruso, F. (2001). Nanoenginneering of particle surfaces. *Adv. Mater.*, *13*, 11-22.

[45] Hu, Y., Wu, G., Liu, Y. & Du, A. (2012). Resultant exchange bias and coercivity on rotatable antiferromagnetic anisotropy in ellipsoidal core/shell nanoparticles. *Phys. Let. A.*, *376*, 1650-1656.

[46] El moussaoui, H., Mounkachi, O., Masrour, R., Hamedoun, M., Hlil, E. K. & Benyoussef, A. (2013). Synthesis and super-paramagnetic properties of neodymium ferrites Nanorods . *J. Alloys. Compds.*, *581*, 776–781.

[47] Rezlescu, N. & Rezlescu, E. (1993). The influence of Fe substitutions by R ions in a Ni-Zn Ferrite. *Solid State Commun.*, *88*, 139-141.

[48] Kamala, B. K., Markandeyulu, G. & Ramana, C. V. (2011). Structural, Magnetic, Electrical, and Magnetoelectric Properties of Sm- and Ho-Substituted Nickel Ferrites. *J. Phys. Chem. C.*, *115*, 554–560.

[49] Jr, A. F. & e Silva, F. C. (2010). High temperature magnetic properties of cobalt ferrite nanoparticles. *Appl. Phys. Lett.*, *96*, 172505.

In: Hysteresis
Editor: José Carlos Dias
ISBN: 978-1-63321-336-4

Chapter 5

HYSTERETIC BEHAVIOR IN SOME NATURAL PROCESSES

Jenő Takács[1], and Akos Nemcsics[2]
[1]Department of Engineering Science, University of Oxford, Oxford, UK
[2]Institute for Microelectronics and Technology, Óbuda University, Budapest, Hungary
Research Institute for Technical Physics and Materials Science, Budapest, Hungary

ABSTRACT

Hysteretic phenomena are described first as part of the spin valve principle.

Although the Magnetic Resistance (MR) effect has been known as a scientific curiosity since 1857, the discovery of the Giant Magneto-Resistive and the Colossal Magneto-Resistive Effects has revolutionized sensor design. Their operational principles founded on a complex, interacting hysteretic behavior. Their modeling also requires complex mathematical operations and only a few of the known hysteretic models are suitable for describing these complex physical phenomena. The recent discovery and modeling of inverse spin-valve effect allowed us to investigate further some of the new physical effects. In this chapter we give examples for hysteresis in other areas of physics as well. Some detailed description is included in this chapter of the hyperbolic model, which found to be easiest to apply to hysteresis, particularly in non-magnetic cases. In this chapter a number of complicated hysteretic behaviour are shown, which occur in physical processes and can be successfully modeled by using this model. Cases are included from fields of spin-valve technology, semiconductor and surface physics. Free normalization, detailed in the 7th Section, is an essential part of the modeling process, gaining importance in investigations in all fields of the natural sciences. This provides greater flexibility for the researcher using modeling as a research tool.

Keywords: general hysteresis, spin-valve hysteresis, hysteresis in surface physics, hysteresis in semiconductors, modeling, free normalization

INTRODUCTION

Hysteresis is a well-known and well documented, recognized phenomenon in physics. Although people associate hysteresis mostly with magnetism, it occurs in many, if not most of the branches of natural science. One will find hysteresis not only in physics but in other branches of science, like in engineering, bio-processes, surface and semiconductor physics, just to mention a few of the obvious areas. You will also find hysteresis in fields unrelated to natural sciences like fiscal processes and economy. The investigation of the hysteretic processes, which are different in every case, leads mostly to difficult or totally unsolvable mathematical formulations. To make good, often highly accurate approximations and predictions, various mathematical models are used. Most of these hysteresis models are however based on the hysteretic phenomenon in magnetism and directed towards magnetic applications [1-5]. These models usually fall into two categories: The first category of the models is based on the physics of ferro-magnetics, while the second group is founded on the similarity with the sigmoid shape of the major hysteresis curve. We can neglect the first group, because of their specific magnetic oriented application and concentrate on the models based on the geometry or shape of the hysteretic process.

In these, so called geometrical models the one suggested by Ferenc Preisach in 1935 is still the most popular in scientific circles [2]. Although some authors suggested various modifications for specific applications it still remained as a fundamental tool for formulating various hysteretic phenomena in its heuristic, original or modified forms.

Since physical processes (except in magnetism), not subject to a phenomena, similar to internal demagnetization [6, 7] most of the modified models are not suitable to describe hysteresis, except in magnetism. The so called "classical" Preisach model (unfortunate choice of description) does not include internal demagnetization, therefore it describes the intrinsic or theoretical hysteresis curve [2]. In Preisach based models the fundamental difficulty is the determination of the distribution function, which is referred to, in general terms, as "the identification problem" defined from a set of the symmetrical first order reversal loops and formulated in the Everett integral [11, 12]. To solve the problem of this identification, some authors suggested the substitution of the Preisach distribution function with suitable approximations. Mayergoyz suggested power series [3], while Kadar and Della Torre [13-15] used the bilinear product of Gaussian functions for the solution of the integral. For soft magnetic substances, the Gaussian approach failed to deliver acceptable results, therefore Finocchio and his co-workers used Lorentzian distribution [15].

The two latest approximations will allow analytical evaluation of the Everett integral in closed mathematical form.

The bilinear Gaussian distribution is formulated as:

$$G2(x, y) = A\exp\frac{-B(x-a)^2}{2sig^2}A\exp\frac{-B(y-a)^2}{2sig^2} \tag{1}$$

where *A, B, a* and *sig* are constants [4]. The integration of (1) however, leading to the error function, describing the hysteresis loop, which is a hard function, resisting any further mathematical operations. This makes it very difficult to use it in practice.

Takacs in his paper published in 2000, with the introduction of the hyperbolic distribution solved the identification problem by approximating sigmoid-like hysteresis loops and giving mathematical description of the different properties of various magnetic substances [16]. At the time, this model, based also on Langevin's theory, was purely phenomenological and no connection was assumed to the Preisach "classical" approach.

The hyperbolic two bilinear distribution *T2(x,y)* can be described mathematically in its simplest form as

$$T2(x,y) = A\{\sec h[\alpha(x-a)]\}^2 B\{\sec h[\alpha(y-a)]\}^2 \tag{2}$$

here *A and B* represents the amplitudes, *a* is the shift in horizontal position of the function and α controls the steepness of the curve. The two dimensional difference between a typical hyperbolic and a Gaussian distribution curves is shown in Figure 1. The difference in the surface contour meshes on the two halfs of 3D plots shows the small difference between the two distributions. The use of T(x) distribution is good enough for most practical case.

We can see that, the hyperbolic distribution is running very close to the Gaussian form in spite of the two curves different mathematical formulation [16]. The close relationship between the two models, intrinsic (classic) Preisach and hyperbolic was only discovered and published in recent time [17]. In spite of the large variation of modified Preisach models [8] all variants have failed to model the complicated variety of hysteresis loops, namely the spin valve and anti-spin valve configurations. Its geometrical representation makes it hard to apply it in complex cases. The best approach so far was, to our knowledge, the analytic, hyperbolic approximation.

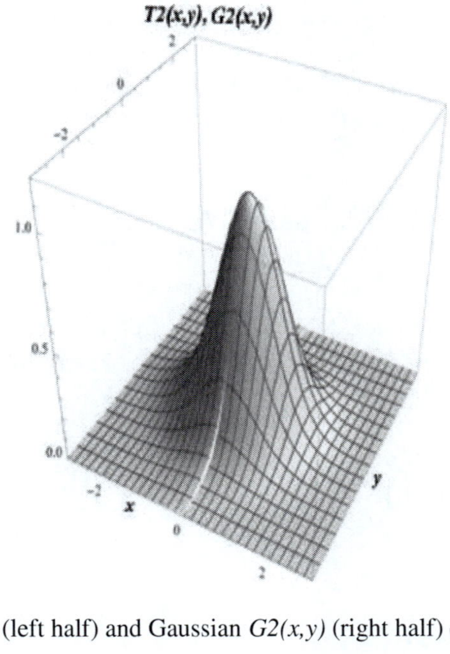

Figure 1. Hyperbolic *T2(x,y)* (left half) and Gaussian *G2(x,y)* (right half) distribution.

The aim of this Chapter is not to repeat the description of sigmoid like hysteresis, well known and well documented from ferro-magnetism, but demonstrate the hysteretic processes taking shapes different from the classical form. Although some of these processes far removed from the magnetic phenomenon, their behavior is strikingly similar to those and their modeling can be done by some of the classic models developed for ferro-magnetic materials. Our intention in this Chapter is to give a tool into the hand of the practical user.

MODEL DESCRIPTION

We assume that hysteresis is not a simple phenomenon but composed of a number of constituents. These simple components, although mixed together, can be separated, modeled individually with their own functions and added together based on Maxwell's principle. This analytical process is used in various branches of physics successfully to describe complex physical phenomena.

When hysteresis is present in a process, the describing curve will differ for forward and backward directions. Based on the first derivative, the two dimensional version of the function given in (2), the hysteresis can be modeled with the following equations [17-19].

$$F_u = A\ \{tanh[\alpha(X - X_c)] + F_0(X_{\mathrm{m}})\} \tag{3}$$

$$F_d = A\ \{tanh[\alpha(X + X_c)] - F_0(X_{\mathrm{m}})\} \tag{4}$$

here F_u and F_d are the output functions for positive and negative going directions, X_c is the measure of hysteresis, called coercitivity in Magnetism, X_m is the maximum value of the horizontal variable and α is the angle at $X = X_c$ therefore the angle of the tangent of the hysteresis loop at the point where it crosses the field axis. The F_0 integration constant can be calculated from the condition, that at X_m, the first return point X_u and X_d must be equal, for all minor and major loops [14, 15]; therefore, by definition:

$$F_0(H_{\mathrm{m}}) = \frac{1}{2}[\tanh\alpha(H_m + H_c) - \tanh\alpha(H_{\mathrm{m}} - H_c)] \tag{5}$$

All components in the complex loop are concurrently subjected to the same physical conditions; therefore the model mathematical equations, describing the process takes up the following normalized form [17]:

$$f_u = \sum_{k=1}^{n}[a_k(f_{uk} + f_{0k}(x_{\mathrm{m}}))] \tag{6}$$

$$f_d = \sum_{k=1}^{n}[a_k(f_{dk} - f_{0k}(x_{\mathrm{m}}))] \tag{7}$$

$$f_{uk,dk} = \tanh[\alpha_k(x \mp x_{ck})] \tag{8}$$

Each and every components or phases in the substance are represented by a separate and different integral of (4). The sum of these integrals completes the modeled hysteresis loop.

Here, all lower case letters represent the appropriate normalized quantities, while the corresponding capital letters are representing the physical quantities in appropriate physical units. In this chapter, wherever possible, the normalized forms will be used, that is

$$f_{0k} = \frac{1}{2}\{\tanh[\alpha_k (x_m + x_{ck})] - \tanh[\alpha_k (x_m - x_{ck})]\} \tag{9}$$

f_{0k} is the normalized form of $F_0(X_m)$. Here n is the total number of processes present and k is the running variable. Experience shows that n is normally between 1 and 3.

By definition $f_{10,}$ the middle line, the ideal loci of the first return points of the loops (major and minor), which belongs to both the ascending and the descending set of branches and has no hysteresis (unhysteretic) can be formulated as:

$$f_{10} = (f_d + f_u)/2 = \sum_{k=1}^{n} \frac{a_k}{2}\{\tanh[\alpha_k (x + x_{ck})] + \tanh[\alpha_k (x - x_{ck})]\} \tag{10}$$

In magnetism this is an entirely theoretical concept, assuming no internal demagnetization [7, 17]. Although experimentally it cannot be realized it is still useful in investigations in theoretical magnetism.

When internal demagnetization is present in a hysteretic process or not negligible unlike in case of non- magnetic phenomena or not very soft magnetic materials the hyperbolic model from (8), (9) and (10) is transformed into the following form, where n_D is the normalized demagnetization coefficient:

$$f_{uk,dk} = -\sqrt{a_k}\,\frac{Arc\tanh\left[\dfrac{\sqrt{a_k n_D}\,tanh[\alpha_k (x \pm x_{ck})]}{\sqrt{-1 + a_k n_D}}\right]}{\sqrt{n_D(-1 + a_k n_D)}} \tag{11}$$

$$f_{0k} = \sum_{k=1}^{n} \frac{1}{2}\left\{\sqrt{a_k}\,\frac{Arc\tanh\left[\dfrac{\sqrt{a_k n_D}\,tanh[\alpha_k (x - x_{ck})]}{\sqrt{-1 + a_k n_D}}\right]}{\sqrt{n_D(-1 + a_k n_D)}}\right.$$

$$\left. -\sqrt{a_k}\,\frac{Arc\tanh\left[\dfrac{\sqrt{a_k n_D}\,tanh[\alpha_k (x + x_{ck})]}{\sqrt{-1 + a_k n_D}}\right]}{\sqrt{n_D(-1 + a_k n_D)}}\right\} \tag{12}$$

$$f_{10} = \sum_{k=1}^{n} \frac{1}{2} \left\{ -\sqrt{a_k} \frac{Arc\tanh[\frac{\sqrt{a_k n_D} tanh[\alpha_k (x - x_{ck})]}{\sqrt{-1 + a_k n_D}}]}{\sqrt{n_D(-1 + a_k n_D)}} \right.$$

$$\left. -\sqrt{a_k} \frac{Arc\tanh[\frac{\sqrt{a_k n_D} tanh[\alpha_k (x + x_{ck})]}{\sqrt{-1 + a_k n_D}}]}{\sqrt{n_D(-1 + a_k n_D)}} \right\} \qquad (13)$$

This transformation is derived from the first derivative of the function and based on the Weiss effective field theory [20-23].

SPIN-VALVE LIKE HYSTERETIC CHARACTER

The aim of this chapter is not to inform the reader about the intricate details of the spin-valve construction and its operation, but to describe and analyze its hysteretic character. Therefore, we only give the reader an overview of the basic operational principle of the device, which is enough to explain its unusual hysteretic behavior. This will lead finally to its mathematical model. The history of spin-valves is dating back to 1857, when Lord Kelvin observed a change of resistance of an iron bar, when subjected to magnetic field [24, 25]. This phenomenon called anisotropic magneto-resistance (MRE) remained a scientific curiosity for well over a century, until 1970, when Hunt discovered the magneto-resistive effect and used it in spin-valve magnetic recording heads [26, 27]. The discovery of the Giant Magneto-Resistive Effect (GMR) [28, 29] and later the Colossal Magneto-Resistive Effect (CMR) in 1994 [30, 31], made the spin-valve formation a popular candidate for applications as hard disc readers, various sensors and Random Access Memories (MRAM), recording heads etc. [32]. The operation of the spin valves based on a quantum property of the electrons called spin. In a magnetic layer subjected to a magnetic field the carrier electrons align their spins to the external magnetic field. When voltage is applied across the spin valve, the polarized electrons move freely across the layer because their spin is aligned with the field. When magnetic field is reversed, the electrons have to flip spins. This requires extra energy which causes the layer to have a higher resistance. The state of the magnetic layers could be "up" or "down" depending on the direction of the applied magnetic field [32]. In simple practical terms, a spin-valve consists of two layers of magnetic materials with a few mono-layer thick magnetic insulating layer between them. The two magnetic layers differ in relative coercivity, one has large coercivity (hard layer) the other one has small (soft layer) one. The soft layer changes polarity at small magnetic excitation, while the hard layer stays in the same state until high magnetic field is applied, as shown in Figure 3.

As a result two states can exist, one is when the two magnetization vectors pointing parallel (Figure 3b) and another one when the vectors are antiparallel (see Figure 3c). The combination of this loops results in a complex hysteresis loop as shown for arbitrary parameters.

Figure 3 shows an arbitrary hysteresis loop of a symmetrically arranged spin valve (SV), modeled by using (6), (7) and (8).

Figure 4 shows the measured loop of an asymmetrically constructed SV, as reported by Dodrill and Kelley [33] and its model, calculated by using the same equations.

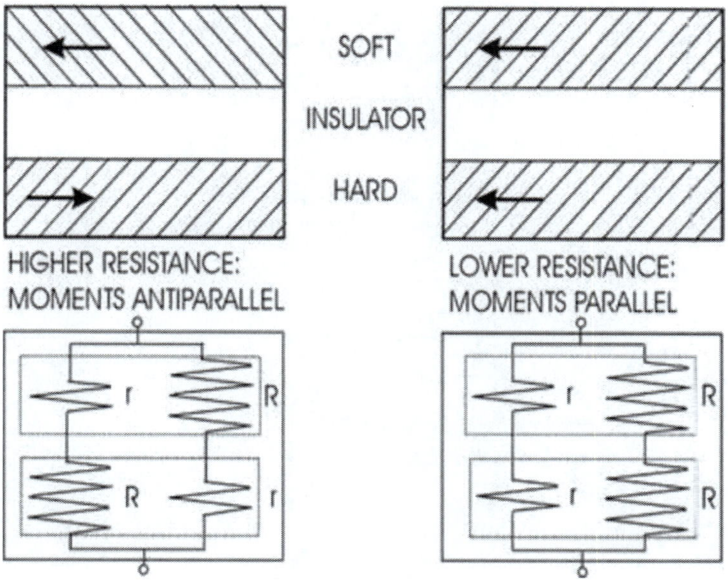

Figure 2. The spin valve acts as a filter for spin polarization, which can be switched on and off by external field.

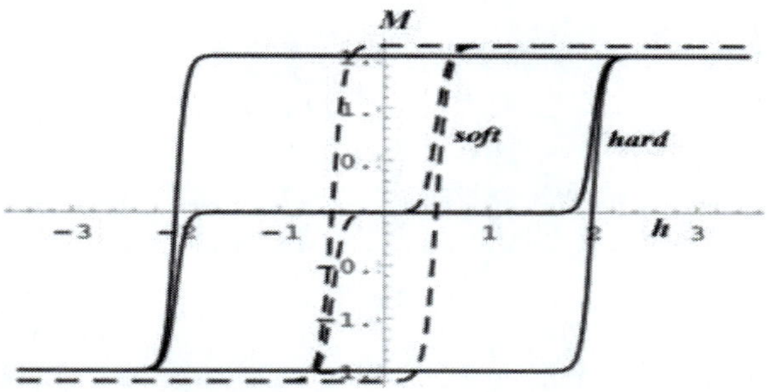

a)

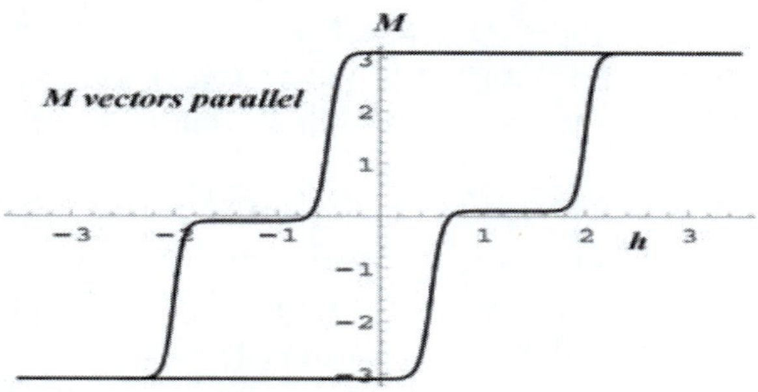

b)

Figure 3. (Continued).

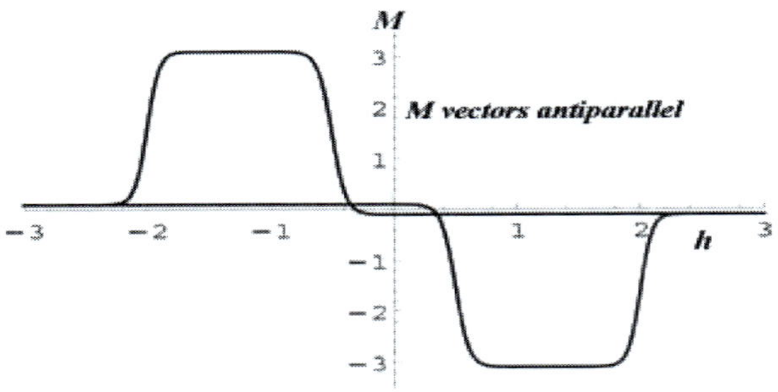

c)

Figure 3. The development of an arbitrary spin valve hysteresis loop: a) Hard and the soft hysteresis loop. b) Magnetization vectors parallel. c) Magnetization vectors antiparallel.

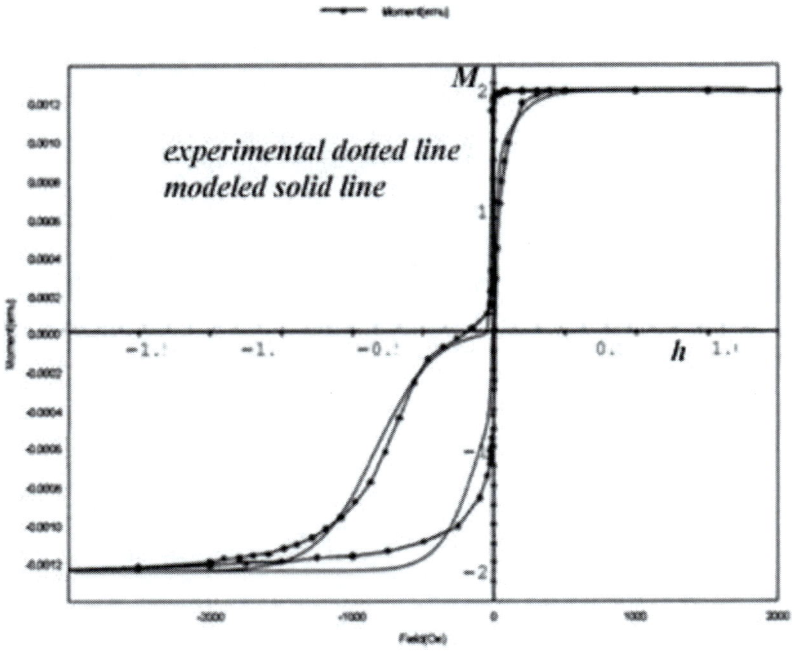

Figure 4. Experimental and its modeled hysteresis loop of an asymmetrical spin-valve.

The magnetic insulator between the magnetic layers is normally a non-magnetic metal like a few Angstrom thick chromium (Cr), copper (Cu), gold (Au), or silver (Ag) [33, 34].

Devices based on the spin-valve principle can manifest an unusual property of negative coercivity. This can produce a combination of three sub loops in the mayor hysteresis loop as shown in Figure 5 schematically.

This phenomenon was discovered in the mayor hysteretic loop of an instrument, developed for temperature monitoring, based on the spin-valve principle [35-37].

The shape, the size and the nature of the hysteresis is the function of the construction, the purpose of the device and the materials used in the design.

Chang reported on a measured loop of a Cu/Co/Cu/Fe multilayer structure with the character, similar to that shown in Figure 6. in heavier line.

Figure 6 also depicts a group of possible hysteresis loops of various layered structures.

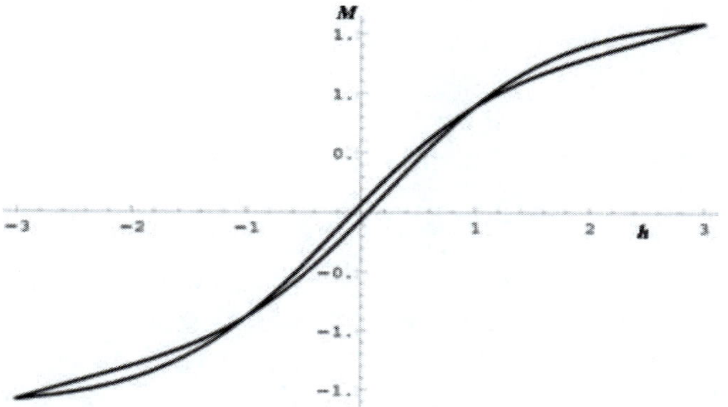

Figure 5. Hysteresis loop, when the coercivity goes negative.

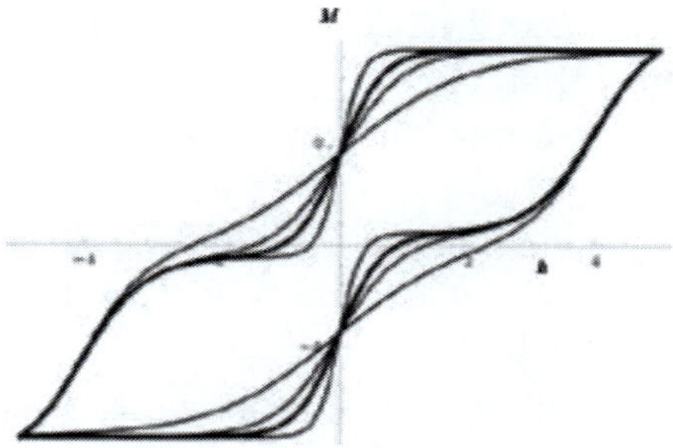

Figure 6. Possible hysteresis loops of layered structures.

HYSTERESIS INDEPENDENT OF FIELD DIRECTION

As we already said, the phenomenon of hysteresis is not restricted to the ferrous substances. In a device certain non-magnetic parameters can show the same effect, although these will not depend on the direction of the field, but only on its magnitude. A typical example is the MRC (Magneto-Resistive Coefficient). The MRC is a representative figure of the quality of the magnetic material used in a device. Instrument sensitivity depends on this number, defined as the ratio (in percentage) between the bulk resistivity of the layered device (ρ_0) and the change in resistivity ($\Delta\rho$) of the magnetic layers between the two states, due to the change in the direction of the field.

$$MRC = \frac{\Delta\rho}{\rho_0} \tag{11}$$

The change in resistivity between the two states of Magneto-Resistance (MRE) is only 2-4%. For the Giant Magneto-Resistance (GMR) this ratio changes to approximately 40%. The Colossal Magneto-Resistance (CMR) has a resistance change about 1300% at room temperature [32]. Modeling the hysteretic characters of MRE cannot be done directly with Equations 6, 7, 8 or 11, 12, 13 due to insensitivity of the device parameter to field orientation. In these cases we have to substitute into the canonic formulae 6 and 7 the following formulae:

$$f_{uk,dk} = \left|\tanh[\alpha_k (x \mp x_{ck})]\right| \tag{12}$$

$$f_{0k} = \left|\frac{1}{2}\{\tanh[\alpha_k (x_m + x_{ck})] - \tanh[\alpha_k (x_m - x_{ck})]\}\right| \tag{13}$$

The nonmagnetic elements, although they often show the hysteresis effects, this is normally independent of the direction of the field and only depends on the absolute value of its magnitude.

In a SV, hysteresis is not restricted to the magnetic layers of the device. The Magneto-Resistive Coefficient (MRC) is also hysteretic as a function of absolute value of the external field excitation; therefore, equations (12) and (13) are applicable. Figure 7 depicts measured and modeled MRC of an experimental device as a function of absolute field strength, as reported by Dodrill and Kelley [33].

Based on magnetic effect, a new transistor similar to an equivalent electronic component has been developed, called spin (-valve) transistor. The original proposal was put forward by Datta and Das in 1990 [38], while the common transistor was invented in 1940.

Both collector and the emitter are made of silicon, with the electrically conducting magnetic insulator layer forming the base.

This magnetic transistor, like other spin-valve design, is based on the two spin states of the electrons fundamentally to store information. The "up" and "down" state of the electrons can be altered without the use of electric current, which makes the spin-transistor small and theoretically more effective and more sensitive than the conventional transistors.

It is operational at a wide range of temperature. These spin transistors are still in a development stage [39]. Figure 8 is a schematic depiction of the collector current versus field characteristic of a spin transistor. The model of Figure 8 also applied formulae 12 and 13 for obvious reason [40].

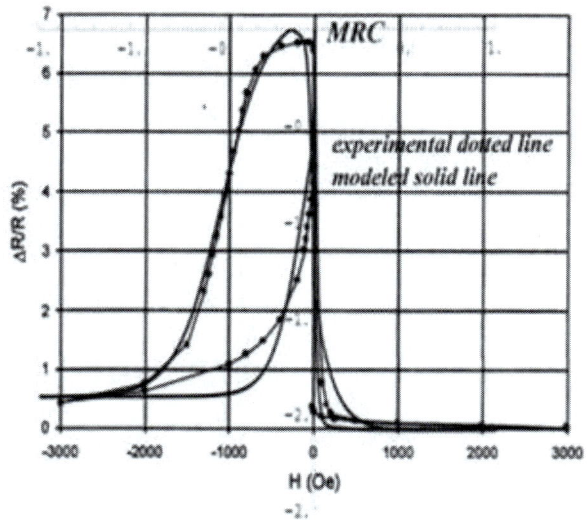

Figure 7. Experimental and modeled Magneto-Resistive Coefficient (MRC).

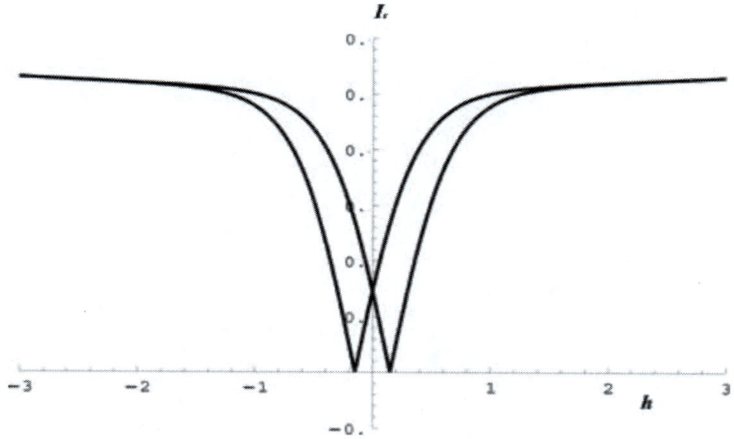

Figure 8. Characteristic shape of collector current versus magnetic field in a spintronic transistor.

HYSTERESIS IN SURFACE PHYSICS

Brief Description of the Experiment

In the experiment we investigate the changes in the behavior of GaAs and InAs crystal surfaces against temperature variation, based on the experiments of Yamagushi and Horikoshi [41]. For the growth of low-dimensional nanostructures it is vital to know the growth kinetics of the process. The crystal growth can be observed in-situ by monitoring the surface condition (i.e. the surface reconstruction) by Reflection High-Energy Electron Diffraction (RHEED). The RHEED intensity variation against temperature shows hysteretic behavior. During the experiment the change in temperature was slow, so every point could be regarded as being in the state of thermal equilibrium. As the temperature rises the RHEED intensity gradually diminishes. Both in the case of GaAs and InAs the surface at lower temperature shows surface reconstruction. The change of the direction of the temperature shows, that the process in subject of hysteresis. In case of both semiconductors the observed hysteresis loops fall within approximately a 50°C temperature range. In case of InAs we had two distinct hysteresis loops. At lower temperature the observed wide loop is an indication of smaller intensity variation. At higher temperature however we have seen a narrower loop with large and sudden intensity change. In case of GaAs there is only on real loop at lower temperature and a suspected degenerate pseudo loop at higher temperature. In the second loop the ascending and the descending branches seemingly overlap. The surface behavior of the bulk material is different from that of the surface, because due to the energy minimalization, it forms various surface reconstructions [42, 43]. These reconstructions, functions of the temperature, result in very complicated phase-diagrams. Each of the $I(T)$ graphs is composed of two hysteresis loops (in case of the GaAs we assume that the second loop is degenerate). We can conclude from the model that each of the loops describe one of the separate processes and that these processes are interlinked. An inverse spin-valve shows similar coupled hysteretic character. The upper parts of Figures 9a and 9b depict the coupled nature of the process and the complex functional relationship between the main and other constituent loops for InAs and GaAs respectively.

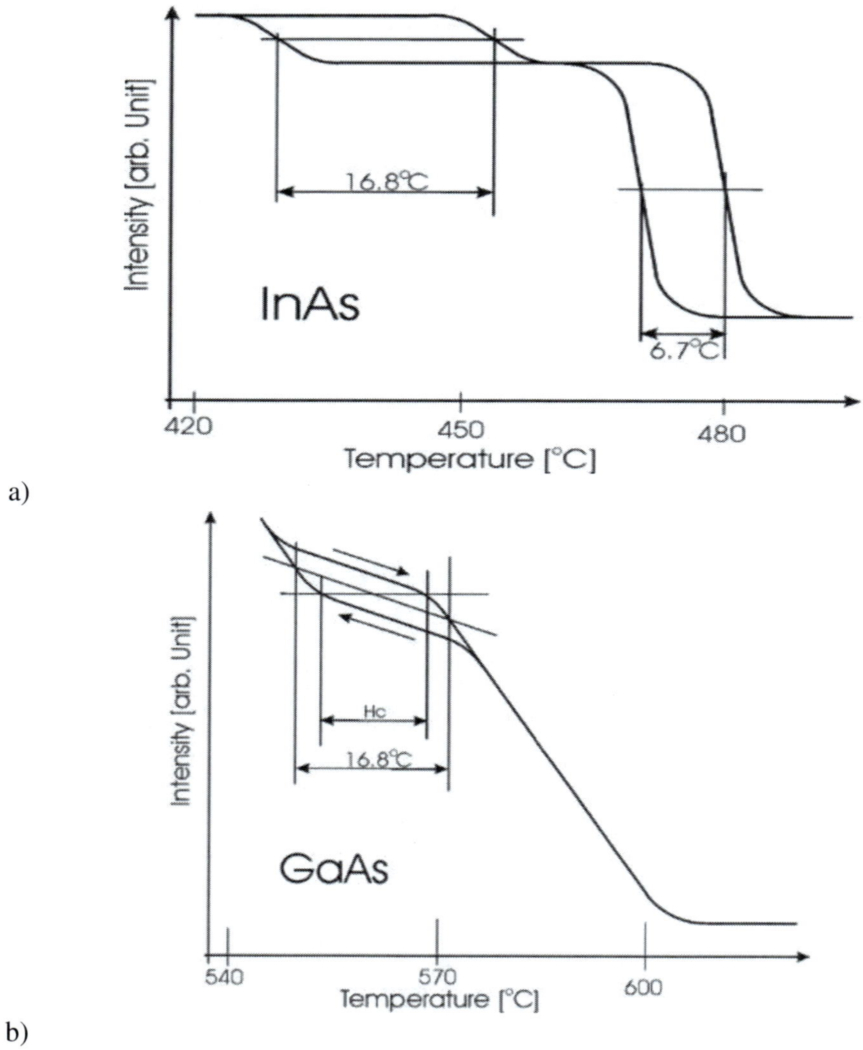

Figure 9. The expected main and its constituent loop for: a) InAs, b) GaAs.

Modeling Hysteretic Phenomenon in Surface Science

The inverse spin-valve effect, as part of the hysteretic processes have only been discovered less than ten years ago. Although the phenomenon described in this paper is far removed from the physical mechanism of spin-valves, its general character is strikingly similar. In general terms, the inverse spin-valve effect involves two coupled hysteretic processes, similarly to the spin-valve and characterized by two hysteresis loops flowing into each other as the excitation varies periodically.

Although there are a number of known models for describing hysteresis, so far there is only one, the hyperbolic model, which can describe this rather complex phenomenon of two coupled hysteretic processes, like in an inverse spin-valve [42, 43].

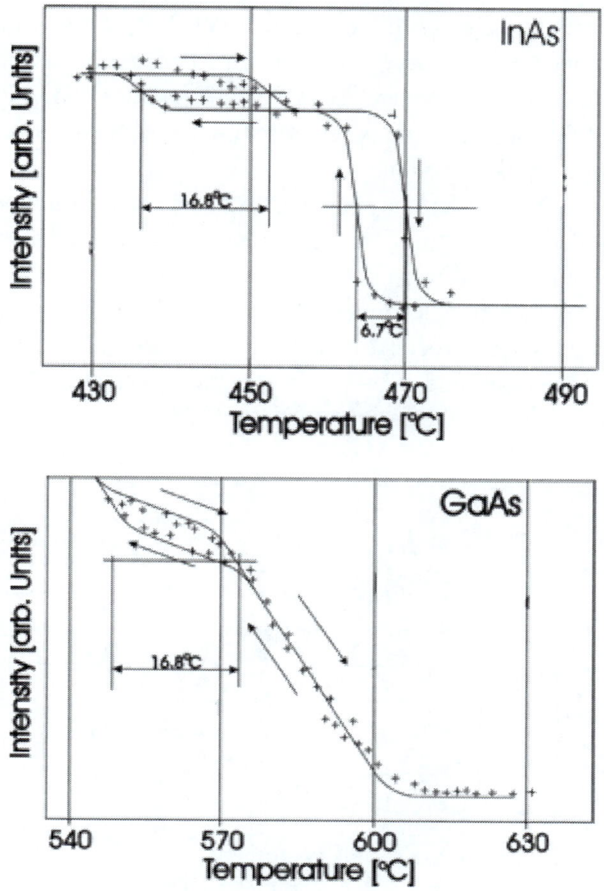

a)

b)

Figure 10. Measured and modeled hysteresis loop for: a) InAs, b) GaAs.

Although the model predicts the presence of a very narrow loop in the GaAs (see Figure 10) intensity plot at higher temperatures, its width is probably within the experimental error. It is interesting to note that while the physical parameters used in modeling are different for the two substances the approximate ratios between the amplitudes (A_n) remained nearly the same.

This is a good indication that although the range of temperature shifted, the surface reactions are the same for the two experimental substances verified by the model as we have initially assumed.

Example for Hysteresis in Physical Parameters of Semiconductor Devices

The introduction of Si into the semiconductor technology led to the development of the MOS field effect transistors, based on the advantageous physical and technological properties of Si [44-47]. For very high frequency applications (GHz) however Si is not suitable because it has low electron mobility.

GaAs, which has an electron mobility ($8x10^3$ cm^2/Vs) [47], 4 to 5 times that of Si, would be a better choice. This however develops high trap density on its surface leading to a hysteresis effect in its Capacity-Voltage (C-V) characteristics. Any improvement in the operation properties of this device requires the investigation of the form and magnitude of the hysteretic properties of its MOS junction.

When bias voltage is applied the thickness of the depletion zone changes, which determines the thickness of the neutral zone, governing the effective distance between the two electrodes, forming a capacitor. When the voltage is sweeping up and down, extra energy is required to free the trapped carriers leading to the effect of hysteresis in the C-V character of the device. The different surface reconstructions follow different C-V hysteresis curves, as shown in Figure 11 for the (2x4) and c(2x4) reconstructions [48, 49]. The two reconstructions are running concurrently between the same voltage limits.

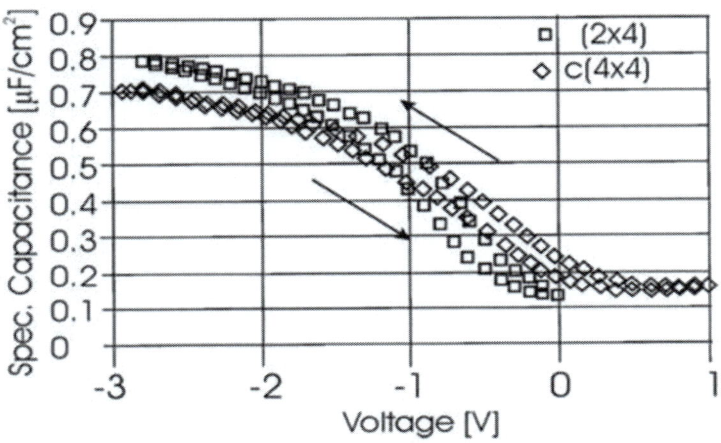

Figure 11. Measured hysteresis loops of (2x4) and c(2x4) surface reconstructions in the C-V characteristic of MOS made of GaAs.

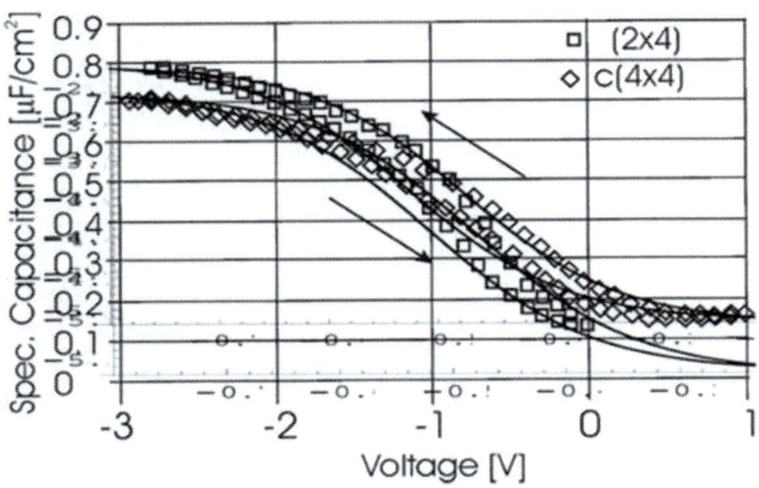

Figure 12. Experimental and modeled hysteresis loops of (2x4) and c(4x4) surface reconstructions in the C-V characteristic of a GaAs MOS device.

Figure 12 depicts the experimental hysterersis loops of the C-V curves for the two different reconstructions with their modeled equivalents. The solid lines are representing the modelled loops.

FREE NORMALIZATION

In the study of hysteresis modeling, it is an imperative to arrive to the correct physical parameters. To choose the most suitable model depends on the nature of the problem in hand. It requires not only mathematical skill, but patience and good judgment to achieve the required result, the best accuracy. Right through this Chapter we used the hyperbolic model with free normalization as the most suitable for solving non-magnetic hysteretic problems. One of the important parts of the process of modeling is normalization. The over-whelming majority of the researchers, in physical and other sciences as well, chose the maximum value of the parameter as unity, relating all other parameters to this peak value. Although widely accepted, it has certain draw backs.

In case of two or more sets of different data, it is difficult to choose which maximum value should be the base of normalization. It often leads also to undesirable fractions, difficult to use in numerical calculations. In free normalization the base is only determined at the end of satisfactory numerical fitting, giving free hand to the researcher in normalization. In the process, only the parameter ratios need to be observed. When the calculated model looks similar to the experimental curve, they are compared and changes should be introduced in the parameters to correct the difference. The same process is repeated until the best fit is achieved.

For the iteration process Matlab, Mathematica or similar mathematical software packages can be used and it can be done manually or assisted by a computer.

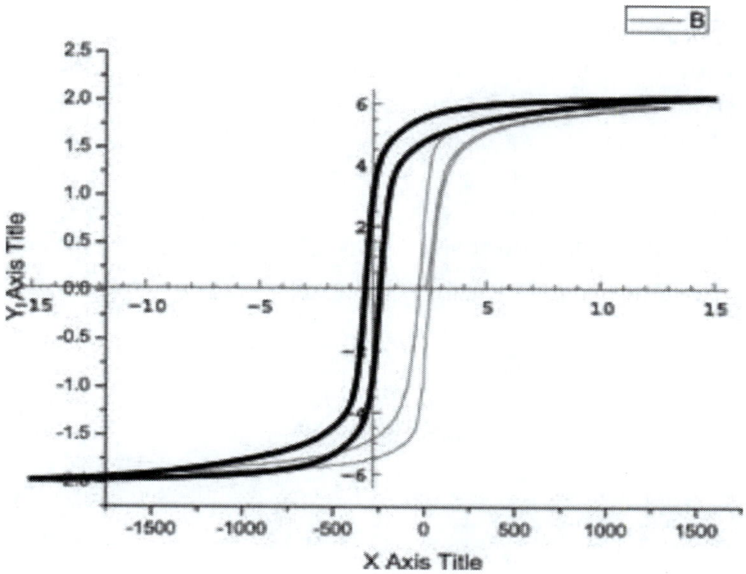

a)

Figure 13. (Continued).

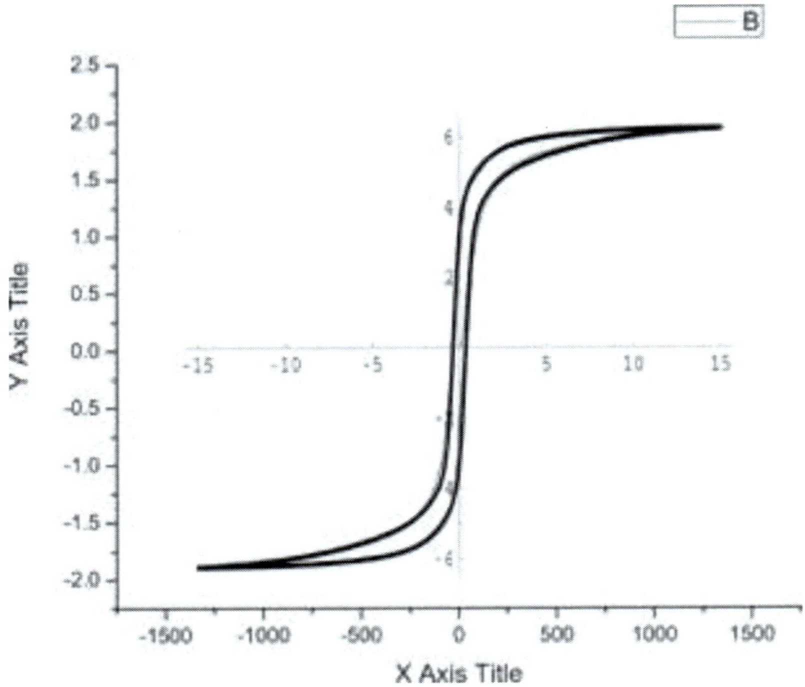

b)

Figure 13. Measured and calculated hysteresis loops a) before; b) after fitting.

At the start of the iteration, both the experimental curve and its calculated model should be saved in any of known transparent formats, like TIFF, JPEG, EPS, PDF, etc. Then the calculated model is copied (see Figure13 a) on to the experimental curve and stretched until the best fit is reached (see Figure13 b) (free or conform transformation). Then to improve the fit, change the parameters, as needed. This iteration is repeated until the best fit between the model and the measured curves is achieved. There are readily available iteration software packages, which can help the user in the final stage of the iteration. When the best fit is reached the physical values of the normalized units can be read from the two coordinate systems. In the given example for instance, on the horizontal axis one normalized unit represents 88 A/m physical units and on the vertical axis one equals 0.22 Tesla (see Figure 13 b). At start we must remember that the sum of the "remanence" figures (curve intersections with the vertical axis) of the constituent loops gives the total "remanence" (called in magnetism), while the maximum magnetization is given by the sum of maximum of the component loops.

CONCLUSION

Hysteresis is described as an often occurring phenomenon in various branches of physics. Hysteresis as a phenomenon however is not restricted to physical sciences. Other fields of natural sciences as well as some fields outside of those already mentioned, are subjected to hysteresis. The aim here is not to venture into the other field of sciences like biology, high finance etc. only to sample the fields of physics within the limited frame of this chapter. All examples described in this chapter was taken from various branches of physics, namely from spin–valve technology, semiconductor and surface physics etc. In this chapter hysteresis is described in general terms, not as a magnetic property of the substance, like in classical magnetism, but as a more generalized effect. In this effort a great deal of help was given by the process of mathematical modeling not oriented towards magnetism. With few exceptions (like the classic Preisach's model), models take into account the internal demagnetization, which only present in magnetic substances. For mathematical simplicity, the hyperbolic model was used right through this chapter, which is based on Langevin's theory and closely related to Preisach's unmodified, classic model. The application of the Preisach's model is often difficult in non-magnetic cases, due to its fundamental geometrical structure. One of the unique feature of the hyperbolic model is the free normalization, where after simple mathematical operations, the normalization base, only determined after a successful curve fitting (see Figure 13 b), giving very often more freedom of choice for the user dealing with complex hysteretic phenomena.

REFERENCES

[1] D. Jiles, *Introduction to magnetism and Magnetic Materials*. Chapman and Hall; London, (1998).

[2] F. Preisach, Zeitschrift fur *Physik.* (1935), Vol. 94, p. 227.

[3] I. D. Mayergoyz, *Mathematical Models of Hysteresis and their Applications*, Academic Press, Elsevier, New York, US (2003).

[4] E. Della Torre, *Magnetic hysteresis*. IEEE Press, New York. US. (1999).

[5] A. Ivanyi, *Hysteresis Models in electromagnetic Computation*. Akademiai Kiado, Budapest, Hu. (1997).

[6] J. Takacs (2013) *Appl. Math. Modelling,* Vol. 37, pp. 7315-7318.

[7] G. Bertotti (1998) *Hysteresis in Magnetism*, Academic Press, London.

[8] D. C. Jiles and D. I. Atherton, *IEEE Trans. on Magn*., (1983), Vol. 19. p. 2183.

[9] D. C. Jiles and D. I. Atherton (1986), p. 48, *JMMM*, Vol. 61.

[10] D. C. Jiles, *IEEE Trans. on Magn.* (1994), p. 4326, Vol. 30.

[11] D. H. Everett and W. I. Whitton, *Trans. Faraday. Soc.* (1952) 48, p. 749.

[12] D. H. Everett, *Trans. Faraday Soc.* (1954) 50, p.1077.

[13] Gy. Kadar and E. Della Torre (1988), p. 3001, *J. of Applied Phys*., 63.

[14] Gy. Kadar, E. Kisdi-Koszo, E. L. Kiss, L. Potocky, M. Zatroch and E. Della Torre (1989), p. 3931 *IEEE Trans. on Magn.* 25.

[15] G. Finocchio, M. Carpentieri, E. Cardelli and B. Azzerboni, *JMMM* (2006), 300, pp. 451-470.

[16] J. Takacs, (2000), p. 1002 *COMPEL*, 20, p. 4.

[17] J. Takacs "The Everett integral and its analytical Approximation", In: L. Malkinski, *Advanced Magnetic Materials*. Intech Publication, (2012).

[18] J. Takacs and I. Meszaros, (2008), p. 3137, *Physica B.* 403.

[19] J. Takacs, Gy. Kovacs and L. K. Varga, (2008), p. e1016, *JMMM*, 320.

[20] P. E. Weiss, *Comptes Rendus*, (1906), 143, pp. 1136-1149.

[21] P. E. Weiss, *J. de Phys*. (1907). Vol. VI. p. 661.

[22] J. Takacs, *Apl. Math. Mod.* 37. (2013) pp. 7315-7318.
[23] J. Takacs, *Journal of Metallurgy*, (2012), Vol. 2012, Art. ID 752871, pp. 1-5.
[24] W. Thomson, *Philosophic Magazine* (1858) London.
[25] W. Thomson, *Proc. Royal Soc. London*, Vol. 8, (1856-1857), pp. 546-550.
[26] R. P. Hunt (1970) *A Magneto-resistive Readout Transducer IEEE Trans.* Mag- 6 3 (digest only).
[27] R. P. Hunt (1971) *A Magneto-resistive Readout Transducer IEEE Trans.* Mag-7 1 (digest only).
[28] M. N. Baibich. M. Broto, J. A. Fert, F. Nguyen Van Dau and F. Petroff (1988) *Phys. Rev. Lett.* 61, p. 21.
[29] R. L. White, (1992) *IEEE* Mag-28 p. 5.
[30] A. P. Ramirez, *J. Phys.* (1997) pp. 8171-8199.
[31] M. Samant and S. S. Parkin (2001) *paper HE-03 Intermag Conference.*
[32] J. C. Mallinson, *Magneto-Resistive Spin Valve Heads* (2002) Academic Press N.Y.
[33] B. C. Dodrill and B. J. Kelley, *Internal report*, Lake Shore Cryotonics Ink.
[34] N. Tombros, C. Jozsa, M. Popinciuc, H. T. Jonkman, B. J. van Wees, Electronic spin transport and spin precession in single graphene layers at room temperature. *Nature* 2007, *448*, pp. 571–574.
[35] I. Jedlicska, R. Weiss, *Private communication.*
[36] I. Jedlicska, R. Weiss and R. Weigel (2008), *IEEE Trans. on Ind. El*ectronics, 884.
[37] I. Jedlicska, R. Weiss and R. Wiegel (2010) *IEEE Trans. on Ind. Electronics* 57, 5 p. 1728.
[38] S. Datta, and B. Das (1990). "Electronic analog of the electrooptic modulator". *Applied Physics Letters* 56, pp. 665-667.
[39] M. Gobbi, A. Bedoya-Pinto, F. Golmar, R. Llopis, F. Casanova, and L. E. Hueso, C_{60}-based hot - electron magnetic tunnel transistor, *Appl. Phys. Lett.* Vol. 101. Iss.10, p. 2012.
[40] H. Yamaguchi, Y. Horikoshi: *Surface Structure Transitions on InAs and GaAs (001) Surfaces; PRB,* Vol. 51 (1995), pp. 9836-9854.
[41] A. Nemcsics, J. Takacs and J. Bozsik, *Proc. SISY 2010 Subotica* (Sept. 10 2010), pp. 389-392.
[42] A. Nemcsics and J. Takacs, *Acta Polytechnica Hungarica* Vol. 7. No. 5, (2010), pp. 109-128.
[43] S. M. Sze: *VLSI Technology*; McGraw-Hill Int. Ed., New York (1988).
[44] W. S. Ruska; *Microelectronic processing*; McGraw-Hill Int. Ed., New York (1988).
[45] S. M. Sze: *Physics of Semiconductor Devices*; John Wiley and Sons, New York (1981).
[46] E. S. Yang: *Microelectronic devices*; McGraw-Hill Int. Ed., New York.
[47] I. M. Vitomirov et al.: *J. Electronic Materials* 22, (1993) pp. 309-313.
[48] M. Pristovsek et al.: *J. Crystal Growth* 248, (2003) pp. 254-258.
[49] J. Takacs, *Mathematics of hysteretic phenomena*. Wiley-Vch, Weinheim, FRG, (2003).

In: Hysteresis
Editor: José Carlos Dias

ISBN: 978-1-63321-336-4

Chapter 6

HYSTERESIS PHENOMENA IN AERODYNAMICS

G. Alonso[*] and J. Meseguer
IDR/UPM, E.T.S.I.A.E., Universidad
Politécnica de Madrid, Madrid, Spain

ABSTRACT

Hysteresis phenomena are not strange in aerodynamics, when the fluid effects on a body are dynamically coupled with the structural response of that body. It is the case of some aeroelastic instabilities of bluff bodies, but also of the dynamic stall of aeronautical airfoils.

Transverse galloping is a type of aeroelastic instability characterized by large amplitude, low frequency oscillation of a structure in the direction normal to the mean wind direction. Some elastic bluff bodies under the action of a fluid flow can experience transverse galloping and lose stability if the flow velocity exceeds a critical value. In some applications, transverse galloping is an undesired phenomenon (the well-known cases of ice-coated electrical power lines and maritime cables). But induced transverse galloping can also be interesting for energy harvesting applications, i.e. to extract useful energy from the surrounding flow.

For any or other reason, the study of galloping has focused the attention of researchers during the last decades. One of the main points of interest in the analysis of transverse galloping is the associated hysteresis phenomenon. In this chapter, this analysis is performed for bodies with many different cross sections to show, based on data obtained from wind tunnel testing, that hysteresis in transverse galloping depends on the shape of the aerodynamic force coefficient curve, in particular, on the number and position of the inflection points of that curve. The influence of parameters like the cross-section geometry, the mechanical properties and the incident flow angle of attack is investigated.

On the aeronautical side, dynamic stall is a flow phenomenon which occurs on many applications were rotating wings are involved, such as helicopter rotor blades, as well as on commonly used pitch-regulated wind turbines under yawing conditions and during gust loads.

[*] Corresponding author: G. Alonso. IDR/UPM, E.T.S.I.A.E., Universidad Politécnica de Madrid, E-28040 Madrid, Spain. E-mail: gustavo.alonso@upm.es.

During this process, the fluid mechanics of the stall are generally considerably different from those obtained on an airfoil under static conditions. Wind tunnel tests have been performed to analyse the unsteady pressure distributions on a NACA 0012 airfoil in pitching motion. The pressure distributions were measured during several oscillating cycles at different reduced frequencies and oscillating amplitudes.

The dependence of the hysteresis in the aerodynamic forces on different parameters (damping, reduced frequency, characteristics of the flow) is investigated. Experiments reveal that the dynamic stall process is very sensitive to the character either laminar or turbulent of the airfoil boundary layer.

Keywords: aerodynamics, turbulence, airfoil, aeroelastic, galloping

INTRODUCTION

Hysteresis naturally appears in a large variety of unsteady aerodynamic problems where boundary layer separation and reattachment occur. Typical examples of this behaviour can be found in periodic airfoil stall, and flutter, as well as in galloping when two-dimensional bluff bodies are considered. However, in these two phenomena there are remarkable differences. In the case of airfoils, high angle of attack aerodynamics is characterized by extensive unsteady flow separation with significant hysteresis loops in the aerodynamic coefficients due to dynamic stall phenomenon. In the case of an oscillating airfoil, hysteresis appears at each oscillation cycle, in the sense that aerodynamic forces, even in the case of symmetric airfoils, are different when the angle of attack grows than the ones obtained when the angle of attack decreases.

In the case of the flow past galloping bodies, hysteresis is related to the different behaviour of the amplitude of the oscillations experienced by the body as the flow velocity increases or decreases.

A typical system investigated by many authors is the pitch-plunge airfoil with various types of nonlinearity, which gives rise to dynamic stall.

Hysteresis problem in airfoil aerodynamics are mainly connected with dynamic stall, which is one of the most fundamental problems in the area of unsteady aerodynamics. Dynamic stall is a serious aeroelastic instability problem in rotating machines, such as helicopters, turbine blades and compressors.

For example, during the forward or manoeuvring flight of helicopters, the angle of attack of the rotor blade sections periodically changes and reaches large values as the blades rotate and pass through the so-called retreating blade position. In contrast to steady-state airfoil stall, on an airfoil whose incidence is increasing rapidly, the onset of the stall can be delayed to incidences which may exceed the static stall angle by a significant amount. However, once dynamic stall occurs, the aerodynamic loads are generally more severe compared to steady stall and may cause significant increases in the blade stresses and the control system loads.

Dynamic stall also occurs on horizontal axis wind turbines, where rather complex blade incidence variations are generated by the unsteady inflow conditions due to atmospheric turbulence and during operation in yaw [1].

The phenomenon of dynamic stall on airfoils and lifting surfaces in unsteady flow environments has been studied for many years. The effects of unsteady motion on aircraft stall characteristics have been recognized and received considerable attention.

Numerous experimental and computational investigations have shown that the unsteady flow can be separating or reattaching over a large portion of the top surface of the airfoil, and that the predominant feature of dynamic stall is the formation and shedding and convection over the upper surface of the airfoil of an energetic vortex-like disturbance from the leading edge of the airfoil, which induces a nonlinearly fluctuating pressure field and produces transient variations in forces and moments that are fundamentally different from their steady-state counterparts [2] After the leading-edge vortex passes the airfoil trailing edge and goes into the wake, the flow progresses to a state of full separation over the upper surface. This is accompanied by a sudden loss of lift and a decrease in pitching moment. Furthermore, when the angle of attack becomes low enough, the flow will finally reattach again from the leading edge. An excellent review on unsteady airfoils is given in [3], and additional reviews on experimental techniques and results of dynamic stall investigations have been presented in [4, 5].

As already stated, dynamic stall occurs on a lifting airfoil when it is subjected to an increasing angle of attack, which subsequently exceeds the normal static stall angle at some significant pitch rate. During this process, the fluid mechanics of the stall are generally considerably different from those obtained on an airfoil under static conditions [6]. The dominant feature characterizing dynamic stall is a strong vortex, which begins near the leading edge, enlarges, and then travels downstream [3, 7].

Much basic research has been devoted to this problem [8-10] aiming to control this phenomenon, and even to cancel it. Boundary-layer control has been employed to have effect on the dynamic stall phenomenon and includes methods such as blowing, suction and pulsed blowing, as well as control of boundary layer transition [11-15]. Modifications of the airfoil geometry such as leading edge slats [16-18], leading-edge droop, rotation of the leading edge, as well as dynamically deforming geometries [19] have been investigated.

On the other hand, the phenomenon known as galloping is characterized by structural cross-sections that are aerodynamically unstable, so that small amplitude vibrations generate forces which increase the amplitudes to large values. Galloping is defined as a typical instability of slender structures [20-24]. It is a relatively low-frequency oscillatory phenomenon of elongated bluff bodies acted upon by a wind stream. The frequency at which the bluff object responds is much lower than the frequency of vortex shedding. Galloping was explained theoretically by Den Hartog [20]. Thereafter, researchers such as [25-27], amongst others, investigated the galloping of bodies of various cross-section shapes for a wide range of flow conditions.

There are two types of galloping: cross-wind and wake. Cross-wind galloping-induced oscillations are caused by forces which act on a structural element as it is subjected to periodic variations in the angle of attack of the wind flow. Usually the periodically varying angle of attack is generated by across-wind oscillation of the structure. The frequency of the oscillations is approximately the same as the natural frequency of the structure, provided that the density of the structure is much higher than the flow density. Wake galloping occurs when two bodies are present, where one is upstream, producing a wake, and one downstream, within that wake, and the cylinders are separated by a distance of few diameters. In this type of flow the downstream body is subjected to galloping oscillations induced by the wake of the upstream cylinder. Consequently, the upstream cylinder tends to rotate clockwise while the downstream cylinder rotates anti-clockwise, inducing torsional oscillations [28-31]; this last type of galloping instability is not considered in the following.

In the last decades galloping has focused the attention of many researches, and large efforts have been devoted to study the galloping features of many bodies having different cross-sections. Many researchers [22, 25, 32-36] have studied the instability mechanism in the flow over a square cylinder, which gives rise to galloping vibration. Besides, the influence of relevant parameters like the incident turbulence [26, 37-41], Reynolds number [42-44], or the hysteresis phenomenon [31, 34, 35, 45-48] have been treated. Concerning hysteresis phenomena, it must be pointed out that the quasi-steady theory was first presented in [33] where a fifth order polynomial damping nonlinearity was introduced to describe the aerodynamic force. This formulation was later developed in [34] by extending the approxima-tion up to seventh order. This new approximation allows for an accurate representation of the point of inflection in the forcing aerodynamic term. The relationship between point of inflection and the existence of a hysteretic loop was proven by [28].

Although most of the effort in galloping research has been concentrated in bodies whose cross section geometry is a square or rectangular cross-section, prismatic bodies with other cross-sectional shapes have been also considered [22, 27, 49-61], a relevant effort being devoted to ice-coated power-lines [62-64], and recently a large amount of work is being concentrated on the galloping behaviour of cables used in bridge structures [65-73], and to the galloping of the bridge decks [74, 75].

The aim of this chapter is to present the relevant features of the hysteresis phenomenon which may appear linked to the separation and re-adherence of the boundary layer on the upper or lower surfaces of the body, by considering different body configurations. This chapter is organized as follows: the influence of the boundary layer regime on the hysteresis of pitching symmetric airfoils is described in the next section, and the linear mathematical model that describes the onset of transverse galloping and the methods to analyse the galloping hysteresis are presented in the third section, whereas in the fourth section the results obtained related to hysteresis from experimental measurements performed at IDR/UPM are shown. Finally, conclusions are drawn in the last section.

Effects of Turbulence on the Hysteresis of an Oscillating Airfoil at Low Reynolds Numbers

In this section the unsteady aerodynamic behaviour of an airfoil oscillating in pitch at subsonic speed is analysed. The analysis has been performed experimentally by considering a wide range of operational parameters such the upstream flow velocity (thus the Reynolds number, Re), the reduced frequency, ω, and the amplitude of the oscillation, α_a, around given mean values, α_m, of the angle of attack α. Besides, the influence of the boundary layer regime, either laminar or turbulent, on the hysteresis is studied. To perform this analysis a passive boundary layer control device, which consisted of a wire 2 mm in diameter placed a distance of 3 mm in front of the airfoil leading edge and parallel to it.

Tests were carried out in the A4C wind tunnel of IDR/UPM (Instituto Universitario de Microgravedad “Ignacio Da Riva”, Universidad Politécnica de Madrid). This wind tunnel is an open circuit wind tunnel whose test chamber 0.20 m width, 1.80 m height and 2.00 m long. Upstream turbulence intensity can be modified by changing the porosity of wire meshes placed at the contraction entrance.

Wind velocity profile at the model test section is uniform within 1%, the turbulence intensity being around 3%. The maximum velocity of the stream at the test section is 23 m/s, which provides a Reynolds number based on the airfoil chord, c =0.25 m, of Re= 4×10^5 [76].

Several NACA 0012 airfoil models were manufactured for the test program. In all cases the model chord was 25 cm. It must be pointed out that as the model front area is 6.5 per cent of the test section area (even at the highest angles of attack of the model), no provisions for blockage corrections of the measured loads have been undertaken.

The models are 190 mm span and the gap between airfoil tips and the lateral walls of the test chamber is 5 mm, which is 2.6 % of the span.

The models are made of Necuron®, which is a low density plastic material. The different airfoil models were machined in two parts (an upper and a lower shell) to facilitate further instrumentation. To make pressure taps, holes are drilled on the airfoil surface and brass tubes are fitted to the holes and mounted flush with the surface. Flexible plastic tubes are connected at the ends of the brass tubes. The plastic tubes leave the airfoil through a hollow suspension axis, which is supported on by two bearings. Pressure taps are connected to a pressure scanner, model ZOC 33/64PxX2 from Scanivalve. The dynamic pressure in the test chamber is measured by a Pitot tube Air Flow 048 connected to a Schaewitz Lucas pressure transducer.

All the experimental sequences, settling the different values of the angle of attack, reduced frequency and pressure signals measurement and archiving, were controlled by a PC.

For the static tests, the models are mounted on a rotating platform ESP100/RV80PP from Newport. During the experiments the angle of attack was swept from 0° to 23° at 0.5° and 1° steps (from 0 to 0.4 radian). For the dynamic tests a simple oscillation device (crank-shaft mechanism) is used. The pitch rotation axis is fixed at the airfoil quarter chord (aerodynamic centre). The model angle of attack can be varied as $\alpha = \alpha_m + \alpha_a \sin\omega t$, where α_m is the mean value of the angle of attack, α_a the oscillating amplitude, ω the circular frequency of oscillation and t the time. Data are acquired from surface pressure taps, the wind tunnel reference pressure transducer, and the angle of attack encoder by means of a 12 bit, 64 channels with a maximum sampling rate of 6 kHz. Data are digitally filtered using a low pass filter with various cut-off and transition frequencies. All oscillatory data presented here are an average of several cycles.

Preliminary experiments were performed to ensure both that the gap does not affect the flow around the airfoil and that the model behaves as a two-dimensional airfoil.

Three NACA 0012 airfoil test models equipped with pressure taps are used. One of the models was equipped with 45 pressure taps distributed in three sections, located at y/b = 0.10, 0.25 and 0.50, where y is the distance from one of the model tip and b the model span. There are 10 pressure taps at y/b = 0.10 section, another 10 pressure taps at y/b = 0.25 section, and the remaining 25 pressure taps are located on the y/b = 0.50 section. This model is used to evaluate the two-dimensional character of the flow, as it is demonstrated from the comparison of measured pressure distributions at different span locations, where no significant differences are found [77]. A second airfoil model, was equipped with 25 pressure taps as in the y/b = 0.25 section of the first model. The main difference with the first one is that here the pressure taps are located a plane which forms an angle of 20 degrees with respect to the model middle section. The purpose of this second model was to evaluate there was not any disturbances between adjacent taps [78, 79] by comparing pressure distributions obtained with this model with those measured on the first one.

Based on the results measured with above described models, a third NACA 0012 model, equipped with 39 pressure taps was built up. Pressure taps are distributed all along the y/b = 0.50 section (19 pressure taps at the upper side, another 19 pressure taps at the lower side, plus an additional tap just at the leading edge). This third airfoil model can incorporate a turbulence generator consisting of a wire 2 mm in diameter and 190 mm long placed just in front of the airfoil leading edge a distance x_w = 3 mm apart.

The clean configuration, without any wire, was labelled as NACA 0012-C (the label C stands for clean configuration), whereas the one equipped with the turbulence generator was labelled as NACA 0012-T. This model was used for a test campaign where several testing conditions were considered, changing the Reynolds number, the oscillation amplitude as well as the reduced frequency.

Sketches of the different airfoil models are shown in figure 1. The different tests performed are summarised in table 1.

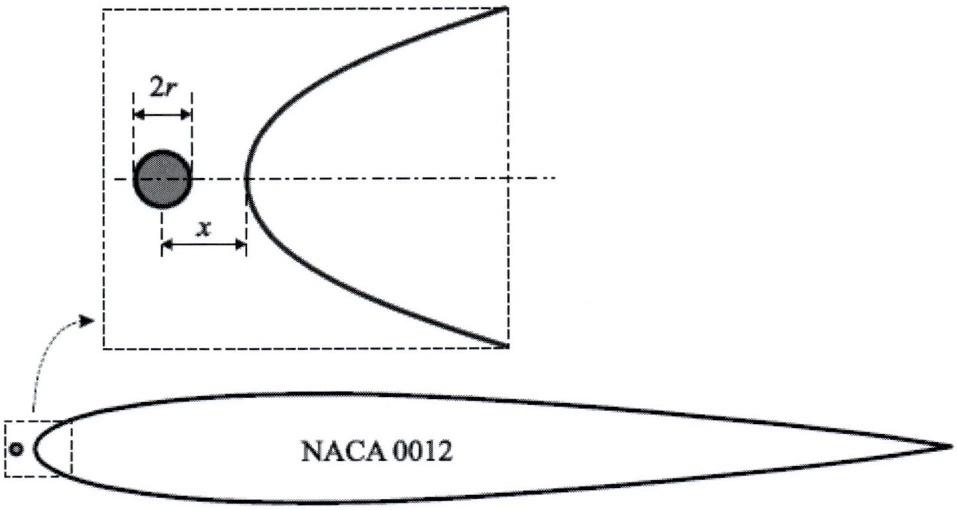

Figure 1. Sketches of the clean airfoil model and of the model with the leading edge wire acting as a turbulence generator.

Table 1. NACA 0012 airfoil configurations (C means clean airfoil, and T with a turbulence generator at the leading edge) and testing cases considered in the wind tunnel tests related to pitching airfoil hysteresis

Configuration	Re×10^{-5}	α [radian]		k	
C, T	1.3	0-0.4		0	
	1.7	0-0.4		0	
	2.5	0-0.4		0	
	4.0	0-0.4		0	
	1.7	0.07sinωt	0.3	0.7	1.0
	1.7	0.4sinωt	0.3	0.7	1.0

Reynolds number, Re = cU_∞/ν; angle of attack, α; and reduced frequency, $k = \omega c/(2U_\infty)$. In these expressions, c is the airfoil chord, U_∞ the incoming air velocity, ν the air kinematic viscosity, and ω the circular frequency of the oscillation.

EXPERIMENTAL RESULTS

First of all, series of static tests were performed before dynamic tests to determine the static stall characteristics of the models. Pressure distributions were measured at static conditions ranging the angle of attack range from 0° to 23° (close to the stall region a 0.5° step was used). Four values of the Reynolds number were considered, namely 1.3×10^5, 1.7×10^5, 2.5×10^5 and 4.0×10^5. These static tests were carried out by changing the angle of attack from 0° to 23° with the appropriate steps and after form 23° to 0° (obviously no differences were found between measurements made with increasing angle of attack and those made with decreasing angle of attack). At each angle of attack pressure data were acquired during 15 s at 100 Hz sampling rate.

As it is well known, Reynolds number drives the stalling characteristics of lifting airfoils, which in turn governs the maximum lift coefficient. In medium thickness airfoils, as is the case of NACA 0012, at low Reynolds numbers (10^4 - 10^5), boundary layer separation occurs at the leading edge, where the boundary layer is still laminar, then a short recirculation bubble is formed, the reattached boundary layer behind the bubble being now turbulent. If the angle of attack increases, the adverse pressure gradient increases accordingly, and the whole boundary layer separates once the angle of attack becomes larger than some threshold value [80]. The influence of Reynolds number on static lift coefficient curve is shown in figure 2.

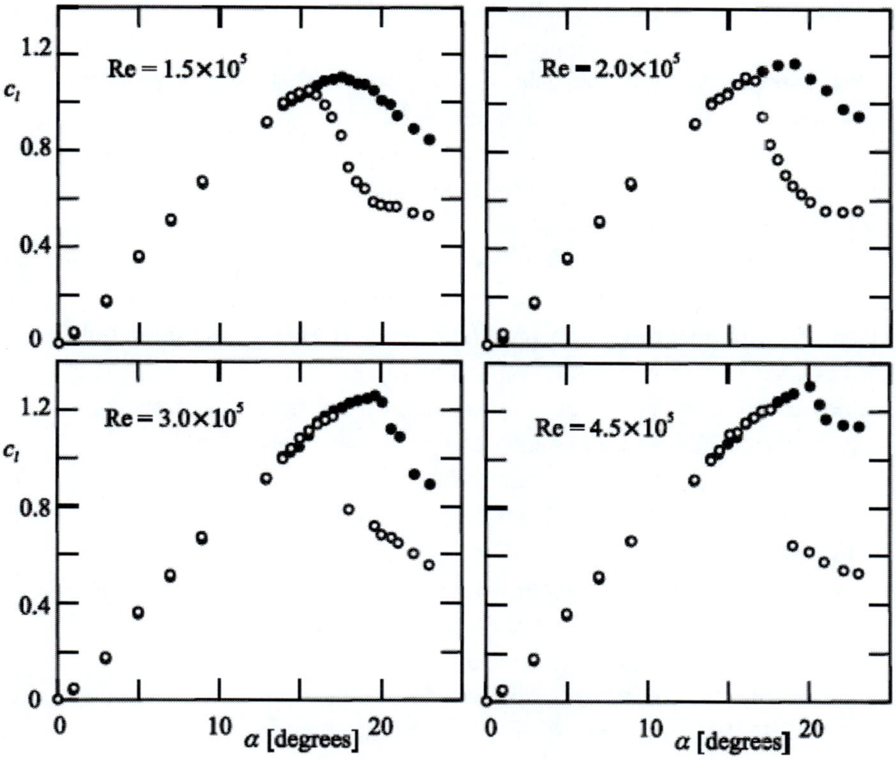

Figure 2. Variation with the angle of attack, α, of the lift coefficient, c_l, of the clean NACA 0012 airfoil (white symbols, configuration C) and the airfoil equipped with a turbulence generator located at 3 mm in front of the leading edge (black symbols, configuration T), for different values of the Reynolds number.

With a cylindrical wire placed some distance apart from the leading edge of the airfoil, even at low Reynolds numbers, the wire boundary layer is separated, and a turbulent, or at least unsteady, wake is formed behind it. When such an unsteady wake impinges on the airfoil, it introduces a remarkable perturbation on the airfoil boundary layer, triggering the transition to the turbulent regime of the existing boundary layers either at lower or the upper side of the airfoil depending on the angle of attack of the airfoil. If the turbulent wake of the wire causes the transition of the boundary layer in the airfoil upper side, boundary layer separation at the leading edge is delayed [81]. In this way, the airfoil angle of attack can be increased without boundary layer separation, and consequently, safe airfoil operation can be extended beyond the maximum angle of attack corresponding to the stalling limit of the laminar boundary layer case, although aerodynamic drag increases accordingly.

Concerning the dynamic tests, according to the information included in table 1, different values of the Reynolds number, of the amplitude of the angle of attack oscillation and of the reduced frequency were considered. The NACA 0012 model was tested without the leading edge wire (C configuration) and with the turbulence generator at the leading edge (T configuration).

The first aspect to be pointed out about these dynamic tests is that, contrary to the static stall cases, the dynamic stall characteristics of this airfoil were found to be only weakly dependent on Reynolds number. Therefore, in the following only data corresponding to a Reynolds number Re = 1.7×10^5 are reported. Pressure distributions were measured, and recorded, at least 125 times during each pitch cycle. The lift and pitching moment about the quarter-chord were obtained by numerically integrating the pressure distributions along the chord. For each airfoil position, the values of the coefficients used to present experimental results are the averaged values of those obtained for at least ten pitch cycles.

As stated in table 1, the effect of oscillation amplitude, reduced frequency and a turbulence generator of an oscillating airfoil was analysed in two cases: in the first one (un-stalled case) the maximum amplitude of the angle of attack was smaller than the static angle of stall (α_a = 4° = 0.07 radian), whereas in the second one (stalled case) the maximum amplitude of the angle of attack (α_a = 23° = 0.4 radian) was larger than the static angle of stall. Note that, according to the results presented above (figure 2), the static stall angle of this airfoil is close to 16°.

The values of reduced frequency, k, were chosen taking into account the definition of quasi-steady flow, unsteady flow and highly unsteady flow given in [6].

In figure 3 measured results are compared with theoretical ones provided by Theodorsen's potential flow theory [82] for the un-stalled case (α_a = 4°) and two values of the reduced frequency, k =0.03 and k =0.1.

As it can be observed the agreement between experiments and theoretical predictions is noteworthy (note that Theodorsen's method applies to very thin airfoils, flat plates, whereas the relative thickness of the airfoil here considered is 12%).

The effects on the lift coefficient versus angle of attack curves of the reduced frequency (k =0.03, 0.07 y 0.1), the oscillation amplitude, and the presence of the turbulence generator for the C and T configurations of a NACA 0012 airfoil, are presented in figure 4 (un-stalled case, α_a = 4°) and in figure 5 (stalled case, α_a = 23°).

From data presented in figures 4 and 5 the influence of the main oscillation parameters on the hysteresis cycle can be outlined.

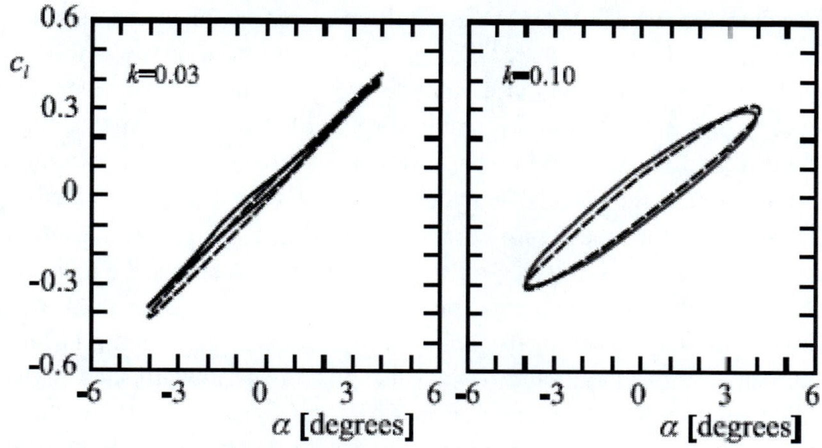

Labels indicate the values of the reduced frequency, $k = \omega c/(2U_\infty)$.

Figure 3. Variation with the angle of attack, α, of the lift coefficient, c_l, of a NACA 0012 airfoil whose angle of attack changes periodically according to the law $\alpha(t) = 0.07\sin\omega t$; solid lines represent experimental data, whereas dashed lines correspond to results from potential flow theory [82].

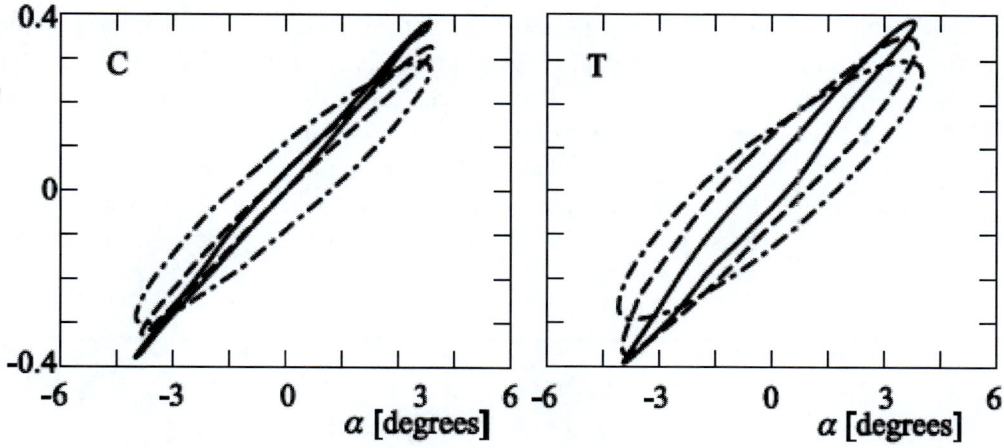

Figure 4. Variation with the angle of attack, α, of the lift coefficient, c_l, of a NACA 0012 airfoil whose angle of attack changes periodically according to the law $\alpha(t) = 0.07\sin\omega t$; experimental results obtained for the indicated values of the reduced frequencies $k = \omega c/(2U_\infty)$: $k = 0.03$ (solid lines); $k = 0.07$ (dashed lines); $k = 0.10$ (dot-dashed lines). C) results corresponding to the clean airfoil, T) the same airfoil with a leading edge turbulence generator.

Concerning the reduced frequency, from the results corresponding to the un-stalled case ($\alpha_a = 4°$, figure 4), it can be concluded that the width of the hysteresis loop increases as the reduced frequency does. This means that, as in the case of plunging airfoils [78] by increasing the reduced frequency, the differences between dynamic pressure coefficients measured in the nose-up part of the pitching movement and those of the nose-down part increase (in the plunging oscillation the corresponding phases are the upstroke portion of the loop and downstroke portion). In addition, from the comparison of the results obtained with a turbulence generator at the leading edge with those obtained with the clean airfoil, it can be concluded also that the width of the hysteresis loop increases when the airfoil is equipped with a turbulence generator which forces boundary layer transition on the airfoil.

For large values of the oscillation amplitude, figure 5 shows the influence of both the reduced frequency and the airfoil configuration on the lift coefficient versus angle of attack curve. These plots give the variation of the instantaneous lift coefficient c_l with the instant-neous angle of attack α of the airfoil. There are three loops in each figure and each loop is for one reduced frequency k. In the figure corresponding to the clean airfoil configuration, the measured static lift curve is plotted as a reference. For large values of the reduced frequency the loops behave as in the un-stalled case, but as k decreases, at larger values of the angle of attack the lift plots evolve into counter-clockwise hysteresis loops with the instantaneous lift lower than the steady un-stalled values when the pitching is upward, and higher than the steady un-stalled values when the pitching is downward. Obviously, to force the transition of the boundary layer on the airfoil to a turbulent one makes this boundary layer less dependent on the reduced frequency (here counter-clockwise hysteresis loops only appear for intermediate values of the reduced frequency, say $k \cong 0.07$).

If the frequency is sufficiently high then it appears that flow separation does not have time to develop. Under these conditions, the onset of stall occurs at the maximum angle of attack in the pitch cycle. This delay in stall onset is, in part, because of the so-called kinematic induced camber effect that is associated with pitch rate. This effect progressively alleviates the leading-edge pressure and pressure gradients for a given value of lift and thus delays the onset of flow separation to a higher angle of attack [6]. This phenomenon is associated with the kinematics of change of angle of attack which produces a normal velocity distribution along the airfoil chord that induces a distribution of curvature along it [2].

Furthermore, the rate of increase varies with increasing the reduced frequencies. As the reduced frequency increases, the rate of increase of both maximum c_l and dynamic stall angle decreases.

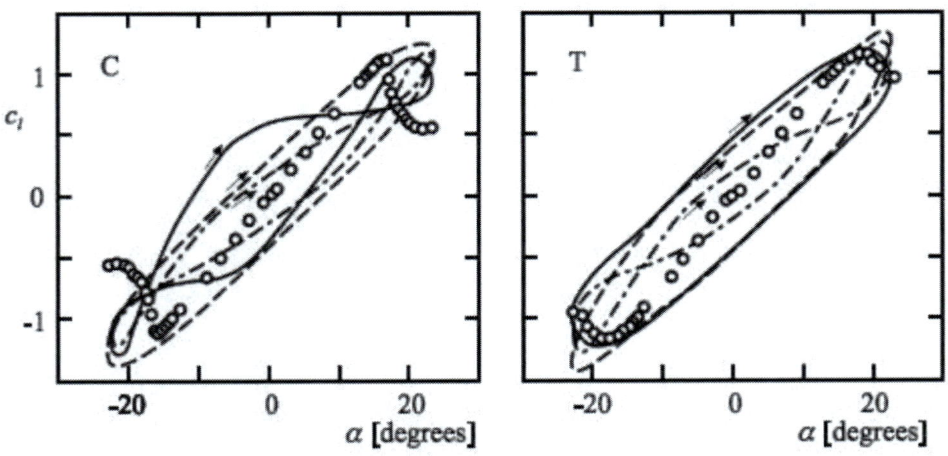

The symbols represent static results, $k = 0$.

Figure 5. Variation with the angle of attack, α, of the lift coefficient, c_l, of a NACA 0012 airfoil whose angle of attack changes periodically according to the law $\alpha(t) = 0.4\sin\omega t$; experimental results obtained for the indicated values of the reduced frequencies $k = \omega c/(2U_\infty)$: $k = 0.03$ (solid lines); $k = 0.07$ (dashed lines); $k = 0.10$ (dot-dashed lines). C) results corresponding to the clean airfoil, T) the same airfoil with a leading edge turbulence generator.

From the measured pressure coefficients, the variation with the angle of attack, α, of the static pitching moment coefficient at the quarter chord location, c_m, have been calculated also. The results obtained are depicted in figure 6 for several values of the Reynolds number.

Similarly to the case of the lift coefficient, the variation of the instantaneous aerodynamic moment coefficient c_m with the instantaneous angle of attack α of the airfoil in an oscillation cycle are represented in figure 7 for both configurations: clean airfoil and airfoil with a leading edge wire.

Note that, when compared with the clear case, the turbulence generator diminishes the magnitude of the pitching-moment coefficient. Note that pitching-moment coefficient curves reproduce the same behaviour as the lift coefficient ones. Note also that the different c_m-α hysteresis loops have a positive total aerodynamic damping.

To get additional insight on the behaviour of an oscillating airfoil, the influence of the leading edge wire on the pressure distribution for a typical configuration is shown in figure 8.

Due to adverse pressure gradients, the laminar boundary layer separates near the leading edge. However, because of the transition to turbulent of the separated laminar flow, the shear layer reattaches on the airfoil upper side and the recirculation bubble is formed near the leading edge [84, 85]. The laminar recirculation bubble where the boundary layer is separated clearly appears as a small but intense suction peak in the pressure distribution on the body surface.

However, when the angle of attack increases beyond a threshold value, boundary layer separates and then the lift coefficient decreases.

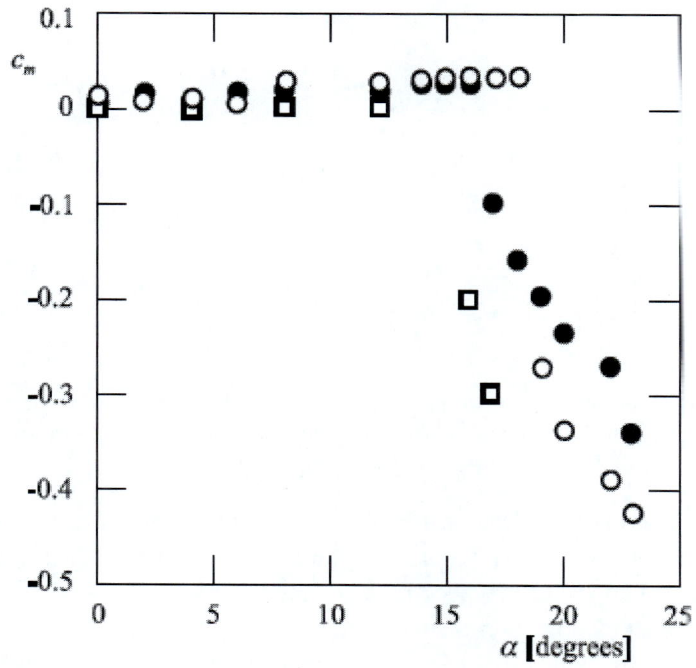

Result for R = 6×10[6] are from [83].

Figure 6. Variation with the angle of attack α of the steady pitching moment coefficient at the aerodynamic center, c_m, for several values of the Reynolds number: Re = 1.7×10^5 (white circles), Re = 4×10^5 (black circles) and Re = 6×10^6 (squares).

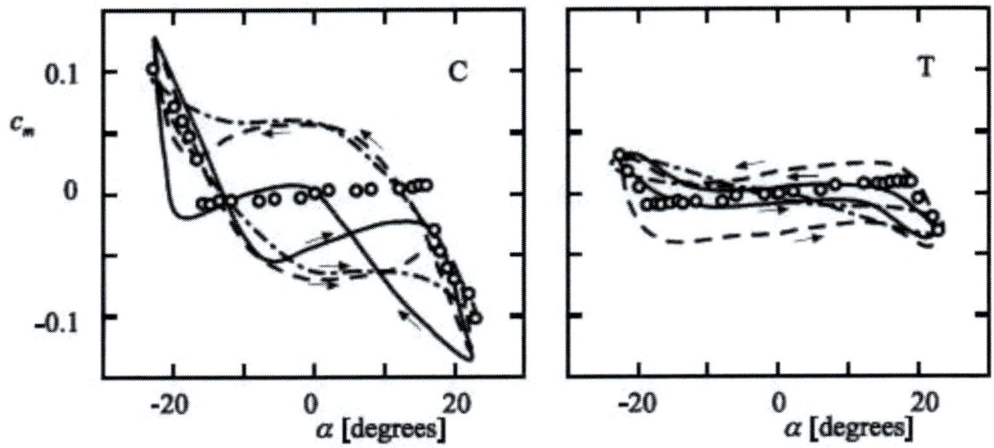

The symbols represent static results, $k = 0$.

Figure 7. Variation with the angle of attack, α, of the pitching moment coefficient at the aerodynamic center, c_m, of a NACA 0012 airfoil whose angle of attack changes periodically according to the law $\alpha(t) = 0.4\sin\omega t$; experimental results obtained for the indicated values of the reduced frequencies $k = \omega c/(2U_\infty)$: $k = 0.03$ (solid lines); $k = 0.07$ (dashed lines); $k = 0.10$ (dot-dashed lines). C) results corresponding to the clean airfoil, T) the same airfoil with a leading edge turbulence generator.

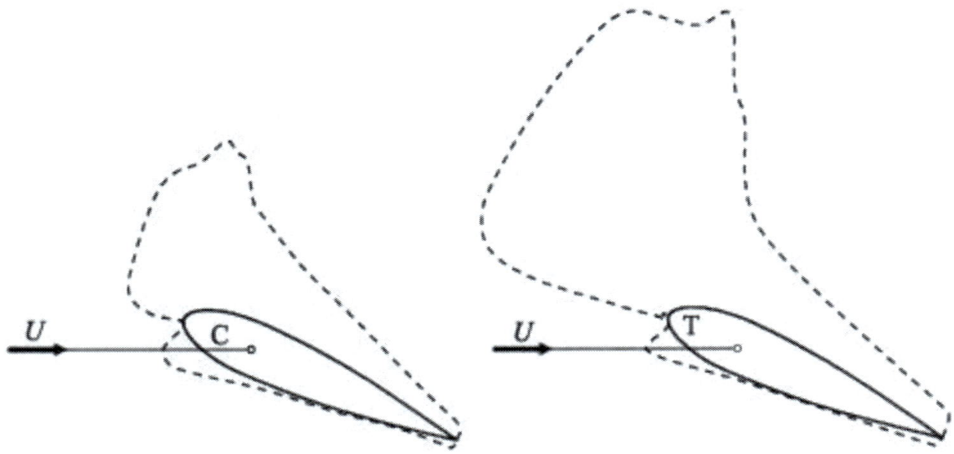

Figure 8. Comparison of the pressure distribution on the surface of a NACA 0012 airfoil. Results correspond to the clean airfoil (C) and the airfoil equipped with a leading edge turbulence wire 3 mm in front of the leading edge (T). In both plots the Reynolds number is $Re = 1.7\times10^5$.

Therefore, for the cases under consideration (medium-thickness airfoils that stall at the leading edge), the boundary layer separation starts close to the leading edge and quickly spread until the whole upper surface becomes separated.

This results in an almost instantaneous increase in the nose-down pitching moment which, together with the hysteresis caused by dynamic stall delay, causes large values of negative damping. An attempt to improve the damping characteristics is made by using a passive boundary layer control device, such as a vortex generator (leading edge wire), on the NACA 0012 airfoil.

The goals being to improve the maximum lift coefficient as well as to reduce, or to eliminate the negative damping which appears in the experimental configuration here considered.

In conclusion, an extensive experimental investigation has been performed on an oscillating airfoil in a pitch mode over a range of reduced frequencies, $k = 0, 0.03, 0.07$ and 0.1, from low to high angles of attack. The unsteady aerodynamic loads are calculated by integrating surface pressure measurements. Variations of the aerodynamic coefficients with the angle of attack show a remarkable sensitivity to the reduced frequency and boundary layer turbulence. By using a simple element like a wire placed at the leading edge, noticeable lift coefficient increments are obtained by forcing the transition of the laminar boundary to a turbulent regime. Pitching moment coefficients are also sensitive to both reduced frequency and boundary layer turbulence.

The measured data show that the incorporation of a simple leading-edge turbulence generator like a wire modifies the aerodynamic damping of the airfoil.

Galloping Phenomena. Linear Approach to the Onset of Galloping

The aeroelastic galloping problem can be modelled as a two-dimensional body immersed in a uniform flow with constant velocity U and fluid density ρ, as shown in figure 9. The two-dimensional body mass per unit length is m, and the cross-section height is b. The body is elastically mounted on a support with damping characterized by a damping coefficient ζ and a stiffness $m\omega_n^2$ (where ω_n is the angular natural frequency, $\omega_n = 2\pi n_n$).

Theoretical foundations of galloping are well established and understood, very often supported by experiments. Galloping can be explained by taking into account that although the incident velocity U is uniform and constant, because of the lateral oscillation of the body, in a body reference system, the total velocity changes both magnitude and direction with time.

Therefore, the body angle of attack also changes with time, hence the aerodynamic forces acting on it (figure 9). The body at rest is oriented at a given angle of attack α_0 with respect to the incident flow (α_0 is defined with respect to some representative direction anchored to the body, as, for instance, its zero lift direction).

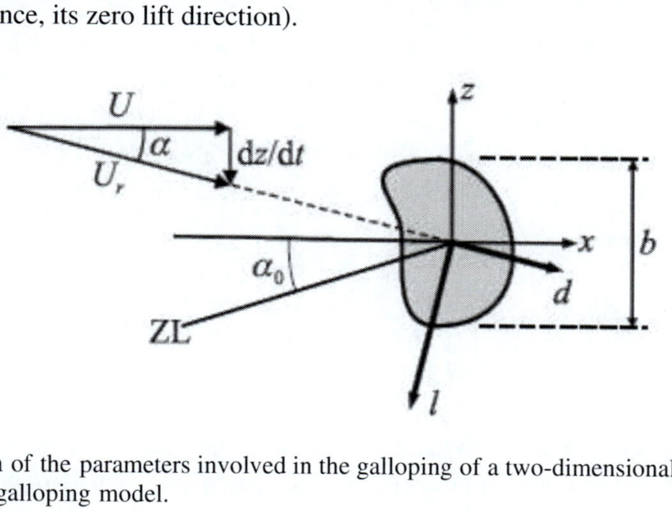

Figure 9. Definition of the parameters involved in the galloping of a two-dimensional body in the one degree of freedom galloping model.

Assuming the structure oscillating along z axis direction within the uniform flow, the relative velocity between the fluid and the body is

$$U_r = \sqrt{U^2 + \left(\frac{\mathrm{d}z}{\mathrm{d}t}\right)^2}, \tag{0.1}$$

and the angle of attack due to oscillation is, in a first approximation (and because α is a small angle), $\tan\alpha \cong \alpha = (\mathrm{d}z/\mathrm{d}t)/U$, where α is the angle between the unperturbed flow velocity U_r and the x axis. Therefore, the aerodynamic forces, drag $d(\alpha_0+\alpha)$ and lift $l(\alpha_0+\alpha)$, are

$$\left.\begin{aligned} d(\alpha_0+\alpha) &= \frac{1}{2}\rho U_r^2 b c_d(\alpha_0+\alpha) \\ l(\alpha_0+\alpha) &= \frac{1}{2}\rho U_r^2 b c_l(\alpha_0+\alpha) \end{aligned}\right\}, \tag{0.2}$$

where, as said above, b is a characteristic length normal to incident flow, and c_d and c_l stand for the drag coefficient and for the lift coefficient, respectively. The projection of those forces in z axis direction is

$$f_z(\alpha) = -d(\alpha_0+\alpha)\sin\alpha - l(\alpha_0+\alpha)\cos\alpha. \tag{0.3}$$

On the other hand, the equation of the movement of the body is:

$$m\left(\frac{\mathrm{d}^2z}{\mathrm{d}t^2} + 2\zeta\omega_n\frac{\mathrm{d}z}{\mathrm{d}t} + \omega_n^2 z\right) = f_z(\alpha) =$$

$$= -\frac{1}{2}\rho U_r^2 b\big(c_d(\alpha_0+\alpha)\sin\alpha + c_l(\alpha_0+\alpha)\cos\alpha\big). \tag{0.4}$$

Considering the movement to be quasi-steady, and assuming the angles of attack due to oscillation to be small enough ($\alpha \ll 1$), from the expression (0.4) the Glauert-Den Hartog criterion for galloping instability is derived [20, 86]. In effect, the introduction of the following expansions: $U_r = U + O(\alpha^2)$, $\sin\alpha = \alpha + O(\alpha^3)$, $\cos\alpha = 1 + O(\alpha^2)$, $c_d(\alpha_0+\alpha) = c_d(\alpha_0) + O(\alpha)$, and $c_l(\alpha_0+\alpha) = c_l(\alpha_0) + (\mathrm{d}c_l/\mathrm{d}\alpha)_{\alpha 0}\alpha + O(\alpha^2) = (\mathrm{d}c_l/\mathrm{d}\alpha)_{\alpha 0}\alpha + O(\alpha^2)$, remember that the definition of α_0 implies $c_l(\alpha_0) = 0$, into expression (0.4) gives

$$\frac{\mathrm{d}^2z}{\mathrm{d}t^2} + 2\zeta\omega_n\frac{\mathrm{d}z}{\mathrm{d}t} + \omega_n^2 z = -\frac{\rho U^2 b}{2m}\left[\left(c_d(\alpha_0) + \left.\frac{\mathrm{d}c_l}{\mathrm{d}\alpha}\right|_{\alpha 0}\right)\alpha + O(\alpha^2)\right] \tag{0.5}$$

and then, neglecting $O(\alpha^2)$ terms, since $\alpha = (\mathrm{d}z/\mathrm{d}t)/U$, if the aerodynamic force (proportional in this case to $\mathrm{d}z/\mathrm{d}t$) is considered as a contribution to the total damping of the system (aerodynamic damping), the total damping coefficient is:

$$\zeta_T = \zeta + \frac{\rho U b}{4 m \omega_n}\left(\frac{\mathrm{d}c_l}{\mathrm{d}\alpha} + c_d\right)\Bigg|_{\alpha=\alpha_0} \tag{0.6}$$

therefore the oscillation will be stable if $\zeta_T > 0$ and unstable if $\zeta_T < 0$. As the mechanical damping ζ is generally positive, instability will only occur if $(\mathrm{d}c_l/\mathrm{d}\alpha + c_d) < 0$, which is a necessary condition for galloping instability. The sufficient condition is $\zeta_T < 0$, or, according to (0.6):

$$\left(\frac{\mathrm{d}c_l}{\mathrm{d}\alpha} + c_d\right)\Bigg|_{\alpha=\alpha_0} < -\frac{4 m \zeta \omega_n}{\rho U b} \tag{0.7}$$

The expression that gives necessary condition for galloping, $(\mathrm{d}c_l/\mathrm{d}\alpha + c_d) < 0$, is known as Glauert-Den Hartog criterion, or simply Den Hartog criterion, and it is pertinent to cite here the footprint of page 21 of Païdousis et al. [24]: "It is both curious and significant that Scruton [33, 49] never cites DenHartog's work, attributing the galloping criterion to Glauert instead; it is even more curious that Den Hartog does not mention Glauert's work, even though he was familiar with the ARC Reports and Memoranda. Parkinson and Smith [34] cite both, making it clear that Glauert's work has precedence".

It must be pointed out that in (0.7) the second member tends to zero when the velocity increases, which means that the possibility of galloping increases as velocity increases; another conclusion that can be derived from (0.7) is that, as the second member tends to infinity when the velocity tends to zero, galloping will only be possible beyond a minimum threshold value of the incident wind velocity.

To use the Glauert-Den Hartog criterion for galloping instability one clearly needs to know the aerodynamic coefficients c_l and c_d as a function of the angle of attack associated to the body oscillation, α. For a given section shape, this data may be obtained from static wind-tunnel measurements, in which the orientation of the section is varied relative to the wind, generating results such as those in figure 10, where the variation with the body reference angle of attack, α_0, of the aerodynamic lift and drag coefficients of triangular isosceles cross section cylinders is shown. For the application of the Glauert-Den Hartog criterion for galloping instability the quasi-steady assumption must be invoked, according to which c_l and c_d in the course of the oscillation around a given value of α_0 are the same at each angle of attack as the values measured at the corresponding steady angle of attack α_0 in wind-tunnel experiments.

As already stated, galloping has been assumed a quasi-steady phenomenon, so that the time evolution of the body is modelled as a succession of static equilibrium states. From the point of view of fluid dynamics, such a concept is applicable if the motion is slow enough, which implies that the characteristic time associated to body oscillation (the oscillation period) is large enough when compared to the time spent by a fluid particle to pass across the neighbourhood of the body (the residence time). According to [24], two different criteria have been proposed concerning the applicability of quasi-steady fluid dynamics hypothesis: the first one was formulated by Fung [87], and the other by Blevins [22].

Fung [87] states that any disturbance experienced by the oscillating body at a certain point in its oscillatory motion must be swept downstream sufficiently far, by the time the body comes back to that same point (one period later), for the disturbance to no longer affect the flow around the body. If disturbances are swept downstream with a velocity equal to the free-stream velocity U, then during a period of oscillation disturbances move a distance U/f_n, f_n being the frequency of the oscillation. In [87] it is proposed that this distance should be at least ten times the body characteristic length b, that is $U/(f_n b)\geq 10$.

In [22], the criterion is that the frequency of shed vortices (the disturbance to the mean flow) must be at least twice as large as the oscillation frequency f_n. Assuming a typical value of the Strouhal number $S = f_S b/U = 0.2$, where f_S stands for the shedding vortices frequency, it follows that should be $U/(f_n b) \geq U/(f_S b) = 2/S \cong 10$. Thus, as pointed out in [24], it is interesting to note that, although different physical reasoning was employed, the same end result was obtained by Blevins and Fung for quasi-steady fluid dynamics to be applicable.

THEORETICAL FOUNDATIONS OF THE HYSTERESIS IN GALLOPING

The description of the behaviour of a structure under the action of an incident flow is an extremely complex problem, whose theoretical analysis requires the settling of simplified models for making the analysis more affordable.

In the derivation of such a simplified mathematical model some simplifying assumptions are needed, the ones giving to the simplest model being that the structure is described as a linear oscillator of one degree of freedom, that the structure is sufficiently slender to consider two-dimensional flow, and finally that the incident flow is free of turbulence.

Assume a structure that at rest is oriented at given angle of attack α_0 with respect to the incident flow. If the structure is oscillating along z axis direction within an uniform flow with velocity U (figure 9), then, by following a procedure similar to the one stated in previous subsection, the relative velocity between the fluid and the body is $U_r = [U^2 + (\mathrm{d}z/\mathrm{d}t)^2]^{1/2}$, and the angle of attack due to oscillation is $\Delta\alpha = \mathrm{atan}((\mathrm{d}z/\mathrm{d}t)/U_r)$. Therefore, drag $d(\alpha_0+\Delta\alpha)$ and lift $l(\alpha_0+\Delta\alpha)$ are:

$$\left.\begin{aligned} d(\alpha_0+\Delta\alpha) &= \frac{1}{2}\rho U_r^2 b c_d(\alpha_0+\Delta\alpha) \\ l(\alpha_0+\Delta\alpha) &= \frac{1}{2}\rho U_r^2 b c_l(\alpha_0+\Delta\alpha) \end{aligned}\right\}, \tag{0.8}$$

respectively, where the different symbols are already defined in the previous subsection. The projection of these forces in z axis direction is

$$f_z(\alpha_0+\Delta\alpha) = -\frac{1}{2}\rho U_r^2 b\left[c_d(\alpha_0+\Delta\alpha)\sin\Delta\alpha + c_l(\alpha_0+\Delta\alpha)\cos\Delta\alpha\right], \tag{0.9}$$

thence, since $U_r = [U^2 + (\mathrm{d}z\mathrm{d}t)^2]^{1/2} = U^2[1+(\Delta\alpha)^2]$, the dimensionless coefficient of the aerodynamic force in the normal direction to the incident flow becomes

$$c_f(\alpha_0+\Delta\alpha)=-\left[1+(\Delta\alpha)^2\right]\left[c_d(\alpha_0+\Delta\alpha)\sin\Delta\alpha+c_l(\alpha_0+\Delta\alpha)\cos\Delta\alpha\right] \tag{0.10}$$

Therefore, the equation of the movement of the body, assuming that the mass per unit length is m, and that it is elastically mounted on a support with lineal damping characterized by a dimensionless structural damping coefficient ζ_η and an undamped natural frequency ω_η, in dimensionless variables reads:

$$\frac{d^2\eta}{d\tau^2}+2\zeta_\eta\frac{d\eta}{d\tau}+\eta=-\mu U_d^2 c_f(\alpha_0+\Delta\alpha), \tag{0.11}$$

where η denotes the dimensionless vertical position, $\eta = z/b$, τ is the dimensionless time, $\tau = \omega_\eta t$, μ is the dimensionless mass of the structure per unit length, $\mu = \rho b^2/(2m)$, and U_d is the reduced velocity, $U_d = U/(\omega_\eta b)$.

From equation (0.11) the main features concerning the dynamics of galloping can be derived. Assuming the angles of attack due to oscillation to be small enough ($\Delta\alpha << 1$), the normal aerodynamic force coefficient can be expanded in powers of $\Delta\alpha$ (which is proportionnal to $d\eta/d\tau$):

$$c_f=\sum_{n=0}^{\infty}a_n(\Delta\alpha)^n. \tag{0.12}$$

where, obviously, the linear term of the expansion (0.12) can be considered as a contribution to the total damping of the system (aerodynamic damping). Then, since the mechanical damping is generally positive, instability can only occur if the aerodynamic damping is negative, thus the well-known Den Hartog stability criterion for the potential susceptibility of a structure to galloping starting from a given equilibrium position is obtained: $(dc_l(\alpha)/d\alpha)|_{\alpha 0}$ + $c_d(\alpha_0) < 0$, as obtained in the previous section.

Once the function $c_f(\alpha_0)$ is known, equation (0.11) can be solved either numerically or by asymptotic methods if the nonlinear aerodynamic term is small. In the case that both aerodynamic and damping forces are small compared with inertia and stiffness forces, solutions of equation (0.11) tend to a limit cycle of quasi-harmonic oscillations (this behaviour of the structure is quite usual because their mean densities are much higher than the density of the surrounding fluid). Depending on the shape of the curve $c_f(\alpha_0)$ some hysteresis phenomena can appear, the hysteresis being characterized by the existence of a solution that can alternately reach different cycle limits (the amplitude of the cycle limit of the body oscillation can be different, depending on whether the reduced velocity is increasing or decreasing).

Leaving apart the early paper of Parkinson and Smith [35], recently the dynamic behaviour of galloping bodies has been analytically analysed by using a perturbation method known as Krylov–Bogoliubov method [42, 46], and the solutions corresponding to a variety of $c_f(\alpha_0)$ curves typical of bluff bodies were studied in order to clarify the correlation between the number of inflection points in the $c_f(\alpha_0)$ curves and the hysteresis that can appear in galloping oscillations, the main conclusion being that the number and position of the inflection points determine the regions in the c_f-α_0 plane prone to hysteresis in the transversal galloping oscillation. The same conclusion results by analysing the galloping dynamics through method of multiple scales [24]. Thus, since $\Delta\alpha$ is small, in a first attempt it is $c_f(\alpha_0) \approx c_l(\alpha_0)$, and then the existence of an inflection point implies $d^2c_l(\alpha)/d\alpha^2 = 0$ (note that in this last expression, α instead α_0 of has been written).

Approximate solutions of the Krylov–Bogoliubov method for galloping are obtained by assuming that the aerodynamic and damping forces are much smaller than the inertial and stiffness ones.

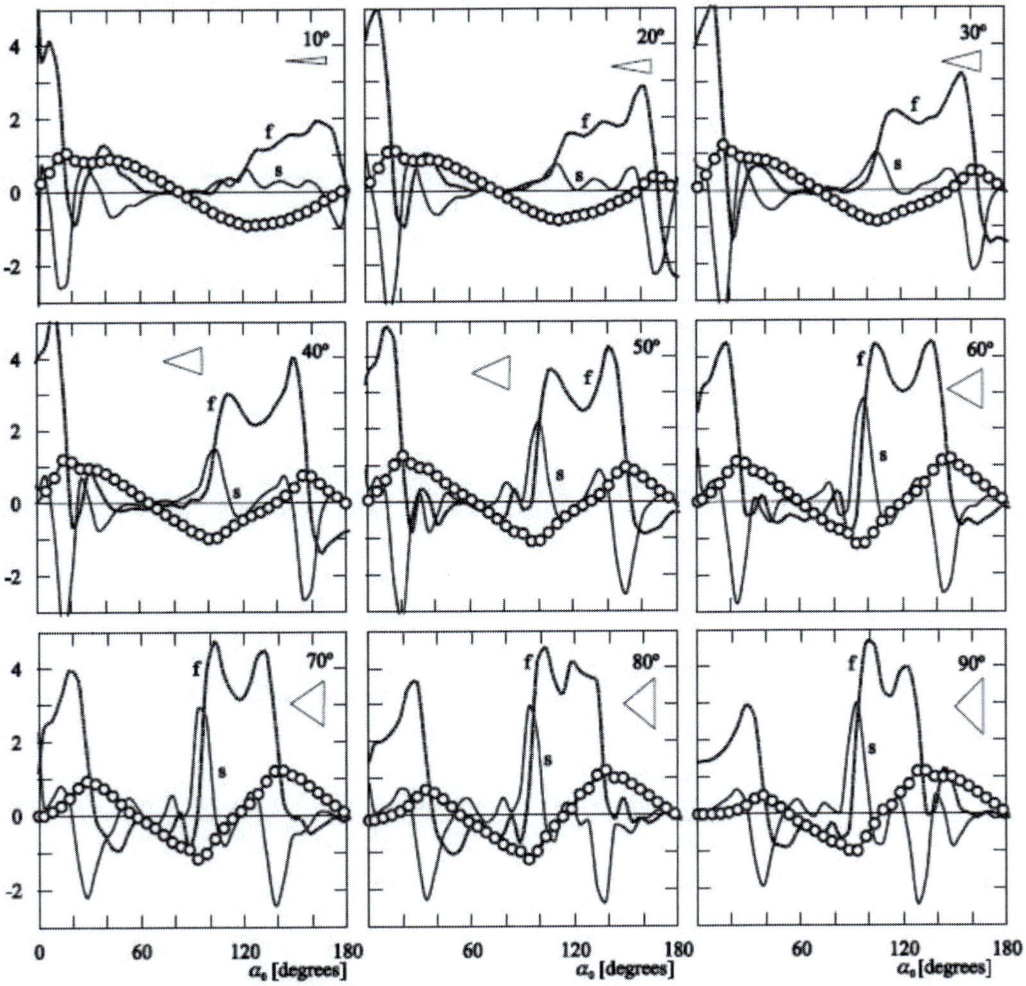

Experimental results are from [48, 54].

Figure 10. Variation with the angle of attack, α_0, of the lift coefficient, c_l (white circles), the Glauert-Den Hartog function, $dc_l/d\alpha + c_d$ (dashed line, label f), and the function $(d^2c_l/d\alpha^2)/10$ (solid line, label s). Note that, due to scale constraints, $(d^2c_l/d\alpha^2)/10$ instead of $d^2c_l/d\alpha^2$ has been represented. Experimental results correspond to triangular cross-section bodies with main vertex angle β ranging from 10° to 60°, as indicated in the plots. Note that the curve $c_l(\alpha)$ as well as $d^2c_l/d\alpha^2$ are anti-symmetric with respect to α = 180°.

From the analysis of the approximate solution, since it is clear the existence of a link between the hysteresis phenomenon and the number of inflection points at the aerodynamic force coefficient curve, the problem reduces to the analysis of the number of terms which must be retained in expression (0.12) when the aerodynamic force coefficient curve is expanded in the neighbourhood of a given value of the angle of attack, because, obviously, the dynamic response of the structure depends on the same number of terms, as explained below.

Transverse galloping accounts for low frequency self-excited oscillations, and it can be developed only in structures whose cross-sections are non-circular. When the density of the structure is much higher than the air density, the frequency of the oscillations is approximately the same as the natural frequency of the structure.

Transverse galloping can be considered as a quasi-steady phenomenon, which allow the introduction of drastic simplifications hypothesis to describe the linearized aerodynamic forces. Since galloping occurs at high flow velocities, the quasi-steady assumption is justified by comparing the characteristic time and the residence time. The residence time of the flow, $t_r = L/U$, must be small compared to the characteristic time, which is related with the changes in the boundary conditions, $t_c = 1/\omega_0$ (where L and U are the characteristic length and velocity of the flow, and ω_0 the natural frequency of vibration of the structure).

Therefore, the condition for quasi-steady assumption becomes $t_r/t_c = \omega_0 L/U << 1$. Note that the fulfilment of this inequality implies that the aerodynamic force acting on the oscillating structure is approximately equal to the static aerodynamic force evaluated at the instantaneous angle of attack of the incident flow.

However, the linearized model is not appropriate for describing the time evolution of the structure once galloping is started, and to perform the dynamic analysis, post-critical, nonlinear terms must be retained. These nonlinear terms introduce a limit in the amplitude of the oscillations, and drive the system towards a limit cycle of oscillation. Once the limit cycle oscillation has been reached, the energy input to the structure by the flow equals the energy dissipated by structural damping averaged along a cycle [42].

In [35], the amplitude of the limit cycle oscillations as a function of the flow velocity is shown, and some galloping experiments on the hysteresis effect observed in galloping are described. In those experiments it is found that after the onset of galloping, multiple solutions for the amplitude of oscillations can appear for a range of flow velocity. In addition, depending on whether the reduced velocity is increasing or decreasing in the experimental sequences, the amplitude of the limit cycle oscillations can be different. As already stated, the galloping hysteresis is related to the existence of inflection points in the curve of the static aerodynamic coefficient $c_z(\alpha)$, where α is the angle between the incident flow and a reference direction measured in the static equilibrium position of the structure.

The presence of an inflection point is due to the reattachment of the flow on either the lower or the upper side of the body, as can be deduced from the visualization experiments carried out Luo et al. [45], concerning the hysteresis phenomenon in the galloping oscillation of square cylinders.

Assuming small values of the velocity ratio $(\mathrm{d}z/\mathrm{d}t)/U_r$, the static normal force coefficient can be expanded in a Taylor series of the angle of attack, expression (0.12), and then, equation (0.11) can be rewritten as a nonlinear ordinary differential equation:

$$\frac{d^2\eta}{d\tau^2}+2\zeta_\eta\frac{d\eta}{d\tau}+\eta=\mu U_d^2\sum_{n=0}^{\infty}a_n(\Delta\alpha)^n, \qquad (0.13)$$

expression (0.13) being valid in a small range of values of the angle of attack close to the static reference angle $\alpha = \alpha_0$.

Obviously, if only the first term of the expansion (0.12) is retained, Glauert-Den Hartog's result is recovered and the critical reduced velocity of galloping is given by $U_g = 2\xi_\eta/(\mu a_1)$. Once the function $c_z(\alpha)$ is known, equation (0.13) can be solved either numerically or by asymptotic methods if the nonlinear term is small. In the case that both aerodynamic and damping forces are small compared with inertia and stiffness forces (of order of unity in the dimensionless equation), solutions of equation (0.13) tend to a limit cycle of quasi-harmonic oscillations (which is a quite usual behaviour if the structure mean density is much higher than that of the surrounding fluid).

Another consequence that can be derived from the Taylor series expansion around a given value $\alpha = \alpha_0$ of the angle of attack is that only the antisymmetric part of $c_z(\alpha)$ plays a relevant role in the dynamics. In effect, the aerodynamic work in an oscillation cycle can be expressed as the integral in a cycle of the product of the lateral force by the lateral velocity:

$$\int_0^T F_z\frac{dz}{dt}dt=\frac{1}{2}\rho U_r^2 b\int_0^T c_z\frac{dz}{dt}dt \qquad (0.14)$$

Obviously, under harmonic oscillations, only the terms with odd powers in the polynomial $c_z(\alpha)$ yield nonzero contributions to the integral (0.14). Thus, in the following, only weak aerodynamic forces are considered and only odd terms in the polynomial expression for $c_z(\alpha)$ are retained.

Since the hysteresis region is characterized by the existence of a solution that can alternately reach different cycles limits, the characteristic polynomial that provides the amplitude of the limit cycle oscillation must have more than a real positive solution. A standard procedure to analyse the behaviour of the solution of (0.13) after the introduction of the series expansion (0.12) is to study the solutions corresponding to a variety of $c_z(\alpha)$ curves typical of bluff bodies in order to clarify the correlation between the number of inflection points and the hysteresis. According to [46, 47], if a polynomial with only terms up to $(\Delta\alpha)^3$ is fitted to experimental data on $c_z(\alpha)$, the result which is obtained from this antisymmetric polynomial approximation (the variation of the dimensionless amplitude of the oscillation limit cycle, A, with the reduced velocity U_R) does not exhibit inflection points and then hysteresis does not appears. The minimum order of an antisymmetric polynomial which exhibits one inflection point is five. In this case, the dimensionless average amplitude as well as the influence of the position of the point of inflection in the size of the hysteresis region can be determined. Seven is the minimum order of an antisymmetric polynomial to get two inflection points, and then the existence of three different solutions separated by two saddle–node bifurcation points can be demonstrated. The response of the system with two inflection points is given in figure 11.

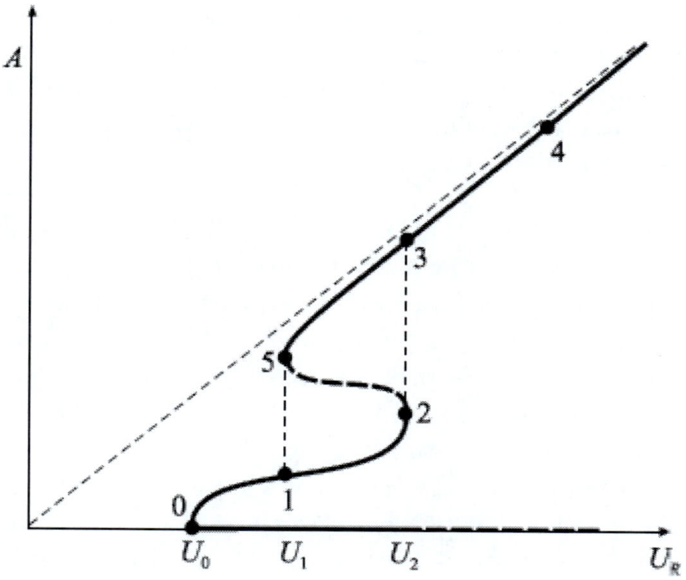

Figure 11. Galloping amplitude, A, versus reduced flow velocity, U_R, of a body whose lateral force curve has two inflection points.

Observe that the appearance of two inflection points brings a hysteresis phenomenon with three limit cycle oscillations. Obviously, the shape of the $c_z(\alpha)$ curve depends on the geometry of the body and the initial orientation of that body with respect to the incident flow (i.e. the initial angle of attack before oscillations occur). For some interesting geometries from the point of view of their potential application to actual engineering problems, the variation of $c_l(\alpha)$ and $c_d(\alpha)$ for the whole range $0 \le \alpha \le \pi$ has been obtained experimentally: triangular cross section bodies were considered in [54-56], biconvex and rhomboidal sections in [57], elliptical sections in [58], and H and Z shaped beams in [59] and [60], respectively.

Experimental coefficients were used to determine galloping stability regions according to the Glauert-Den Hartog criterion. But also from those aerodynamic coefficients it is easy to determine the $c_z(\alpha)$ curve from any given value of the angle of attack. Plotting the antisymmetric part of that curve, and looking at the number and position of the inflection points, the analysis presented in [46, 47], can be applied to determine regions of hysteresis in the galloping phenomenon for the respective bodies and initial angles of attack.

The analysis also allows the determination of the qualitative behaviour of the oscillating response by a simple inspection of the static aerodynamic data. For example, if there is only one inflection point at the $c_z(\alpha)$ curve, it can be automatically deduced that the oscillator may present hysteresis via a subcritical bifurcation. Moreover, some information on the magnitude of the hysteresis region can be obtained from the location of the inflection point. It should also be emphasized that the qualitative behaviour of the response to galloping and the number of limit cycles are independent of the number of terms of the polynomial used to fit the aerodynamic data whenever the former reproduces all the inflection points of the latter curve.

Returning to equation with seven odd terms and possible real roots (see figure 11, where velocities U_1 and U_2 are identified.), the following cases can occur [24],

1 for $U < U_0$ the system is stable;

2 for $U_0 < U < U_1$ and for $U > U_2$ two roots are found, one of which corresponds to a stable limit cycle, whereas the trivial solution is now unstable;
3 for $U_1 < U < U_2$ four roots are found, involving (a) the unstable source at the origin of the phase plane, (b) a stable limit cycle nesting within (c) an unstable one, which in turn nests within (d).

Note that, according to figure 11, the bifurcation is supercritical, but one can have (a) a single, small amplitude, stable limit cycle; (b) two, one with small and the other with large amplitude, with an unstable one in-between; or (c) only a large amplitude stable limit cycle.

Parkinson and Smith [35] also conducted experiments for square prisms, with variable ζ. By plotting the results in reduced variables: $\mu a_1 A/(2\zeta)$ versus $\mu A_1 U/(2\zeta)$, where μ is the dimensionless mass defined in equation (0.11), A is the galloping amplitude, and a_1 is related to the first term of the Taylor series (0.12), the results measured for various values of ζ collapse nicely onto effectively a single curve, as shown in figure 12. The experimental results agree remarkably well with the theoretical curve.

As the flow velocity is increased to beyond U_2 in figure 11, the galloping amplitude should jump from point 2 to point 3, and then continue on to point 4; whereas starting from point 3 and reducing U to less than U_2, the galloping follows the path from 3 to 5, and then jumps down to point 1. Remarkably, even these jumps, indeed the whole hysteresis phenomenon, are well followed by the experimental data.

Experimental Results Concerning Galloping

In this section experimental results concerning the hysteresis of galloping isosceles triangular cross-section prisms are presented.

As already stated, hysteresis requires that for a given angle of attack, α, and triangle geometry (identified by the main vertex angle, β) the body must be prone to gallop (that means that the Glauert-Den Hartog stability criterion must be fulfilled) and that close to the angle of attack under consideration the $c_f(\alpha) \approx c_l(\alpha)$ curve present an inflection point (note that, accordingly with the already used nomenclature, α_0 instead of α should be written, although here the subscript 0 is omitted, because there is not possibility of confusion).

Therefore, there are two conditions to be fulfilled simultaneously: the Glauert-Den Hartog criterion, which implies $\mathrm{d}c_l(\alpha)/\mathrm{d}\alpha + c_d(\alpha) < 0$, and the existence of a inflection point, which roughly speaking implies $\mathrm{d}^2 c_l(\alpha)/\mathrm{d}\alpha^2 = 0$.

The hysteresis phenomenon of two dimensional bodies having isosceles triangular cross-sections (figure 13) when subjected to cross-flow translational galloping vibration has been analysed through wind tunnel experiments. Triangular cross-section bodies have been chosen because these bodies have been widely studied in the past [54-56, 88], so that a large amount of information is now available. In order to clarify the influence of the geometry on the hysteresis, six different triangular cross-sections have been considered, so that the hysteresis map in the angle of attack - triangle main vertex angle (α, β) has been experimentally studied.

According to the results already published [54, 55, 57, 58], the stability to transverse translational galloping of triangular cross-section cylinders are both cross-sectional geometry and angle of attack dependent.

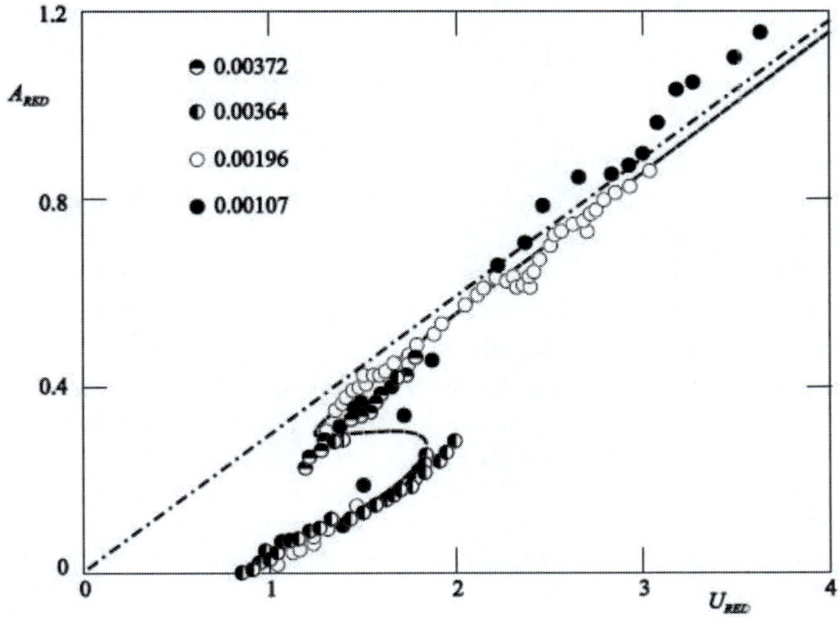

After [35].

Figure 12. Variation with the reduced flow velocity $U_{RED} = nA_1U/(2\zeta)$, of the reduced galloping amplitude $A_{RED} = na_1A/(2\zeta)$ of square prisms. Symbols identify experimental data points according to the values of the dumping coefficient as indicated in the insert. In experiments the Reynolds number ranges from 4×10^3 to 20×10^3; the dimensionless mass is $\mu = 4.3\times10^{-4}$.

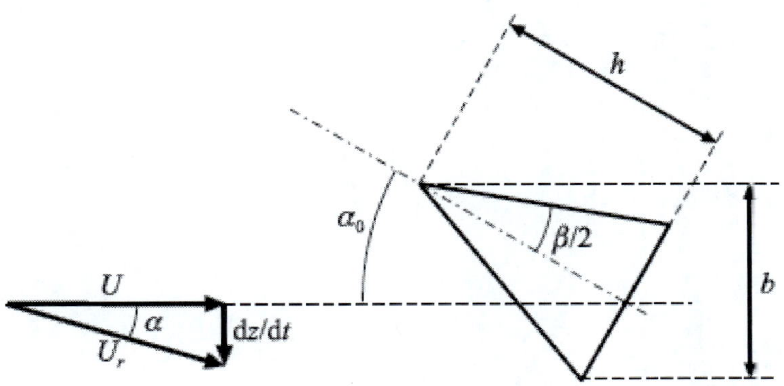

Figure 13. Definition of the geometry of the isosceles triangular bodies, and of the parameters involved in the one degree of freedom galloping model.

Available experimental results also show that configurations which are stable at a certain reduced incident wind velocity become unstable when that velocity is increased beyond a critical value that depends on the configuration (cross-section geometry and angle of attack). Isosceles triangular cross-section prisms may present galloping instability (if the appropriate critical incident wind velocity is reached) in configurations where the main vertex of the triangle is almost pointing the incident flow (at angles of attack higher than the ones where the triangular profile stalls, around $\alpha = 20°$) or in configurations with the base of the triangle facing the wind flow ($\alpha \approx 180°$, the flow approaching clearly a bluff body), covering the angles of attack interval $\alpha \approx 180^{\circ} \pm \beta/2$, basically when the wind flow is aligned with the lateral sides of the triangle [55], Another less critical instability region appears for intermediate values of α, between 50º and 90º, provided the vertex angle β is high enough (figure 14).

However, as already stated this is not the only criterion to be considered for hysteresis, in addition the $c_l(\alpha)$ curve must present an inflection point within the range of values of the angle of attack where galloping can takes place. Figure 10 shows the variation with the angle of attack α of the functions $c_l(\alpha)$, $dc_l(\alpha)/d\alpha + c_d(\alpha)$ and $d^2c_l(\alpha)/d\alpha^2$ corresponding to the six triangle geometries under consideration (β = 10º, 20º, 30º, 40º, 50º and 60º). In these plots inflection points are identified by the condition $d^2c_l(\alpha)/d\alpha^2 = 0$, thus only angles of attack close to these inflection points lying within galloping instability regions have been tested. These points where hysteresis can appears are identified with thick striped lines in figure 15.

To perform the experiments reported in [48], an open circuit wind tunnel was used. Wind tunnel test chamber is 0.15 m width, 0.90 m high and 1.20 m long. Wind velocity profile at the model test section is uniform within ±1%, the turbulence intensity being around 2%. The wind tunnel is powered by an electric motor with variable frequency drive, which provides a better control on the wind velocity.

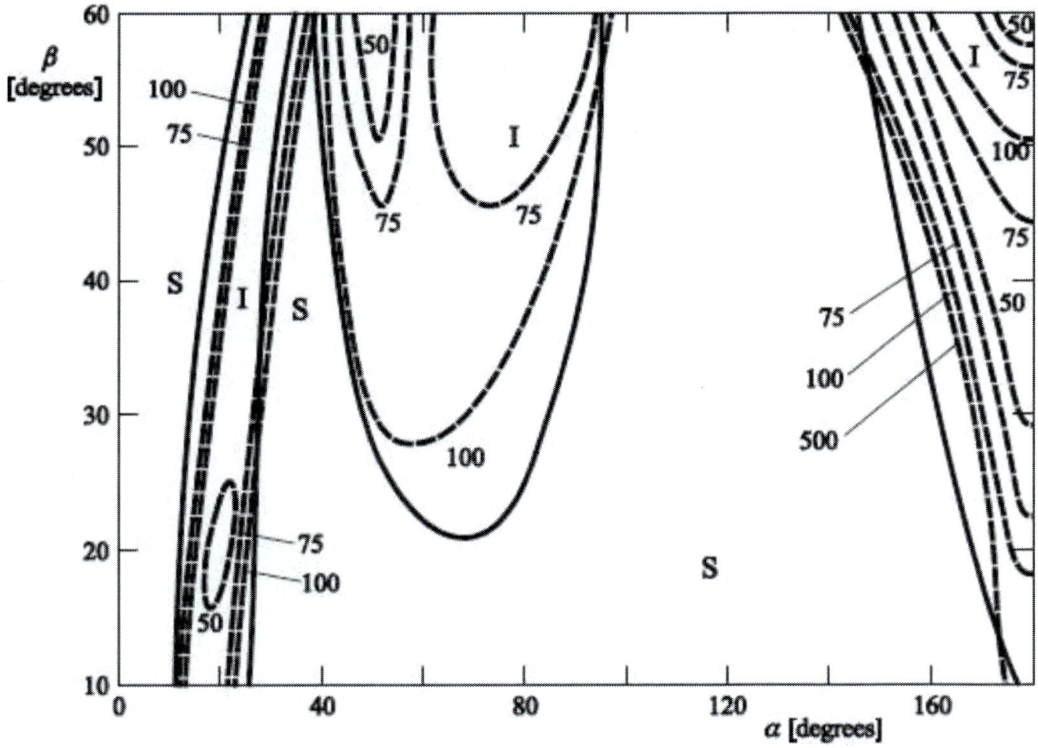

From [54].

Figure 14. Stability diagram in the angle of attack-main vertex angle plane (α, β) based on dynamic test results. Numbers on the curves (dashed lines) indicate the value of the reduced incident wind velocity U_R. Label "I" identifies instability regions, while "S" indicates stability regions in the diagram. Note that the stability diagram is symmetric with respect to α = 180º. Solid lines represent boundaries of the unstable region according to static Glauert-Den Hartog criterion.

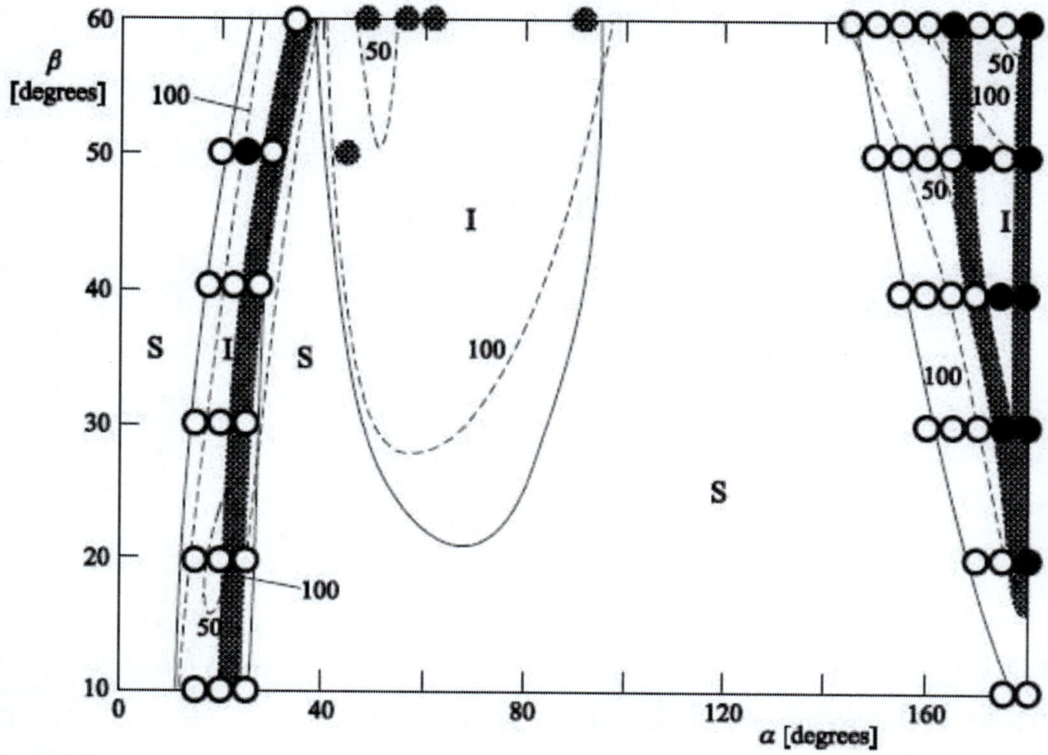

From [48].

Figure 15. Stability diagram in the angle of attack-main vertex angle plane (α, β) based on dynamic test results. Numbers on the curves (dashed lines) indicate the value of the reduced incident wind velocity U_R. Note that the stability diagram is symmetric with respect to α = 180º. Solid line represents the boundaries of the unstable region according to static Glauert-Den Hartog criterion (Alonso et al., 2005). Striped zones indicate the regions where hysteresis can appear. Circles identify the tested configurations: a black circle indicates that hysteresis has been detected, and a white one that this phenomenon has not been observed.

The wind velocity of the stream at the test section of the wind tunnel was in the range from 7 to 26 m/s, which provides Reynolds numbers ranging from 10^4 to 2.6×10^5, based on the model characteristic chord h (figure 13).

Six isosceles triangular cross-section prisms were tested, the main vertex ranging from β = 10º to β = 60º in 10º steps. All triangles share identical maximum characteristic length h = 0.10 m, the base of the triangle being $2h\tan(\beta/2)$. The model span is 0.145 m.

As sketched in figure 16, the selected triangular prism (A) is attached to a rod 8 mm in diameter through an appropriated screw; the rod is fixed to the extreme side of a mechanism (C), in such a way that the prism angle of attack, α, can be set with ±0.5º accuracy (D). The mechanism consists of two long parallel beams which are fixed to a reference frame through ball bearings. At the opposite ends of the long parallel beams is located the vertical small beam that supports the prisms, this small beam being joined to the long ones through ball bearings also. This mechanism allows the vertical (normal to wind) motion of the prisms.

In addition, there are two springs (F), allowing thus to control the frequency of the oscillation in the normal to flow direction. Note that with this testing device oscillation is purely translational.

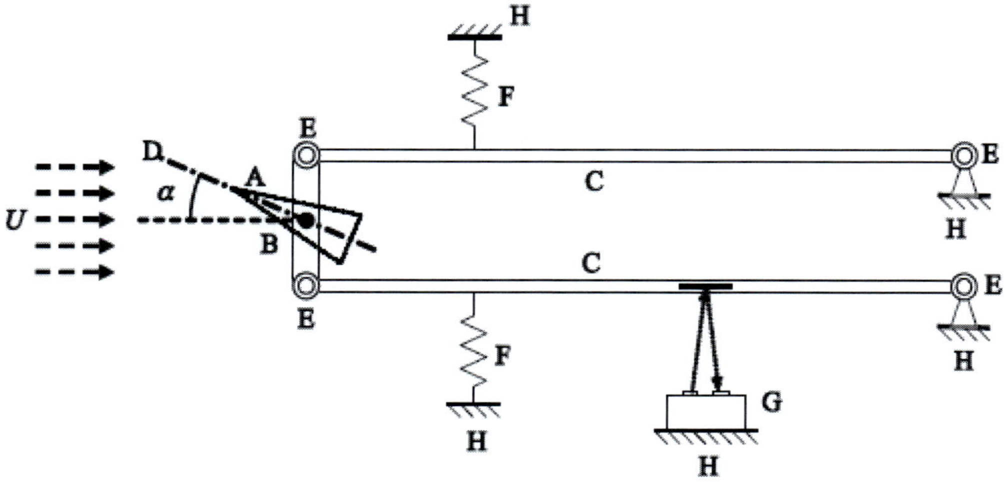

Figure 16. Sketch of the dynamic test lay-out: prism model (A), supporting rod (B), oscillating parallel beams (C), angle of attack scale (D), rotating ball bearings (E), springs (F), laser vibrometer (G), and reference frame (H). All the elements are placed in a tight chamber outside the wind tunnel test, except the prism, which is inside the test chamber, and the supporting rod that joint the prism to the oscillating mechanism, which is partly inside the test chamber and partly outside, passing through a vertical slot made at the test chamber wall.

The normal to flow displacement of the prisms are recorded using a laser vibrometer M7/L100 from MEL Instrument, with a range of 100 mm, a resolution of 6 μm and a sampling frequency of 54 kHz. This equipment is anchored to the fixed frame, the laser beam being directed toward a reflecting surface placed on the lower long beam. Measurements were registered at 800 Hz sampling rate during 25 s with a 12 bit A/D board (from Keithley instruments). Post-processing of the recorded signals gives plots of normal to flow displacement versus time.

Experimental sequence is as follows: once a selected model is fixed to the supporting rod of the holding system, and the chosen reference angle of attack α is set, the experimental sequence runs almost automatically. The wind tunnel is switched on at a small velocity and the output from the laser vibrometer is recorded during 25 s, after a given period of time (typically around 30 second), to let the oscillations reach the limit cycle. Then the wind tunnel velocity is increased and, after waiting again 30 s to overpass the transient period, the new output from the laser vibrometer is recorded. Preliminary tests were performed to determine that a time of 30 s after changing the incident wind velocity was enough to let the configuration reach the limit cycle, with a maximum deviation of 5% in the amplitude of the limit cycle during the following 25 s. The procedure described before is repeated until the entire range of wind velocities is covered (from 7 m/s to 26 m/s at small steps, typically 1 m/s), and then the process is repeated again but now with decreasing wind tunnel velocities (from 26 m/s to 7 m/s, also at small steps). Static and dynamic pressures inside test chamber were measured with an Air Flow 048 Pitot tube connected to a Schaewitz Lucas P-3061-2WD pressure transducer.

The natural frequency ω_η and the structural damping coefficient ζ_η of each one of the models tested have been measured using the same experimental configuration, by letting the model vibrate freely in no wind conditions.

Corresponding values are given in Table 2, together with the mass and the Scruton number, defined as $\mathrm{Sc} = 2m\zeta/\rho b^2$. It must be pointed out that the stiffness of the models has been chosen low enough in order to find critical galloping velocities within the range of velocities allowed by the wind tunnel used in the tests, and also to get natural frequencies below the vortex shedding frequencies (decoupling therefore both phenomena, galloping and vortex shedding). For the same reason, the testing device has been designed with a structural damping as low as possible, since it is not the purpose of this chapter to evaluate the influence of the mechanical properties in the galloping stability.

Once the whole range of velocities has been tested (both in the increasing velocity sense and in the decreasing one), a new value of the angle of attack of reference is set, and the process is repeated again. The same procedure being applied to each one of the triangular cross-section bodies under consideration. The selection on the reference angles of attack has been done taking into account the existing knowledge on the galloping behaviour of these bodies (figure 14).

From the temporal series of normal to wind displacements, $a(t)$, corresponding to each wind velocity U and tests configuration, identified by the angles β and α, the dimensionless root mean square of the lateral displacement $A = a_{\mathrm{rms}}/h$ results, where a_{rms} is the root mean square value of the $a(t)$ series and h the characteristic length defined in figure 13. Besides, the reduced velocity $U_R = U/(\omega_\eta h)$ is also calculated.

The results obtained, A versus U_R, for some selected configurations are depicted in figures 17 and 18, and a summary of the hysteresis behaviour of tested triangular bodies is presented in figure 15. The dependence of the hysteresis on the angle of attack can be observed in figure 15. This phenomenon appears only around the values of α where there is an inflection point.

A total of 31 different configurations, spread on the plane α-β, in the areas of interest (where hysteresis can appear), were tested. These areas are angles of attack close to $\alpha = 30^\circ$ and close to $\alpha = 180^\circ$. The values of α were selected with a difference of 5° between two consecutive values, in order to clearly separate the dynamic responses of two adjacent configurations.

Table 2. Measured mechanical properties of the triangular cross-section prisms: structural damping coefficient and natural frequency

Main vertex angle, β	Structural damping coefficient, ζ	Natural frequency, ω_n [Hz]
10°	0.008	2.5
20°	0.008	2.5
30°	0.008	2.3
40°	0.010	2.1
50°	0.014	2.0
60°	0.014	2.0

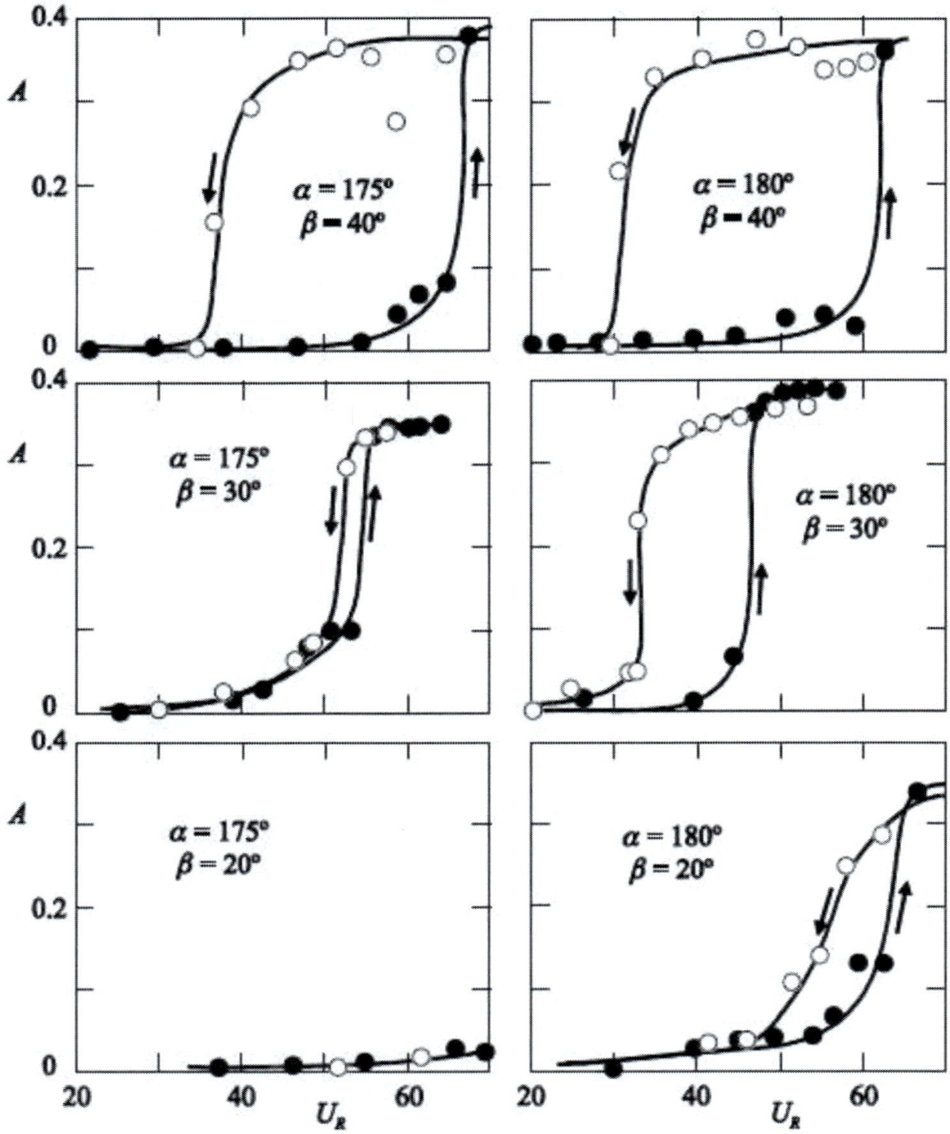

From [47].

Figure 17. Variation with the reduced velocity, U_R, of the dimensionless root mean square oscillation amplitude, A. Experimental results correspond to triangular cross-section bodies with main vertex angles β and angles of attack α, as indicated in the plots. Black circles indicate results obtained by increasing the reduced velocity, whereas white ones indicate results obtained by decreasing the reduced velocity.

Note that in the experimental procedure there are many parameters involved, and although all along the experimentation these parameters have been carefully controlled, the hysteresis itself is a noisy phenomenon influenced by many other variables which are extremely difficult to control, such as small variations in the incident flow conditions (mean velocity and turbulence intensity) or mechanical inaccuracies in the experimental apparatus.

It must be stated that the existence of hysteresis on the dynamic response of a given configuration has been considered to exist only when the A versus U_R curves obtained during the increasing incident flow velocity phase and the decreasing one are clearly different.

It seems, however, that this 5° separation into two consecutive values of the angle of attack is a too large step in tests with low values of α (in the region close to α = 30°). This would explain why no hysteresis has been detected in this area, except for triangles with β = 50°, see figure 15.

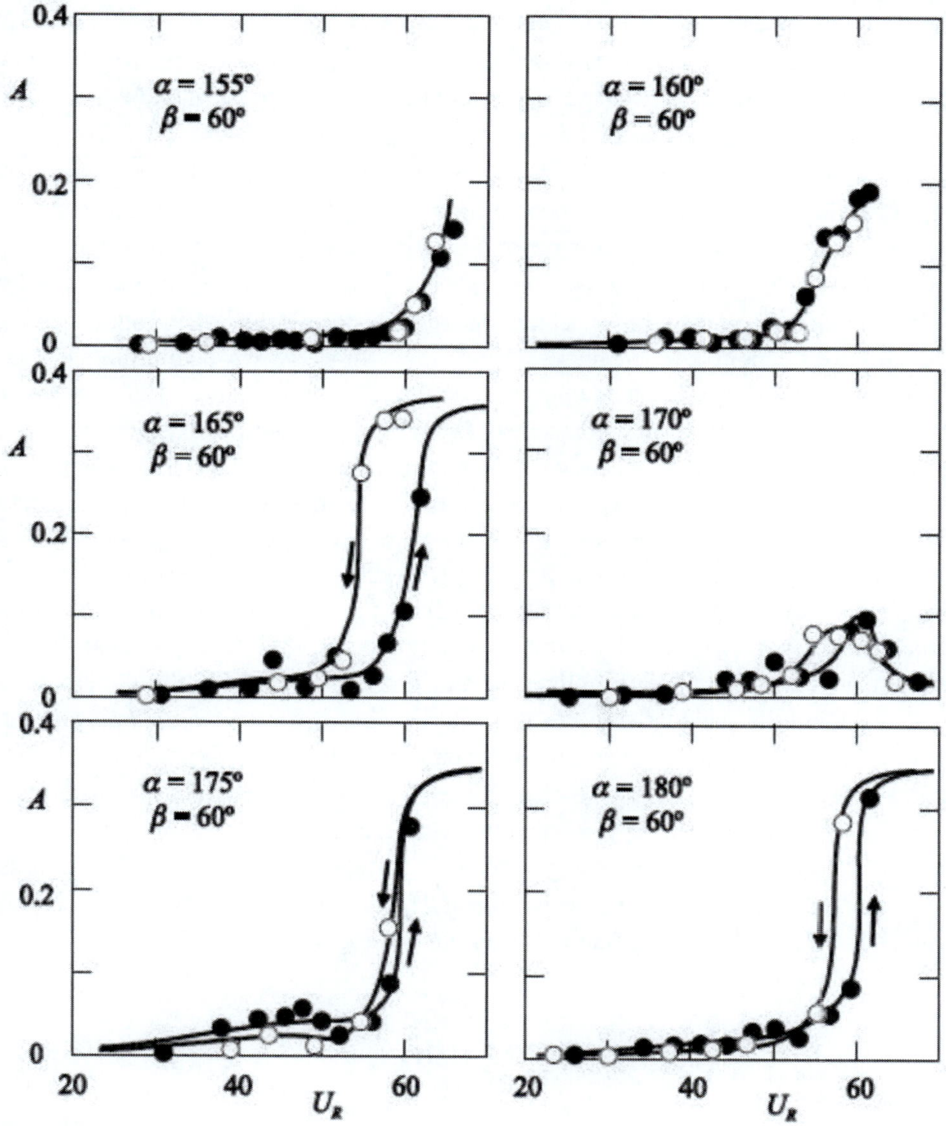

From [47].

Figure 18. Variation with the reduced velocity, U_R, of the dimensionless root mean square oscillation amplitude, A. Experimental results correspond to triangular cross-section bodies with main vertex angle β =60° and angles of attack α, as indicated in the plots. Black circles indicate results obtained by increasing the reduced velocity, whereas white ones indicate results obtained by decreasing the reduced velocity.

There is another explanation based also on the limitations of the test setup. In effect, as it can be observed in figure 15, in this region ($\alpha \approx 30°$) galloping instability is associated with large values of the reduced velocity U_R. Therefore, since $U = U_R\omega_n h$, for a given value of the reduced velocity, the wind speed at the test chamber increases as the natural frequency ω_n does or, according to data shown in Table 2, as the triangle main vertex angle decreases.

Thence, in some configurations the maximum wind tunnel air speed was reached before hysteresis appears.

The situation is much clearer in the region close to $\alpha = 180°$, because for this value of the angle of attack all the configurations tested are prone to show hysteresis provided that $\beta > 10°$. However, it must be stressed that $\alpha = 180°$ configurations become stable when the vertex angle β becomes high enough, so that no galloping occurs at $\alpha = 180°$ for $\beta > 70°$ [54]. This behaviour can be observed in figure 15: triangles with $\beta = 10°$ do not show hysteresis at $\alpha = 180°$, the magnitude of the increasing-decreasing velocity loop growing as the vertex angle grows. The hysteresis becomes maximum for $\beta \approx 40°$, and then it decreases as the value of β grows until it disappears for $\beta \approx 70°$ (see the response of $\beta = 60°$ triangles close to $\alpha = 180°$ in figure 18). Obviously, it can be expected that the hysteresis be less pronounced as the distance to the inflection point increases. This behaviour is shown in the plot included in figure 18, where the response of equilateral triangles ($\beta = 60°$) at angles of attack ranging from $\alpha = 155°$ to $\alpha = 180°$ are represented (according to figure 10 these triangles have $d^2c_l/d\alpha^2 = 0$ at $\alpha \approx 165°$ and at $\alpha = 180°$). Finally, the influence of the free stream turbulence is not considered relevant for this type of geometries with sharp edges, because at any angle of attack the incident flow separates at any of the vertex of the triangle, provided that these vertexes are sharp enough, After separation at a vertex, the flow becomes turbulent, even if the incident flow were initially laminar. This effect was proved by the authors [56] for these same triangles. Experimental results show a good agreement with the theoretical predictions of those simple galloping criteria for angles of attack close to $\alpha = 180°$, when the base of the triangles are facing the incident flow. In the other region of transverse galloping instability where hysteresis should appear, α around 30°, being the size of the region of the same order of magnitude of the interval chosen to change the angle of attack in the test set-up, the tests reproduce the theoretical result only in one of the configurations.

Therefore it can be concluded that the absence of hysteresis in the $\alpha \approx 30°$ region is due to the mechanical limitations of the experimental facility rather than to a real absence of hysteresis (which probably will appear at higher wind velocities than those tested).

CONCLUSION

An investigation on hysteresis phenomena in aerodynamics has been presented in this Chapter. Hysteresis is relatively common in fluid-structure interaction problems, both in civil engineering (bluff body aerodynamics) and aeronautical applications.

The case of hysteresis in the transverse galloping aeroelastic instability has been analysed. Certain bluff bodies under the action of an incident fluid flow can experience transverse oscillations that become unstable when the fluid velocity reaches a certain value. In some applications these fluid-induced oscillations are an unwanted phenomenon that has to be avoided, but in some other applications transverse galloping can be used as an energy harvesting mechanism. The performed research allows us to conclude that the hysteresis phenomenon in transverse galloping is dependent on the characteristics of the aerodynamic force coefficient curve, in particular, on the number of inflection points and their location. Of course, other parameters which are relevant for the galloping instability of a given bluff body, such as the cross-section geometry, the mechanical properties and the incident flow angle of attack have also been investigated.

On aeronautical applications, a relatively common aerodynamic problem is dynamic stall, i.e. the type of stall that appears for instance in helicopter rotor blades or in pitch-regulated wind turbines under yawing conditions and during gust loads. With the analysis of the results obtained from the experiments performed in wind tunnel tests it is possible to conclude that the dynamic stall process is very sensitive to the character either laminar or turbulent of the airfoil boundary layer, and that the hysteresis of the aerodynamic forces depends on parameters both of the structure and the fluid, such as the damping, the reduced frequency, and the characteristics of the flow.

Acknowledgment

The authors are indebted to Ms. A. Ruótolo, who performed most of the tests concerning pitching airfoils.

References

[1] Ekaterinaris, J. A. and Platzerh, M. F., Computational prediction of airfoil dynamic stall, *Progress in Aerospace Sciences* 33 (1997) 759-846.

[2] Lee, T. and Gerontakos, P. Investigation of flow over an oscillating airfoil, *Journal of Fluid Mechanics*, 512 (2004) 313–341.

[3] McCroskey, W. J., Unsteady airfoils, *Annual Review of Fluid Mechanics* 14 (1982) 285-311.

[4] Ericsson, L. E. and Reding, J. P., Fluid dynamics of unsteady separated flow. Part II. Lifting surfaces, *Progress in Aerospace Sciences* 24 (1987) 249-356.

[5] Carr, L. W. and Chandrasekhara, M. S., Compressibility effects on dynamic stall, *Progress in Aerospace Sciences* 32 (1996) 523-573.

[6] Leishman, J., *Principles of Helicopter Aerodynamics*. Cambridge Aerospace Series, Cambridge University Press, New York, US, 2000.

[7] Kuo, C-H. and Hsieh, J. K., Unsteady flow structure and vorticity convection over the airfoil oscillating at high reduced frequency, *Experimental Thermal and Fluid Science* 24 (2001) 117-129.

[8] Lee, B. H. K., Price, S. J. and Wong, Y. S., Nonlinear aeroelastic analysis of airfoils: bifurcation and chaos, *Progress in Aerospace Sciences* 35 (1999) 205-334.

[9] Amiralaei, M. R., Alighanbari, H. and Hashemi, S. M., An investigation into the effects of unsteady parameters on the aerodynamics of a low Reynolds number pitching airfoil, *Journal of Fluids and Structures* 26 (2010) 979-993.

[10] Ghoreyshi, M. and Cummings, R. M., Challenges in the aerodynamics modeling of an oscillating and translating airfoil at large incidence angles, *Aerospace Science and Technology* 28 (2013) 176-190.

[11] Liiva, J. and Davenport, F. J., Dynamic stall of airfoil sections for high-speed rotors, *Journal of the American Helicopter Society* 14 (1969) 26–33.

[12] McCroskey, W. J., McAlister, K. W. and Carr, L. W., Dynamic stall experiments on oscillating airfoils, *AIAA Journal* 14 (1976) 57-63.

[13] Chandrasekhara, M. S., Wilder, M. C. and Carr, L. W., Boundary layer tripping studies of compressible dynamic stall flow, *AIAA Journal* 34 (1996) 96-103.

[14] Green, R. and Galbraith, R., *An investigation of dynamic stall through the application of leading edge roughness,* Department of Aerospace Engineering, University of Glasgow, 1994.

[15] Geissler, W. and Haselmeyer, H., Investigation of dynamic stall onset. *Aerospace Science and Technology* 10 (2006) 590–600.

[16] Greenblatt, D., Nishri, B., Darabi, A., and Wygngagnsky, I., Some factors affecting stall control with particular emphasis on dynamic stall, *AIAA Paper 99-3504*, 1999.

[17] Carr, L. W., Chandrasekhara, M. S., Wilder, L. W., and Noonan, K. W. Effect of compressibility on suppression of dynamic stall using a slotted airfoil, *Journal of Aircraft* 38 (2001) 296-309.

[18] Chandrasekhara, M. S., Wilder, M. C. and Carr, L. W., Compressible dynamic stall control using dynamic shape adaptation, *AIAA Journal* 39 (2001) 2021-2024.

[19] Wandon, J., Bo-sung, L., Kwanjung, Y., and Dong-ho, L., Combining control passive method for dynamic stall control, *Journal of Aircraft* 43 (2006) 1120-1128.

[20] Den Hartog, J. P., Transmission line vibration due to sleet, *Transactions of the American Institute of Electrical Engineers* 51 (1932) 1074-1076.

[21] Den Hartog, J. P., *Mechanical Vibrations*, 4th ed. McGraw-Hill, New York, US, 1956.

[22] Blevins, R. D., *Flow-Induced Vibration*, 2nd edition, Van Nostrand Reinhold, New York, US, 1990.

[23] Simiu, E. and Scanlan, R. H., Wind Effects on Structures. *Fundamentals and Applications to Design,* John Wiley and Sons, Inc., New York, 1996.

[24] Païdoussis, M., Price, S. and de Langre, E., *Fluid–Structure Interactions: Cross-Flow-Induced Instabilities,* Cambridge University Press, UK, 2011.

[25] Novak, M., Aeroelastic galloping of prismatic bodies, *ASCE Journal of Engineering Mechanics Division* 96 (1969) 115–142.

[26] Novak, M. and Tanaka, H., Effect of turbulence on galloping instability, *ASCE Journal Engineering Mechanics Division* 100 (1974) 27-47.

[27] Naudascher, E. and Rockwell, D., *Flow Induced Vibrations – An Engineering Guide,* A. A. Balkema, Rotterdam, Netherlands, 1994.

[28] Luo, S. C., Li, L. L. and Shah, D. A., Aerodynamic stability of the downstream of two tandem square-section cylinders, *Journal of Wind Engineering and Industrial Aerodynamics* 79 (1999) 79-103.

[29] Brika, D. and Laneville, A., The flow interaction between a stationary cylinder and a downstream flexible cylinder, *Journal of Fluids and Structures* 13 (1999) 579-606.

[30] Hover, F. S. and Triantafyllou, M. S., Galloping response of a cylinder with upstream wake interference, *Journal of Fluids and Structures* 15 (2001) 503-512.

[31] Vio, G. A., Dimitriadis, G. and Cooper, J. E., Bifurcation analysis and limit cycle oscillation amplitude prediction methods applied to the aeroelastic galloping problem, *Journal of Fluids and Structures* 23 (2007) 983-1011.

[32] Novak, M., Galloping oscillations of prismatic structures, *ASCE Journal of Engineering Mechanics Division* 98 (1972) 27–46.

[33] Scruton, C., Use of wind tunnels in industrial aerodynamic research. *AGARD Report* 309, 1960.

[34] Parkinson, G. and Brooks, N. P. H., On the aeroelastic instability of bluff cylinders, *Journal of Applied Mechanics* 28 (1961) 225–258.

[35] Parkinson, G. and Smith, J., The square cylinder as an aeroelastic non-linear oscillator, *Quarterly Journal of Mechanics and Applied Mathematics* 17 (1964) 225-239.

[36] Bearman, P. W. and Luo, S. C., Investigation of the aerodynamic instability of a square-section cylinder by forced oscillation, *Journal of Fluids and Structures* 2 (1988) 161-176.

[37] Chadha, J. and Jaster, W., Influence of turbulence on the galloping instability of iced conductors, *IEEE Transactions on Power Apparatus and Systems* 94 (1975) 1489-1499.

[38] Li, Q. S., Fang, J. Q. and Jeary, A. P., Evaluation of 2D coupled galloping oscillations of slender structures, *Computers and Structures* 6 (1998) 513-523.

[39] Hémon, P., Santi, F., Schnoerringer, B., and Wojciechowski, J., Influence of free stream turbulence on the movement-induced vibrations of an elongated rectangular cylinder in cross flow, *Journal of Wind Engineering and Industrial Aerodynamics* 89 (2001) 1383-1395.

[40] Ziller, C. and Ruscheweyh, H., A new approach for determining the onset velocity of galloping instability taking into account the nonlinearity of the aerodynamic damping characteristic, *Journal of Wind Engineering and Industrial Aerodynamics* 69-71 (1997) 303-314.

[41] Hémon, P. and Santi, F., On the aeroelastic behaviour of rectangular cylinders in cross-flow, *Journal of Fluids and Structures* 16 (2002) 855-889.

[42] Barrero-Gil, A., Sanz-Andrés, A. and Roura, M., Transverse galloping at low Reynolds numbers, *Journal of Fluids and Structures* 25 (2009) 1236-1242.

[43] Singh, A. P., De, A. K., Carpenter, V. K., Eswaran, V., and Muralidhar, K., Flow past a transversely oscillating square cylinder in free stream at low Reynolds numbers. *International Journal for Numerical Methods in Fluids* 61 (2009) 658-682.

[44] Joly, A., Etienne, S. and Pelletier, D., Galloping of square cylinders in cross-flow at low Reynolds numbers, *Journal of Fluids and Structures* 28 (2012) 232-243.

[45] Luo, S. C., Chew, Y. T. and Ng, Y. T., Hysteresis phenomenon in the galloping oscilation of a square cylinder, *Journal of Fluids and Structures* 18 (2003) 103-118.

[46] Ng, Y. T., Luo, S. C. and Chew, Y. T., On using high-order polynomial curve fits in the quasi-steady theory for square-cylinder galloping, *Journal of Fluids and Structures* 20 (2005) 141-146.

[47] Barrero-Gil, A., Sanz-Andrés, A. and Alonso, G., Hysteresis in transverse galloping: The role of the inflection points, *Journal of Fluids and Structures* 25 (2009)1007-1020.

[48] Alonso, G., Sanz-Lobera, A. and Meseguer, J., Hysteresis phenomena in transverse galloping of triangular cross-section bodies, *Journal of Fluids and Structures* 33 (2012) 243-251.

[49] Scruton, C., Wind effects on structures. James Clayton Lecture, *Proceedings Institution of Mechanical Engineers* 185 (1971) 301–317.

[50] Ruecheweyh, H., Hortmanns, M. and Schnakenberg, C., Vortex-excited vibrations and galloping of slender elements, *Journal of Wind Engineering and Industrial Aerodynamics* 65 (1996) 347-352.

[51] Zdravkovich, M. M. and Carelas, E., Aerodynamics of a covered pedestrian bridge of a trapezoidal section, *Journal of Wind Engineering and Industrial Aerodynamics* 66 (1997) 141-153.

[52] Kawai, H., Effect of corner modifications on aeroelastic instabilities of tall buildings, *Journal of Wind Engineering and Industrial Aerodynamics* 74-76 (1998) 719-729.

[53] Luo, S. C., Chew, Y. T., Lee, T. S., and Yazdani, M. G., Stability to translational galloping vibration of cylinders at different mean angles of attack, *Journal of Sound and Vibration* 215 (1998)1183-1194.

[54] Alonso, G. and Meseguer, J., A parametric study of the galloping instability of triangular cross-section bodies, *Journal of Wind Engineering and Industrial Aerodynamics* 94 (2006) 241-253.

[55] Alonso, G., Meseguer, J. and Pérez-Grande, I., Galloping oscillations of two-dimensional triangular cross-sectional bodies, *Experiments in Fluids* 38 (2005) 789-795.

[56] Alonso, G., Meseguer, J. and Pérez-Grande, I., Galloping stability of triangular cross-sectional bodies: a systematic approach, *Journal of Wind Engineering and Industrial Aerodynamics* 95 (2007) 928-940.

[57] Alonso, G., Valero, E. and Meseguer, J., An analysis on the dependence on cross section geometry of galloping stability of two-dimensional bodies having either biconvex or rhomboidal cross sections, *European Journal of Mechanics B Fluids* 28 (2009) 328-334.

[58] Alonso, G., Meseguer, J., Sanz-Andrés, A., and Valero, E., On the galloping instability of two-dimensional bodies having elliptical cross sections, *Journal of Wind Engineering and Industrial Aerodynamics* 98 (2010) 438-448.

[59] Gandía, F., Meseguer, J. and Sanz-Andrés, A., Experimental study of the galloping stability of prismatic beams, *EUCASS, 5th European Conference for Aeronautics and Space Sciences,* Munich, Germany, July 1-5, 2013.

[60] Alonso, G., Pérez-Grande, I. and Meseguer, J., Galloping instabilities of Z-shaped shading louvers, *Indoor and Built Environment* submitted (2014).

[61] Ibarra, D., Sorribes, F., Alonso, G., and Meseguer, J., Transverse galloping of two-dimensional bodies having a rhombic cross-section, *Journal of Sound and Vibration* submitted (2014).

[62] Chabart, O. and Lilien, J. L., Galloping of electrical lines in wind tunnel facilities, *Journal of Wind Engineering and Industrial Aerodynamics* 74-76 (1998): 967-976.

[63] McComber, P. and Paradis, A., A cable galloping model for thin ice accretions, *Atmospheric Research* 46 (1998) 13-25.

[64] Gjelstrup, H. and Georgakis, C. T., A quasi-steady 3 degree-of-freedom model for the determination of the onset of bluff body galloping instability, *Journal of Fluids and Structures* 27 (2011) 1021-1034.

[65] Macdonald, J. H. G. and Larose, G. L., A unified approach to aerodynamic damping and drag/lift instabilities, and its application to dry inclined cable galloping, *Journal of Fluids and Structures* 22 (2006) 229-252.

[66] Lemaitre, C., Hémon, P. and de Langre, E., Thin water film around a cable subject to wind, *Journal of Wind Engineering and Industrial Aerodynamics* 95 (2007) 1259-1271.

[67] Cheng, S., Irwin, P. A. and Tanaka, H., Experimental study on the wind-induced vibration of a dry inclined cable - Part II: Proposed mechanisms, *Journal of Wind Engineering and Industrial Aerodynamics* 96 (2008) 2254-2272.

[68] Cheng, S., Larose, G. L., Savage, M. G., Tanaka, H., and Irwin, P. A., Experimental study on the wind-induced vibration of a dry inclined cable - Part I: Phenomena, *Journal of Wind Engineering and Industrial Aerodynamics* 96 (2008) 2231-2253.

[69] Gu, M., On wind–rain induced vibration of cables of cable-stayed bridges based on quasi-steady assumption, *Journal of Wind Engineering and Industrial Aerodynamics* 97 (2009) 381-391.

[70] Matsumoto, M., Yagi, T., Hatsuda, H., Shima, T., Tanaka, M., and Naito, H., Dry galloping characteristics and its mechanism of inclined/yawed cables, *Journal of Wind Engineering and Industrial Aerodynamics* 98 (2011) 317-327.

[71] Kleissl, K. and Georgakis, C. T., Aerodynamic control of bridge cables through shape modification: A preliminary study, *Journal of Fluids and Structures* 27 (2011) 1006-1020.

[72] Kleissl, K. and Georgakis, C. T., Comparison of the aerodynamics of bridge cables with helical fillets and a pattern-indented surface, *Journal of Wind Engineering and Industrial Aerodynamics* 104–106 (2012) 166-175.

[73] Jakobsen, J. B., Andersen, T. L., Macdonald, J. H. G., Nikitas, N., Larose, G. L., Savage, M. G., and McAuliffe, B. R., Wind-induced response and excitation characteristics of an inclined cable model in the critical Reynolds number range, *Journal of Wind Engineering and Industrial Aerodynamics* 110 (2012) 100-112.

[74] Ge, Y. J., Lin, Z. X., Cao, F. C., Pang, J. B., and Xiang, H. F., Investigation and prevention of deck galloping oscillation with computational and experimental techniques, *Journal of Wind Engineering and Industrial Aerodynamics* 90 (2002) 2087-2098.

[75] Ge, Y. J. and Xiang, H. F., Recent development of bridge aerodynamics in China, *Journal of Wind Engineering and Industrial Aerodynamics* 96 (2008) 736-768.

[76] Ruótolo, A. and Meseguer, J., A low cost, low speed wind tunnel for dynamic stall measurement, *European Wind Energy Conference and Exhibition - EWEC 2010,* Warsaw, Poland, 20 - 23 April 2010, P.O. 230, in http://events.ewea.org/ ewec2010/ conference/conference-proceedings/ conference-proceedings-usb/.

[77] Ayuso, L., Sant, R. and Meseguer, J., Influence of leading edge imperfections on the aerodynamic characteristics of NACA 632-215 laminar aerofoils at low Reynolds numbers, *Proc. IMechE, Part G: Journal of Aerospace Engineering* (2013) DOI: 10. 1177/0954410013481418.

[78] Soltani, M. and Rasi Marzabadi, F., Effect of reduced frequency on the aerodynamic behavior of an airfoil oscillating in a plunging motion, *Scientia Iranica* 16 (2009) 40–52, 2009.

[79] Fuglsang, P., Dahl, K. and Antoniou, I., Wind tunnel tests of the risø-a1-18, risøa1- 21 and risø-a1-24 airfoils. *final draft. Risø-R-1112(EN),* Risø National Laboratory, Roskilde, Denmark, 1999.

[80] McCullough, G. and Gault, D. Examples of three representative types of airfoilsection stall at low speed. NACA TN 2502, National Advisory Committee for Aeronautics, Washington, US, 1951.

[81] Meseguer, J., Sanz-Andrés, A., Pérez-Grande, I., Franchini, S., Sanz, J. L., Ortega, F., and Chiappe, L., Use of turbulence generators as a stall-delaying devices in flight at low Reynolds numbers, *Proc. IMechE, Part G: Journal of Aerospace Engineering* 222 (2008) 1007-1013.

[82] Theodorsen, T., General theory of aerodynamic instability and the mechanism of flutter. NACA/TR 496, 1935.

[83] Abbott, I. and von Doenhoff, A., *Theory of Wing Sections*. Dover Publications, Inc., New York, US, 1949.

[84] Charmichael, B., *Low Reynolds number airfoil survey*. CR 165803, NASA, US, 1981.

[85] O'Meara, M. and Mueller, T. Laminar separation bubble characteristics on an airfoil at low Reynolds numbers, *AIAA Journal* 25 (1987) 1033-1041.

[86] Glauert, H., The rotation of an aerofoil about a fixed axis. (British) Advisory Committee for Aeronautics (ARC) R and M No. 595. In: *Technical Report of ARC for* 1918–1919, pp. 443–447. London: H.M.S.O., 1919.

[87] Fung, Y. C., *An Introduction to the Theory of Aeroelasticity*. Wiley, New York, US, 1955.

[88] Iungo, G. V. and Buresti, G., Experimental investigation on the aerodynamic loads and wake flow features of low aspect-ratio triangular prisms at different wind directions, *Journal of Fluids and Structures* 25 (2009) 1119–1135.

In: Hysteresis
Editor: José Carlos Dias

ISBN: 978-1-63321-336-4

Chapter 7

HYSTERESIS MODULATION IN STRUCTURAL CONTROL SYSTEMS AND SEMI-ACTIVE IMPLEMENTATION

Marco Domaneschi*

Politecnico di Milano, Department of Civil
and Environmental Engineering, Milan, Italy

ABSTRACT

Models of hysteresis are employed for simulating control devices in complex structural systems. This chapter is devoted firstly to review the existent literature on passive control system modeling. Subsequently, the exposition is focused on their innovative development represented by the semi-active improvement as one of the most recent solutions in the field of numerical simulation of structural control systems. Some examples of application are finally given.

Keywords: structural control, passive, semi-active, numerical simulation

INTRODUCTION

Phenomenological models of mechanical passive systems with hysteresis can be employed for simulating dampers or isolation devices into structural models, both with the aim of evaluating their effects at a design stage for new buildings or for a supplementary implementation on existing ones. Their mathematical formulation usually allows their application in refined structural finite element models or in simpler closed form analytical approaches.

* Corresponding author: Marco Domaneschi. Piazza Leonardo da Vinci 32, 20133, Milan Italy. E-mail: marco.domaneschi@polimi.it.

A review of existing literature in the field of passive systems modeling is summarized in the first section of the chapter, spreading from recent advances on the dissipative models of hysteresis to numerical solution adopted for reproducing the characteristics of base-isolation systems, with special care to rubber bearings.

An innovative semi-active technology for control systems has been introduced in recent decades as a progress toward more efficient and adaptable solutions, with respect to the classical passive ones. Several examples are reported to the aim of describing the current research trend. Rearranging control reactions simultaneously with the hazardous excitation, provides enhanced structural behavior for improved usefulness and safety. Consequently, current numerical simulation procedures, which proved to be effective for the passive implementation, are expected to include the semi-active description as well.

Along with, the second section of the chapter is focused on existent numerical methodologies for simulating the semi-active modulation of hysteresis components in real time. A recently proposed semi-active developments of a wide employed passive model is discussed. Special attention will be given to the use of the innovative model for design issues and to its adoption in developing new semi-active control laws for driving semi-active dissipative devices. The theoretical formulation of the original settlement is discussed, allowing to focus on the way for tuning the parameters which can be modified on-line for modulating the dissipative component. Therefore, special care is given to the methodology for embedding new algorithms into the analytical formulation of the model and for testing their efficiency.

In the remaining of this chapter, the general capability of the previously discussed numerical approaches is presented with applications to complex structural systems and reference to existing studies from literature. Mitigation of direct dynamic effects (e.g. acceleration, displacements) induced by earthquake and wind loadings represents the most expected purpose of structural control discipline.

However, in conclusion, attention is also given to the positive outcomes related to indirect effects, coming from the dampers implementation on buildings.

OVERVIEW ON STRUCTURAL CONTROL

Passive systems of vibration protection can be presented as the simplest typology of structural control solutions: the control device works without using data collected from other elements as sensors. The passive protection can be considered asymptotically stable.

Active systems are the most effective in the vibration protection of mechanical systems. However, they are complicated by a large number of elements which participate in the input data processing and the formation of the control forces. They are sensors, actuator devices, and other facilities, acting in the amplification and manipulation of data.

The *Semi-active* system can be considered as passive system with smart qualities: they able to change actively its characteristic, driven by an external command, manually or automatically.

Hybrid control systems employ at the same time passive solutions with active or semi-active ones. When active or semi-active systems are considered, the control forces are processed by special algorithms collecting informations on the current conditions from an open or a closed loop data flow.

An open loop control system is a combination in series of elements, from a starting point to another point in a flow scheme. Its transfer function is the result of the product of each single element transfer function. A closed-loop system is a combination of elements with feedback connection [1].

PASSIVE NUMERICAL MODELS OF HYSTERESIS

The Bouc-Wen Model

The most eclectic phenomenological model of hysteretic mechanical systems reported in the present literature is probably the Bouc-Wen (BW) endochronic model [2, 3]. It has been selected by several Authors for the simulation of the hysteretic component for a wide range of control devices. The physical and mathematical consistency of that model [4] and the excellent correspondence between the experimental and numerical results are probably the main reasons which support and justify its frequent implementation. In this section the general approach to the oscillator dynamic equation with the BW model is summarized [5] and a number of results reported in the literature are also summarized.

A hysteretic oscillator can be represented by the following equation of motion [6]:

$$m\ddot{x} + c\dot{x} + \Phi(x, \dot{x}) = f(t) \tag{1}$$

where x is the displacement, the simple superimposed dot means the first time derivative (velocity) and the double superimposed dot the second time derivative (acceleration). In equation (1), $f(t)$ is the force excitation depending on the time parameter t. m is the mass, c the damping coefficient and $\Phi(x, \dot{x})$ is the restoring force expressed by

$$\Phi(x, \dot{x}) = \alpha kx(t) + (1-\alpha)kz(t) \tag{2}$$

$$\dot{z} = A\dot{x} - \beta\dot{x}|z|^{n} - \gamma|\dot{x}|z|z|^{n-1} \tag{3}$$

The auxiliary variable z allows one to introduce a smooth hysteretic behavior; k is the pre-yielding stiffness, α represents the ratio between post and pre-yielding stiffness, managing all the intermediate yielding states. In particular, $\alpha = 1$ provides an elastic response and $\alpha = 0$ an elastic-perfectly-plastic behavior. A, β, γ, n are time invariant parameters defining the amplitude and the shape of the cycles, the linearity in unloading and the smoothness of the transition from the pre- to the post-yield region. In detail, A is related to the initial stiffness, β adjusts the cycle amplitude and, as a result, the dissipation level ($\beta \rightarrow 0$ is employed for low energy dissipation), γ defines the unloading path in the hysteresis, n the smoothness of the transition.

The consequential relationship between the model parameters that appear in equation (3) and the resulting shape of the obtained hysteresis loops is detailed in [7] using a special form of the BW model with normalization.

Division by m leads to a more classical form of the second order differential equation (1) in terms of the circular frequency ω_0 and the damping ratio ζ :

$$\ddot{x} + 2\zeta\omega_0\dot{x} = \frac{f(t) - \Phi(x,\dot{x})}{m} \tag{4}$$

Re-arranging the governing equation (4) in the state space form, the following is obtained:

$$\frac{d}{dt}\begin{bmatrix} z \\ u_1 \\ u_2 \end{bmatrix} = \begin{bmatrix} 0 & 0 & A \\ 0 & 0 & 1 \\ -(1-\alpha)\omega_0^2 & -\alpha\omega_0^2 & -2\zeta\omega_0 \end{bmatrix}\begin{bmatrix} z \\ u_1 \\ u_2 \end{bmatrix} + \begin{bmatrix} -\beta u_2|z|^n - \gamma|u_2|z|z|^{n-1} \\ 0 \\ 0 \end{bmatrix} + \begin{bmatrix} 0 \\ 0 \\ m^{-1} \end{bmatrix} f(t) \tag{5}$$

where $u_1 = x$ and $u_2 = \dot{x}$. The state space formulation is usefully employed for numerical implementation in standard computation environments, also going from single to multi-degree-of-freedom systems.

Several developments of the initial formulation of the BW model can be found in literature, among the others, for reproducing stiffness degradation in hysteretic systems [8], to recover Drucker postulate [9]. Next studies result in comprising additional terms and representing the pinching behavior, as observed in reinforced concrete and masonry structures [10, 11]. The biaxial extension was proposed on reinforced concrete beams [12] and 3D frames [13]. More recently the dynamic stability of the model has been investigated in [14], where the domain limits for the model parameters are fixed, the model passivity, among other general properties, is established in [15]. The nonlinear responses and bifurcations of one-dof BW hysteretic oscillators to harmonic excitations have been also investigated in [16].

Parameters Identification for the BW Model

The identification issue can be approached by considering the physical significance of the parameters which define the BW model of hysteresis. Mass m and pre-yielding circular natural frequency ω_0 of the oscillator are related to pre-yielding stiffness k and the BW parameters α and A by the following equation:

$$\omega_0^2 m = k[\alpha + A(1-\alpha)] \tag{6}$$

Equation (6) is derived from the time derivative of equation (2) where $\dot{z}$ is replaced by its value when z=0. Consequently, equation (6) is the ratio between the restoring force $\Phi(x,\dot{x})$ and relative displacement derivative with respect to time, i.e. $\dot{x}$.

The identification of remaining parameters, namely γ and β, can be fixed by setting to zero the derivative $\dot{z}$ (simplifying by $\dot{x}$) and replacing the expression for z as obtained in equation (2). In light of this comment, the following relation results:

$$\Phi_y = \left[\frac{A}{\gamma+\beta}\right]^{\frac{1}{n}}(1-\alpha)k + \alpha k x \tag{7}$$

From the definition of elastic force and equation (6) one obtains

$$\Phi_y = \left[\frac{A}{\gamma+\beta}\right]^{\frac{1}{n}}(1-\alpha)k + \alpha k \frac{\Phi_y}{k[\alpha + A(1-\alpha)]} \tag{8}$$

With A=1 equation (8) can be rewritten as:

$$\Phi_y = \left[\frac{1}{\gamma+\beta}\right]^{\frac{1}{n}} k \tag{9}$$

and equation (6) becomes

$$\omega_0^2 m = k \tag{10}$$

Equation (9) allows to define different yielding levels of the general BW signature, mainly focusing on the hysteresis cycle amplitude definition rather than on the theoretical elastic limit [8].

The parameters can satisfy the physical and mathematical consistency of the general passive BW formulation in equation (3), verifying some properties from [17] such as bounded stability of input and output, passivity among the others. More details can be also found in [6]. Figure 1 depicts different cycles of hysteresis reproduced by the BW model with essential parameters $A = 1$, $n = 1$, $\alpha = 0.0$, $\beta = \gamma = 40\ \mathrm{m}^{-1}$, $k = 80000$ kN/m, $\Phi_y = 1000$ kN. Influence of their variation has been highlighted in Figure 1.

Lund-Grenoble Friction Model

Friction is a natural phenomenon that is quite complicate to model and it is also an important aspect of many control systems which are able to dissipate mechanical energy by hysteresis cycles, decoupling dynamic motion of the controlled object from its supports, and for high quality servo mechanisms and hydraulic systems as well. Consequently, an accurate model of such behavior results necessary to analyze stability, predict limit cycles and perform simulations among the others.

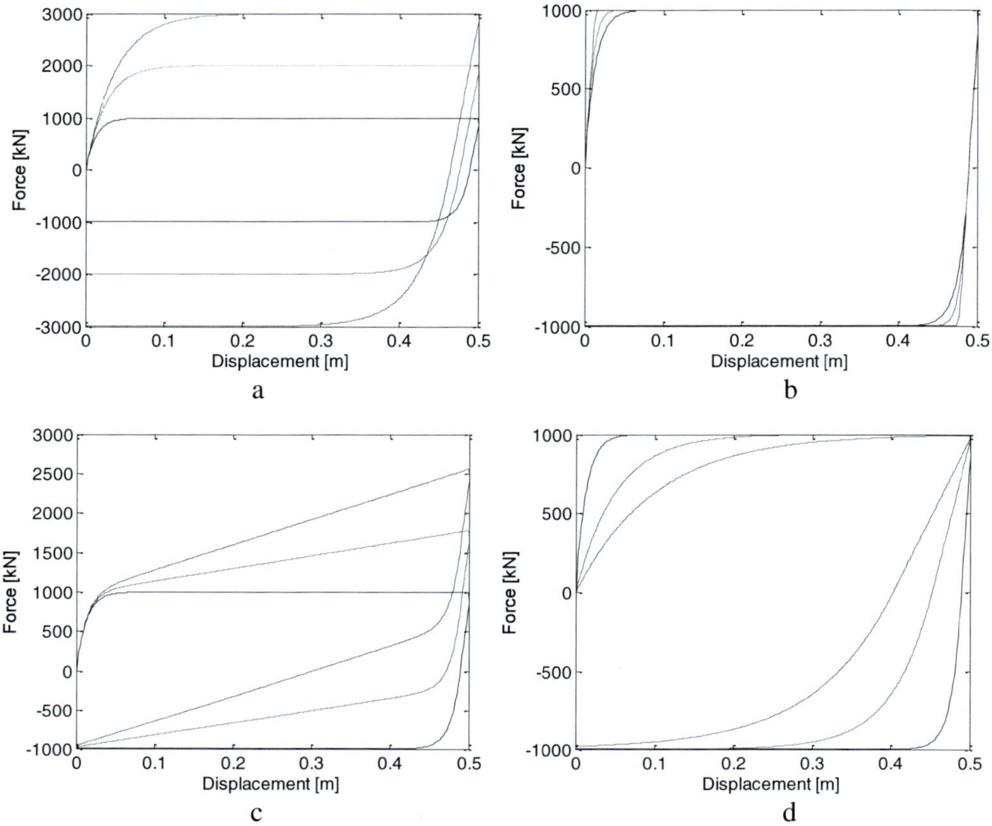

Figure 1. Hysteresis cycles reproduced by BW model: Φ_y = 1000, 2000, 3000 kN (a); n = 1, 2, 10 (b); α = 0.0, 0.02, 0.04 (c); k = 80000, 20000, 10000 kN/m (d).

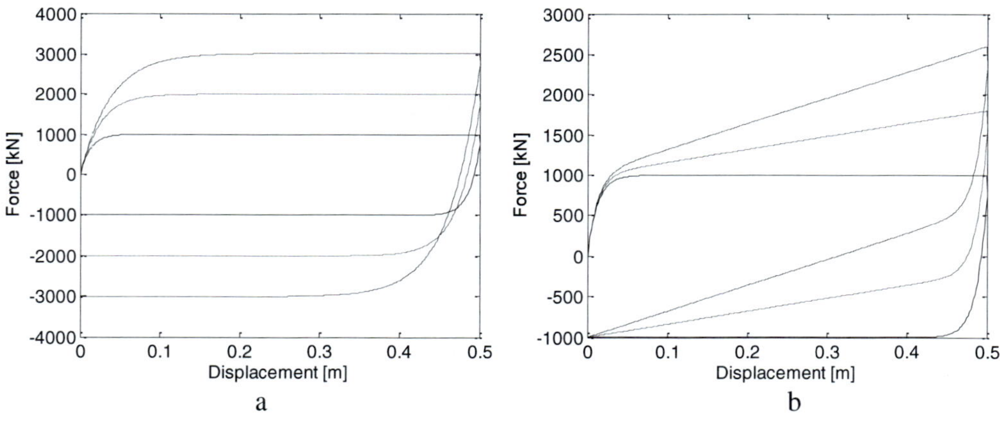

Figure 2. Hysteresis cycles reproduced by LuGre model: (a) with different yielding limits (1000, 2000, 3000 kN respectively through α = 80, 40, 26.5 m^{-1}) and post-yielding behavior k=0, 80000x0.02, 80000x0.04 kN/m.

Classical friction models, such as Coulomb and viscous friction, can be useful in a number of existing numerical schemes. However, in applications with high precision positioning and with low velocity tracking, a more refined model is necessary.

The Lund-Grenoble (LuGre) model [18, 19] allows for a better description of the friction phenomena for low velocities and especially when crossing zero velocity occurs. Efficient simulations of a dynamic systems and control dampers are reported in literature (e.g. [20, 21]). The following relations summarize the LuGre model:

$$\dot{z} = \dot{x} - \alpha|\dot{x}|z \tag{11}$$

$$V = \beta z + \zeta \dot{x} + kx + \varepsilon \dot{z} + V_0 \tag{12}$$

where x is the state variable of the system, z the auxiliary variable, k is the pre-yielding stiffness and ζ the damping coefficient, V is the force and α, β, ε are parameters defining amplitude and shape of the cycles. Figure 2 depicts different cycles of hysteresis reproduced by the LuGre model with basic parameters α = 80 m^{-1}, β =80000 kN/m, $\zeta = k = \varepsilon = V_0 = 0$.

Modeling Base Isolation Systems

Among control systems, seismic isolation ones are probably the most known and adopted with succesful real world applications. Technology spreads from rubber bearings to pendulum systems, even if some peculiar innovations can also be noted (e.g. [22]). Due to their reliable efficient behavior for protecting civil structures from earthquake loadings, isolation system solutions are also studied for special structures such as nuclear power plant buildings with some interesting applications. Furthermore, a consolidated interest on such control solutions covers different fields from research to recently proposed design standards [23].

In this light, as it is well known, high damping rubber bearings (HDRB) show an interesting non-linear mechanical behavior with hysteresis, characterized also by the scragging and Mullins' effects (stiffness and damping degradation) by the horizontal stiffness variation (due to temperature and axial load), by strain-rate dependence and ageing.

Precursor of a large family of numerical models for isoslation systems simulation, Ozdemir and Wen models [24, 25] of smooth hysteresis are probably the most eclectic phenomenological model presented in the present literature. Thanks to their physical and mathematical consistency, with an excellent correspondence between the experimental and numerical results, a number of developments have been established, from their original formulation, including new features and reproducing new technologies [26, 5].

An excellent overview on the present isolation knowledge, including some issues on modeling and analysis options as well, can be found in the research report by Grant et al. [27]. The main characteristics of a seismic isolation system are described, listing several models for the unidirectional numerical simulation. In conclusion, a bidirectional model is also proposed for representing the bearings response to 2D horizontal loading in terms of stiffness, damping and degradation, on the basis of a decomposition of the bearing resisting force as the sum of an elastic component (from Mooney-Rivlin strain energy function) and an hysteretic force.

A non-differential unidirectional model for elastomeric isolation bearings has been proposed by Kikuchi and Aiken [28] as a development from the Fujita one, in order to improve the performance into the large strain range.

The Fujita model has the characteristic of including a procedure to update model parameters, in contrast with other differential models as Ozdemir or Wen one [24, 25]. The modeling approach proposed by Kikuchi and Aiken modifies the Fujita one in order to improve the performance of the modeling outcomes at high shear strain levels; however the effects of strain rate and of variation of axial load on the bearing hysteresis properties are neglected.

Hwang et al. [29] developed an analytical unidirectional model from the proposed approach by Pan and Yang [30]. In this last, the shear force experienced by the isolator is attributed to the sum of a restoring force, describing the skeleton curve, and a damping force, adopted to represent the area of the hysteresis loop. Such model, however, results unable to reproduce the Mullins and describe the scragging effect on the isolation bearings. The development by Hwang et al. model for high damping elastomeric isolation devices is aimed to overcome such limits. Resoring and damping components of the original Pan and Yang model result modified to describe the stiffness degradation and the decrease of loop area trough a special integral term which accounts the energy dissipated by the elastomeric material or bearing during cyclic loading reversals. Nevertheless, the proposed approach neglects axial loads influence, as well as the rubber compounding and vulcanization process, which can also control the hysteresis behavior of elastomeric isolation bearings. The model parameters identification is also not included in this study, which is mainly focused on the prediction of the force-displacement hysteresis.

The unidirectional model for HDRBs by Tsai et al. [31] has been developed by modifying the Wen model to include rate-dependent effects under constant axial loads. Consequently, it has a differential signature. In spite of the good correlation between experimental and numerical results, the model by Tsai et al. does not show the stiffness and damping degradations, which are distinctive features of such devices.

An unidirectional model for HDRBs under constant axial load is proposed by Jankovsky in [32] accounting nonlinear rate dependent effects. I consists in a non-differential model, developed as well from the Pan and Yang solution [30]. The shear stiffness, K, and damping, C, coefficients at a given time, t, are obtained based on the actual values of displacement, $x(t)$, and velocity, $\dot{x}(t)$ as follws:

$$K(x(t),\dot{x}(t)) = a_1 + a_2 x^2(t) + a_3 x^4(t) + \frac{a_4}{\cosh^2(a_5 \dot{x}(t))} + \frac{a_6}{\cosh(a_7 \dot{x}(t))\cosh(a_8 x(t))} \tag{13}$$

$$C(x(t),\dot{x}(t)) = \frac{a_9 + a_{10} x^2(t)}{\sqrt{a^2{}_{11} + \dot{x}^2(t)}} \tag{14}$$

where a_1-a_{11} are parameters to be defined for fitting experimental data (e.g. by least squares method). In particular, a_1 is related to the basic stiffness level which is modified by a_2 and a_3 for higher shear strains, a_4 and a_5 control the stiffness at the maximum displacements; a_6-a_8 increase stiffness value in case of motions with lower amplitudes; $a_9 - a_{11}$ manage damping value so as to obtain appropriate hysteresis loops. It has been proved how cyclic experimental tests are well reproduced by the model, even though its performance gets worse when a seismic signal is applied (see Figure 3 with a_1=212390 N/m, a_2=3431000 N/m^3, a_3=-8767000

N/m^5, a_4=518290 N/m, a_5=4.3595 s/m, a_6=586760 N/m, a_7=3.9664 s/m, a_8=91.482 1/m, a_9=15377 N, a_{10}=408680 N/m^2, a_{11}=0.13985 m/s). The adopted parameters have been collected in an example of application from [32] (Example 2).

Abe et al. [33, 34] proposed differential hysteretic models of laminated rubber bearings, HDRB, lead-rubber bearings (LRB) and natural rubber bearings (NRB), under biaxial and triaxial loading conditions on the basis of experimental results.

Firstly, an unidirectional model is derived by extending the Ozdemir [24] elasto-plastic model; secondly, a bidirectional model of the bearing is developed by including symmetric components of the characteristic matrix.

The restoring force for the device unidirectional model is the sum of three contributions: an elastic-plastic model (F_2 hysteretic component) and two elastic non-linear springs, namely a non-linear elastic spring (F_1) and a hardening spring (F_3).

From an analytical point of view the force-displacement relation for the first non-linear spring reads

$$F_1 = K_1\left\{\beta + (1-\beta)\exp(-\frac{U_{\max}}{\alpha})\right\}U + a\left[1-\exp(-b|U|)\right]\text{sgn}(U) \tag{15}$$

where U is the relative displacement while K_1, a and b are parameters. In Equation (15), the first term reproduces the force linear evolution, while the second one the non-linear behavior.

The hysteretic contribution F_2 is described by the following differential equation

$$\dot{F}_2 = \frac{Y_t}{U_t}\left\{\dot{U} - |\dot{U}|\left|\frac{F_2}{Y_t}\right|^n \text{sgn}(\frac{F_2}{Y_t})\right\} \tag{16}$$

where the Y_t and U_t parameters are in turn defined as

$$Y_t = Y_0(1+\left|\frac{U}{U_H}\right|^p)\quad U_t = U_0(1+\frac{U_{\max}}{U_S}) \tag{17}$$

being Y_0 is the initial yielding force, U_0 the initial yielding displacement, U_H the displacement where hardening starts, U_S a parameter for controlling the degradation of the elastic stiffness of the elastic-plastic spring, U_{max} the maximum displacement experienced during the loading history and p a parameter governing the shape of the hardening branch.

Finally a further non-linear spring is introduced in parallel for capturing the increment of the tangential stiffness experienced by the devices at very high strain levels. This results in the F_3 contribution, defined as

$$F_3 = K_2\left|\frac{U}{U_H}\right|^r U \tag{18}$$

where r is the parameter to prescribe the shape of the hardening curve, K_2 a constant to describe the contribution of the hardening spring to the total stiffness.

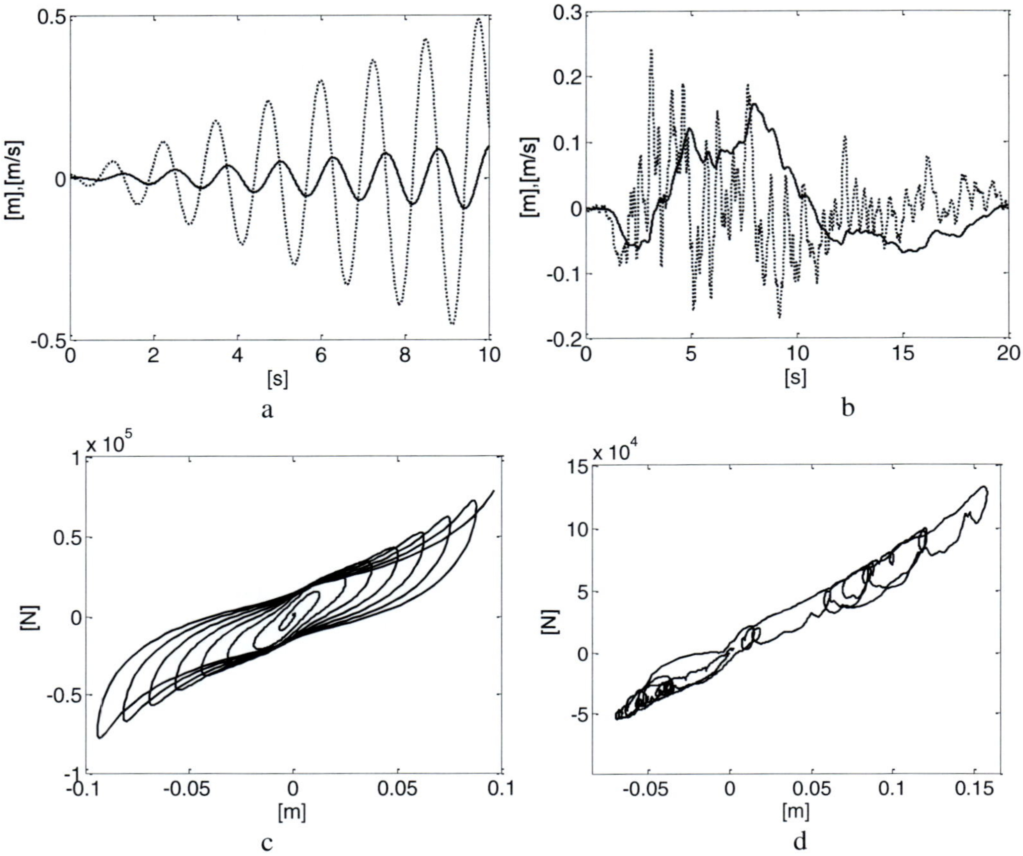

Figure 3. Non-differentical unidirectional Jankovsky model: cyclic loading (a) produces the model response (c). Seismic input (b) the model response (d). Dotted lines in (a) and (b) denote velocity.

Details on these model components can be found also in [23] where it has been designated for isolation system simulation under a nuclear power plant building.

Figure 4 depicts the Abe et al. model response (with K_1=1.51 KN/mm, α=8.97 mm, β=0.301, a=9.5 KN, b=0.0675 1/mm, n=0.451, Y_0=13.7 KN, U_h=52.1 mm, p=4.17, U_0=2.07 mm, U_s=30.4 mm, K_2=0.21 KN/mm, r=1.73) under the same cyclic and seismic condition of Figure 3 for comparison. The adopted parameters have been collected in an example of application from [34] (HDR-D sample). Comparing Figures 3 and 4, even if the models reproduce different devices from literature, the seismic response of the non-differential model is questionable. Besides the tests from Figures 3 and 4, several others have been conducted at different amplitudes: it can be concluded that the differential approach results more accurate in the simulation of the device response under seismic-type loading histories; conversely, both unidirectional models [32] and [33, 34] perform similarly when cyclic loading are considered. Ryan et al. in [35] approach the problem of the influence of the axial load variation on the isolator response. In particular the horizontal stiffness and yielding strength have been deepen. They underlined for both HDRB and LRB isolation devices the following considerations: (i) lateral stiffness decreases with the increasing axial load; (ii) lateral yield strength decreases with decreasing axial load (LRB only); (iii) vertical stiffness decreases with increasing lateral deformation.

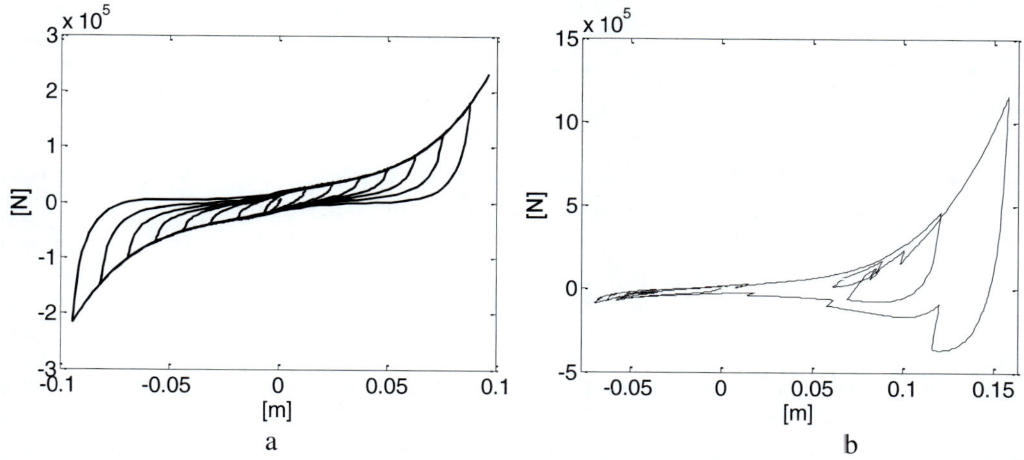

Figure 4. Differential unidirectional Abe et al. model: cyclic loading (Figure 3a) response (a) and seismic (Figure 3b) response (d).

Considerations on HDRB and LRB numerical modeling are also included. Note that the proposed solutions, although an improvement in the isolation knowledge, are declared in [35] as an incomplete representation of the experimental response.

However this literature study can be considered original and of interest for deepening the influence of axial load variation on the isolators response, accounting also different technologies.

Besides the study by Ryan et al., the one by Yamamoto et al. [36] proposes a two-dimensional model for the numerical simulation of seismic isolation bearings including also the influence of axial load. Such model is based on an analytical approach comprising shear and axial springs, having properties which vary with the vertical load.

In particular, the dependence of axial loads is captured by the material nonlinearity formulation of the axial springs and by the transversal geometric nonlinearity of the shear stiffness.

A three dimensional development of this model has been reported in [37]; however, it has to be noted that no evaluation of the model performance is given under three dimensional loading paths (e.g. circular or 8-shaped in the horizontal plane) and/or seismic loads.

Among the approaces herein summarized, the one proposed by Abe et al. [33, 34] seems the most promising for base isolation behavior reproduction in complex structural models. It has been selected in [23] due to its good representation of the experimental response for both cyclic and seismic loading, particularly when bidirectional loading paths in the horizontal plane are considered.

It could also represented an interesting enhancement with respect to simplest bilinear models of isolators, as proposed in existent standards (see [23] for references).

This choice is oriented to eliminate part of uncertainties related to the simulation of the isolation system, being the numerical approach by Abe et al. strictly associated to the actual behavior of the devices.

The Semi-Active Model of Hysteresis

The Semi-Active Choice in the Structural Control Discipline

Semi-active approaches for structural control show several good points in comparison with passive and active solutions, which should be underlined as follow [1, 38, 39]:

1 improvement of passive solutions by online adjustments of the damping or stiffness; this is done according to feedback signals and control commands.
2 simplification of the design and the implementation of control systems in comparison with active solutions; this is also due to their requirements of low power supply and low maintenance costs.

Specifically, passive systems have the advantage of being generally more robust than active systems because they are independent from external power sources and processed commands. They perform with fewer operating costs and require a lower maintenance. Their main disadvantage is their inability to adapt to different intensities of excitation. To overcome this limit, semiactive control schemes can be implemented on structural systems.

Semi-active control strategies share the same advantages as passive systems, requiring only a limited amount of active external power to modify the working parameters of the semiactive control devices. This is contrary to the active control case, in which the external power is used to directly inject mechanical energy in the structure through the reactions at the device ends.

Semi-active systems can also be employed in *decentralized* configurations where the whole control system is subdivided into a number of subsystems. When decentralization is employed, e.g. for space structures as long-span bridges, low-order control laws represent a suitable choice. Such algorithms for control force computation are low-order in the sense that they only use the feedback from the nodes that are directly influenced by the devices [38].

Decentralization of control systems, compared to its centralization, presents several good points, of which two in particular have to be underlined:

1 decentralised control systems perform satisfactorily even under adverse conditions showing interesting robust qualities. When local failures occur, if the system is totally decentralized, none of the devices depends on any other. It is therefore acceptable to foresee the failures of one control device at a time, mantaining the global efficency of the control system by the remaining ones.
2 Furthermore, the advantages of designing decentralized control schemes result from the reduction of transmission costs within the feedback loops, the reduction of the overall computational effort and the possibility of a more effective power supply to the devices.

Another intersting classification related to semi-active control systems, but not exclusively, is related to their *collocated* configuration. It occurs when the processed control forces act along the same degree of freedom along which the monitoring data are collected. On the contrary, the system is termed as *non-collocated*.

Toward the Semi-Active Implementation

Smart semi-active device typologies, as those electro-inductive (EI) and electro-magnet-rheological ones (EMR) can be effectively simulated by numerical models [19, 40]. The existent literature develops appropriate rheological models by connecting together simple elements, as spring and dashpot, with non linear parts, depending on the intrinsic characteristics of the device (e.g. viscous, hysteretic or elastic). Some research works report on the force's hysteretic component of real devices which are efficiently reproduced by differential models: among others, the BW one plays a major role, being able to perform with reasonable efficacy [6]. The overviewed solutions adopted for simulating conventional semi-active dampers, as the EI and EMR devices, are able to reproduce the inherent characteristics of the physical systems, consequently improving their knowledge. However, such devices are also able to modulate their dissipation capability, increasing and decreasing online the hysteretic cycles amplitude by external commands, operating simultaneously with the external excitation and adapting themselves to the structural conditions.

Consequently, the development of an operational method for modifying in real time the BW hysteresis cycles, through suitable semiactive laws, may have consequences of practical interest.

In [6] an innovative analytical approach for controlling, by a modified semi-active version of the Bouc-Wen model, the hysteresis component of simulated semi-active control systems is first presented. The next section is devoted to present it.

A General Approach to Semi-Active Modeling

As introduced in previous section, semi-active devices can be regarded as passive ones able to adjust their reaction on-line with the external excitation, through suitable algorithms. Managing their intrinsic dissipation level or stiffness, removing energy from the system or decoupling the structural motion from the base external excitation, they have been shown to significantly outperform the passive schemes by a moderate energy supply. Their practical installation in a wide variety of structures demonstrates their positive function on alleviation of wind and seismic response of buildings, sometimes showing better performances than active systems, which usually are considered the best performers in terms of internal forces and displacements reduction. Acting simultaneously with the hazardous excitation, they also provide enhanced structural behavior for improved usefulness and safety, making the semi-active techniques a very attractive choice for buildings whose functionality is of paramount relevance [6, 38].

The numerical simulation of control devices allows evaluating their efficiency into complicated structural systems, as high-rise buildings or long-span bridges, under a wide variety of external forces. For improving the normal solutions adopted for simulating conventional semi-active dampers, in [6] an operational method for modifying in real time the hysteresis cycles has been first developed, through suitable semi-active laws embedded into the BW numerical model, with promising consequences for buildings.

When semi-active systems are employed in real applications, the information collected from the monitoring system is processed by suitable algorithms. Subsequently, the damper configuration is modified so as to exert the necessary control reactions.

The same procedure can be translated into the simulation environment, where the BW model parameters are updated to modify, as needed, the hysteresis forces which represent the control actions during the real time mock-up. The methodology in [6], for embedding semi-active algorithms and achieving the resulting semi-active BW expression, considers three different control laws, suitable for managing dissipative devices: the *on/off SkyHook*, the *continuous SkyHook*, the *BangBang*. An innovative one, called *continuous BangBang,* is also introduced. The proposed procedure has been demonstrated effective for managing, in real time, the hysteresis component of semi-active systems defining a semi-active version of the BW model, termed for brevity as DM (DoManeschi) model.

The general formulation of the DM model can be summarized by the following equations:

$$F = \alpha k\left(x_{bc} - x_{\sup}\right) + (1-\alpha)kz + c\left(\dot{x}_{bc} - \dot{x}_{\sup}\right) \tag{19}$$

$$\dot{z} = A\left(\dot{x}_{bc} - \dot{x}_{\sup}\right) - \overline{\beta}\left(\dot{x}_{bc} - \dot{x}_{\sup}\right)\left|z\right|^{n} - \overline{\gamma}\left|\left(\dot{x}_{bc} - \dot{x}_{\sup}\right)\right| z\left|z\right|^{n-1} \tag{20}$$

$$\overline{\gamma} = f(\gamma_{low}, \gamma_{high}, h) \tag{21}$$

$$\overline{\beta} = f(\beta_{low}, \beta_{high}, h) \tag{22}$$

The variable z is the auxiliary BW variable allowing to introduce a smooth hysteretic behavior in the control force F, x_{bc} and x_{sup} are respectively the coordinate of the body mass to be controlled and the support where the device is connected, k the pre-yielding stiffness, α the ratio between post and pre-yielding stiffness (α =1 provides an elastic response and α=0 an elastic-perfectly-plastic behavior), A an n are time invariant parameters, c is the intrinsic viscous component (not considered in the remaining of this chapter).

Parameters $\overline{\gamma}$, $\overline{\beta}$ embeds the general semi-active control law into the BW model for hysteresis modulation defining its semi-active version. They are functions of the BW model parameters γ, β which control the dissipation level, ranging from the high and low limits (γ_{high} β_{high}, γ_{low} β_{low}) which define the semi-active domain, and the adopted h gain [6]. Algorithms can be introduced through equations (21) and (22), guaranteeing mathematical consistency and passivity of the resulting formulation.

Considering the *continuous BangBang* , the following equations replace (21) and (22) ones [6]:

$$\overline{\gamma} = \gamma_{low} + H\left[\left(x_{bc} - x_{\sup}\right)\left(\dot{x}_{bc} - \dot{x}_{\sup}\right)\right]\left\{\min\left[\gamma_{low}, \max\left(\left|h\dot{x}_{bc}\right|, \gamma_{high}\right)\right] - \gamma_{low}\right\} \tag{23}$$

$$\overline{\beta} = \beta_{low} + H\left[\left(x_{bc} - x_{\sup}\right)\left(\dot{x}_{bc} - \dot{x}_{\sup}\right)\right]\left\{\min\left[\beta_{low}, \max\left(\left|h\dot{x}_{bc}\right|, \beta_{high}\right)\right] - \beta_{low}\right\} \tag{24}$$

The device yielding force level is modulated through $\bar{\gamma}, \bar{\beta}$ parameters in equation (19) and equations (23) and (24); *H[.]* is the *Heaviside* step function.

The discrete *BangBang* semi-active configuration of the BW model is defined by the following expressions [6]:

$$\bar{\gamma} = \gamma_{low} - \Delta\gamma H\left[\left(x_{bc} - x_{\sup}\right)\left(\dot{x}_{bc} - \dot{x}_{\sup}\right)\right] \tag{25}$$

$$\bar{\beta} = \beta_{low} - \Delta\gamma H\left[\left(x_{bc} - x_{\sup}\right)\left(\dot{x}_{bc} - \dot{x}_{\sup}\right)\right] \tag{26}$$

Two values can be assumed by $\bar{\gamma}, \bar{\beta}$ parameters, depending whether the high or the low state for damping is required and the increments are defined as $\Delta\gamma = \gamma_{low} - \gamma_{high}$ and $\Delta\beta = \beta_{low} - \beta_{high}$.

In analogy with the last one, the discrete semi-active on/off *SkyHook* ccnfiguration of the BW model results [6]:

$$\bar{\gamma} = \gamma_{low} - \Delta\gamma H\left[\dot{x}_{bc}\left(\dot{x}_{bc} - \dot{x}_{\sup}\right)\right] \tag{27}$$

$$\bar{\beta} = \beta_{low} - \Delta\beta H\left[\dot{x}_{bc}\left(\dot{x}_{bc} - \dot{x}_{\sup}\right)\right] \tag{28}$$

Accordingly to [6], the semi-active *continuous SkyHook* form of the BW model is:

$$\bar{\gamma} = \gamma_{low} + H\left[\dot{x}_{bc}\left(\dot{x}_{bc} - \dot{x}_{\sup}\right)\right]\left\{\min\left[\gamma_{low}, \max\left(\left|h\dot{x}_{bc}\right|, \gamma_{high}\right)\right] - \gamma_{low}\right\} \tag{29}$$

$$\bar{\beta} = \beta_{low} + H\left[\dot{x}_{bc}\left(\dot{x}_{bc} - \dot{x}_{\sup}\right)\right]\left\{\min\left[\beta_{low}, \max\left(\left|h\dot{x}_{bc}\right|, \beta_{high}\right)\right] - \beta_{low}\right\} \tag{30}$$

where $\bar{\gamma}, \bar{\beta}$ parameters can assume continuously variable values between the two high and low limits.

The *h* gain adopted for the continuous semi-active versions (equations (23)-(24) and (29)-(30)) consists in $\gamma_{low} / max(\dot{x}_{body})$, where the denominator is the expected max:mum body mass velocity [6].

When semi-active hysteretic control systems are adopted, their ability in energy dissipation and their aptitude in decoupling the support vibration from the body mass are valuable mitigation resources that should be estimated. These properties are also investigated in [6] for the proposed DM model with different configurations. Such observations may improve the semi-active structural control design perspective.

Figure 5 depicts the mean control force and equivalent damping factor for semi-active dampers with identical sinusoidal input time histories, at increasing amplitudes, for reproducing a standard motion of the dampers connection points.

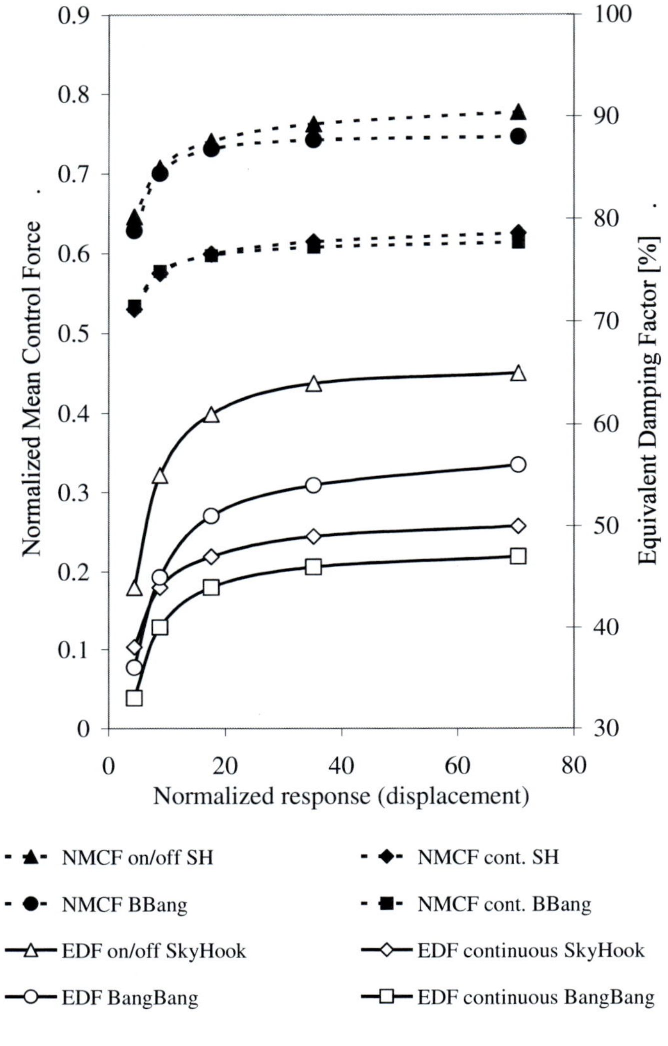

Figure 5. Restrain and dissipation vs. different input cycle amplitudes.

Forces and displacements are here normalized to the device highest yielding level. Four different control algorithms are implemented: the *SkyHook*, the continuous *SkyHook*, the *BangBang* and the continuous *BangBang* law through the following parametes: A=1, n=1, α =0, the yielding forces Φ_y=1000kN and 500kN respectively for the high and low semi-active limit, k=80000kN/m, with $\gamma_{low}=\beta_{low}$ and $\gamma_{high}=\beta_{high}$. A significant decoupling of relative motions between the support and body mass to be controlled, with respect to the discrete algorithms, is exerted by the continuous semi-active laws; on the contrary, highest dissipation levels are provided from the discrete ones.

The BW model in its original formulation, from which the DM one has been derived, has proven to perform with reasonable efficiency in reproducing the force's hysteretic component of real passive device, playing a major role among other differential ones. It can reproduce also the hysteresis signature coming from a wide range of semi-active damper technologies, e.g. an innovative EI one characterized in laboratory for structural control applications on bridges [19]. It represents a fascinating solution for the feasibility of larger passive devices of this type to be installed in long span bridges. This is of interest due to two facts: that they are much shorter than passive hydraulic dampers of identical maximum stroke and that they can easily be converted into the semi-active type, adapting themselves to different seismic intensity levels by using specific control laws. An additional aspect to be underlined of such devices is the self-centering ability after an extreme loading event, realigning the deck with its original axis and the towers.

Figure 6 depicts some input kinematic conditions with standard sinusoidal shapes for the proposed semi-active models of hysteresis. Namely the continuous line is the sinusoidal motion in terms of displacements for the support and the body mass to be controlled (x_{bc} and x_{sup} respectively). Figure 7 reports the hysteresis response of the models. The BW parameters have been set as those in Figure 5 (A=1, n=1, α =0, Φ_y=1000kN and 500kN respectively for the high and low semi-active limit, k=80000kN/m, with $\gamma_{low}=\beta_{low}$ and $\gamma_{high}=\beta_{high}$.).

Several results from literature are intended as a realistic validation of such innovative semi-active technology implementation on complex structures by numerical simulations through finite element models. The next section is devoted to summarize some research works where bridge structures in particular are considered.

However, the numerical simulation by suitable models of hysteresis and the detailed approaches are intended to a wide application at different building typologies.

APPLICATIONS EXAMPLES

The previously discussed numerical approaches have demonstrated to be capable to reproduce control systems with passive and semi-active signature in complex structural models (e.g. finite element ones).

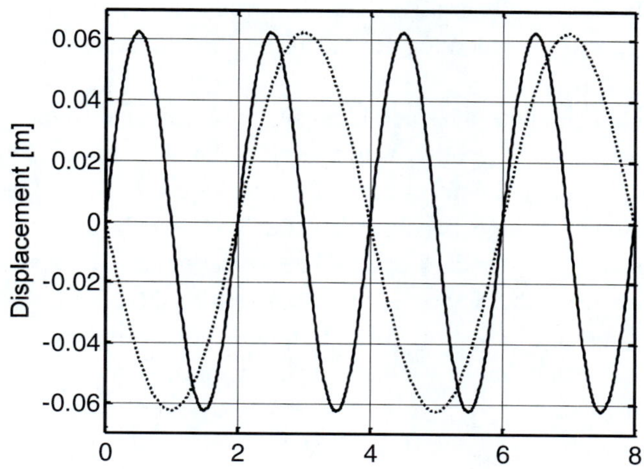

Figure 6. Restrain and dissipation vs. different input cycle amplitudes. Continuous line for the support, dotted for the body mass to be controlled.

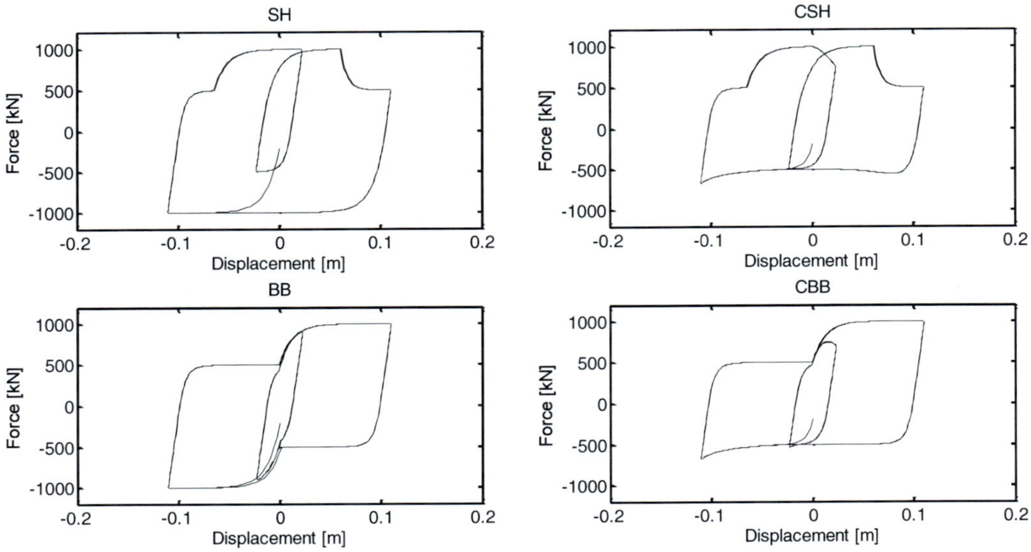

Figure 7. Models response to the sinusoidal input of Figure 6. SH and BB refer to discrete *SkyHook* and *BangBang* models, CSH and CBB respectively to the continuous ones.

Mitigation of direct dynamic effects is traditionally the primary target of control systems (e.g. acceleration and/or displacements mitigation). However, attention is also given to the positive outcomes related to indirect effects coming from the dampers implementation on buildings.

The international control benchmark on cable stayed bridges, attracting in the last years the attention of the control community, has been considered in [38] for studying passive and semi-active decentralized collocated solutions for seismic mitigation. The passive BW and semi-active on/off DM models are employed. Robust feasible solutions are investigated starting from a first step where passive devices are applied to the bridge numerical model between the deck and the piers. Subsequently, an open-loop semi-active improvement of the passive control system is implemented and, finally, an innovative decentralized solution for a semi-active scheme in closed-loop configuration is introduced. The device force displacement relationship fits the results of laboratory tests during the characterization process of an EI device [19].

In [22] the results of simulation tests to identify the mechanical signature of an innovative isolation device known as the Roll-N-Cage (RNC) isolator are presented. The seismic performance of an RNC passive control scheme is subsequently investigated on an updated model of the cable-stayed bridge benchmark. The force–displacement relationship of the device is reproduced using the standard Bouc–Wen model of smooth hysteresis. Subseqently, the numerical assessment of the device efficiency is established through its implementation into a bridge model considering several ground motions as external excitations. It was found that the RNC isolator is promising as a reliable isotropic horizontal isolation device for bridge structures.

The same updated version of the controlled cable-stayed bridges is considered in [41] including new aspects in the knowledge of the relative transversal motion of the main girder with respect to the towers under seismic motions. Earthquake restrainers in the transverse direction are considered in the original benchmark configuration.

The proposed updated bridge model first reproduces this kinematic arrangement, subsequently this restriction has been overcome and a new configuration with transversal releases is studied. The control strategies consist in passive and decentralized semi-active systems working in the horizontal plane, likewise those studied for the original statement in the previous investigation [38]. The proposed arrangement schemes markedly mitigate the transversal structural response of the bridge with sensible effectiveness.

The main goal of the study in [42, 43] consists in the evaluation of the control system performance, for iconic bridges, when local failure occurs in the control system devices. Model of a cable-stayed bridge and a suspension one are developed at the numerical level in a commercial finite element code, starting from original data. The models are used to simulate the structural response under extreme loading conditions, such as seismic excitation and wind. Different types of control systems for the mitigation of the bridge response, implementing passive and semi-active dampers respectively, are employed through the BW and MD approaches from the literature.

Some general observations, by means of a proposal for quantitative robustness indexes in terms of control performance, are also provided. Resilience advantages in bridge structures by means of smart control strategies is an innovative research topic currently under investigation at the Politecnico di Milano.

Collocated and non collocated decentralized control strategies for an existent suspension bridge are studied in [44] against strong wind buffeting adopting a numerical approach. An original procedure for implementing BW and DM models into a multipurpouse finite element code is included by a mixed implicit-explicit step by step formulation. The bridge model results useful to assess the efficacy of passive and discrete semi-active control schemes. Their efficacy in mitigating instant loading effects is shown and the factors contributing to their positive performance both for mitigating instant structural variables and also fatigue damages are highlighted. Protection of composite decks in long span bridges against fatigue damage through decentralized control schemes is an interesting research topic currently under investigation at the Politecnico di Milano.

The numerical model of an existing suspension bridge is used in [45] to reevaluate an optimization procedure for a passive control strategy, already proven effective in [44] with a simplified model of the buffeting wind forces. The BW is used for reproducing the dissipative behavior of the passive control scheme. Such optimization procedure, previously implemented with a quasi-steady model of the buffeting excitation, is reevaluated adopting a more refined version of the wind-structure interaction forces. It is based on the use of indicial function, adopting to reflect coupling with the bridge orientation and motion. The final outcomes show that the previously identified optimal passive configuration withstands.

The research work [23] is focused on the development of a numerical procedure for the computation of seismic fragilities in base isolated Nuclear Power Plants. A special procedure for fragility computation is proposed making use of the Response Surface Methodology to model the influence of the random variables on the dynamic response and the Monte Carlo method is used to compute the failure probability. The nuclear building is equipped with a base isolation system based on the introduction of High Damping Rubber Bearing elements showing a markedly non linear mechanical behavior and simulated by the Abe et al. [33, 34]. The isolation devices result the critical elements in terms of seismic risk and the dynamic behavior of the building, belonging to rigid body motion one, is captured by low-dimensional numerical models.

REFERENCES

[1] Domaneschi, M. (2006). Structural Control of Cable-stayed and Suspended Bridges. *Ph.D. Thesis*. University of Pavia, I. file:///C:/Users/ghisi/Downloads/Domaneschi.pdf.

[2] Bouc, R. (1967). Forced vibrations of mechanical systems with hysteresis. *Proceedings of the 4th International Conference on Nonlinear Oscillations*. Prague, Czechoslovakia.

[3] Wen, Y. K. (1976). Method for random vibration of hysteretic systems. *Journal of the engineering mechanics division ASCE*. 102, 249–263.

[4] Erlicher, S., Point, N. (2004). Thermodynamic admissibility of Bouc–Wen type hysteresis models. *C. R. Mecanique*. 332, 51–57.

[5] Casciati, F., Faravelli, L. (1991). Fragility Analysis of Complex Structural Systems. *Research studies press ltd*. Taunton, Somerset, England.

[6] Domaneschi, M. (2012). Simulation of Controlled Hysteresis by the Semi-active Bouc-Wen Model. *Computers and Structures*. 106–107, 245–257.

[7] Ikhouane, F., Hurtado, J. E., Rodellar, J. (2007). Variation of the hysteresis loop with the Bouc–Wen model parameters. *Nonlinear Dynam*. 48, 361–380.

[8] Baber, T. T., Wen, Y. K. (1981). Random vibration of hysteretic, degrading systems. *Journal of the engineering mechanics division ASCE*. 107, 1069-1087.

[9] Casciati, F. (1989). Stochastic Dynamics of Hysteretic Media. *Structural Safety*. 6, 259-269.

[10] Noori, M., Dimentberg, M., Hou, Z., Christodoulidou, R., Alexandrou, A. (1995). First-passage study and stationary response analysis of a BWB hysteresis model using quasi-conservative stochastic averaging method. *Probabilist. Eng. Mech*. 10, 161-170.

[11] Foliente, C. G., Singh, M. P., Noori, M. N. (1996). Equivalent linearization of generally pinching hysteretic, degrading systems. *Earth. Eng. Struct. D*. 25, 611-629.

[12] Park, J., Wen, Y. K., Ang, A. H. S. (1986). Random vibration of hysteretic systems under bi-directional ground motions. *Earthq. Eng. Struct. D*. 14, 543-557.

[13] Casciati, F., Faravelli, L. (1998). Stochastic equivalent linearization for 3D frames. *J. Eng. Mech. ASCE*. 114, 1760-1771.

[14] Ikhouane, F., Manosa, V., Rodellar, J. (2007). Dynamic properties of the hysteretic Bouc-Wen model. *Syst. Control. Lett*. 56, 197–205.

[15] Ismail, M., Ikhouane, F., Rodellar, J. (2009). The hysteresis bouc-wen model, a survey. *Arch. Comput. Method. E*. 16, 161–188.

[16] Lacarbonara, W., Vestroni, F. (2003). Nonclassical Responses of Oscillators with Hysteresis. *Nonlinear Dynam*. 32, 235–258.

[17] Ismail, M., Ikhouane, F., Rodellar, J. (2009). The hysteresis Bouc-Wen model, a survey. *Arch. Comput. Method. E*. 16, 161–88.

[18] Canudas de Wit, C., Olsson, H., Astrom, K. J., Lischinsky, P. (1995). A New Model for Control of Systems with Friction. *IEEE Transactions on Automatic control*. 40(3), 419-425.

[19] Casciati, F., Domaneschi, M. (2007). Semi-active Electro-inductive Devices: Characterization and Modelling. *Journal of Vibration and Control*. 13(6), 815-838.

[20] Boston, C., Weber, F., Guzzella L. (2010). Modeling of a disc-type magnetorheological damper. *Smart Mater. Struct*. 19, 045005, 12pp.

[21] Do, N., Ferri, A., Buachau, O. A. (2007). Efficient Simulation of a Dynamic System with LuGre Friction. *Journal of Computational and Nonlinear Dynamics*. 2, 281-289.

[22] Ismail, M., Rodellar, J., Carusone, G., Domaneschi, M., Martinelli L. (2013). Characterization, modeling and assessment of Roll-N-Cage isolator using the cable-stayed bridge benchmark. *Acta Mechanica*. 224, 525–547.

[23] Perotti, F., Domaneschi, M., De Grandis, S. (2013). The numerical computation of seismic fragility of base-isolated NPP buildings. *Nuclear Engineering and Design*. 262, 189–200.

[24] Ozdemir, H., 1976. Nonlinear transient dynamic analysis of yielding structures. *Ph.D. Dissertation,* Division of Structural Engineering and Structural Mechanics, Department of Civil Engineering, University of California, Berkeley.

[25] Wen, Y. K., 1976. Method for random vibration of hysteretic systems. *J. eng. mech. ASCE.* 102. 249-263.

[26] Sivaselvan, M. V., Reinhorn, A. M., 2000. Hysteretic models for deteriorating inelastic structures. *J. eng. mech. ASCE.* 126(6), 633-640.

[27] Grant, D. N., Fenves, G. L., Auricchio, F. (2005). Modelling and analysys of high-damping rubber bearings for the seismic protection of bridges. *Research Report No. ROSE-2005/01*, ROSE School, Pavia, I.

[28] Kikuchi, M., Aiken, I. D. (1997). An analytical hysteresis model for elastomeric seismic isolation bearings. *Earthquake Engng. Struct. Dyn*. 26, 215-231.

[29] Hwang, J. S., Wu, J. D., Pan, T. C., Yang, G., 2002. A Mathematical Hysteretic Model for Elastomeric Isolation Bearings. *Earthquake Engng. Struct. Dyn.* 31, 771-789.

[30] Pan, T. C., Yang, G., 1996. Nonlinear analysis of base-isolated MDOF structures. *Proceedings of the 11th World Conference on Earthquake Engineering*, Mexico. Paper No. 1534.

[31] Tsai, C. S., Chiang, T. C., Chen, B. J., Lin, S. B., 2003. An advanced analytical model for high damping rubber bearings. *Earthquake Engng. Struct. Dyn.* 32, 1373-1387.

[32] Jankowski, R., 2003. Nonlinear Rate Dependent Model of High Damping Rubber Bearing. *B Earthq. Eng*. 1, 397-403.

[33] Abe, M., Yoshida, J., Fujino, Y., 2004a. Multiaxial Behaviors of Laminated Rubber Bearings and Their Modeling. I: Experimental Study. *J. Struct. Eng.-ASCE*. 130, 1119-1132.

[34] Abe, M., Yoshida, J., Fujino, Y., 2004b. Multiaxial Behaviors of Laminated Rubber Bearings and Their Modeling. II: Modeling. *J. Struct. Eng.-ASCE*. 130, 1133-1144.

[35] Ryan, K. L., Kelly, J. M., Chopra, A., 2005. Nonlinear Model for Lead-Rubber Bearings Including Axial-Load Effects. *J. Struct. Eng.-ASCE*. 131, 1270-1278.

[36] Yamamoto, S., Kikuchi, M., Ueda, M., Aiken, I. D., 2009. A mechanical model for elastomeric seismic isolation bearings including the influence of axial load. *Earthquake Engng. Struct. Dyn*. 38, 157-180.

[37] Kikuchi, M., Nakamura, T., Aiken, I. D., 2010. Three-dimensional analysis for square seismic isolation bearings under large shear deformations and high axial loads. *Earthquake Engng. Struct. Dyn.* 39, 1513-1531.

[38] Domaneschi, M. (2010). Feasible control solutions of the ASCE benchmark cable-stayed bridge. *Structural Control and Health Monitoring*. 17, 675-693.

[39] Casciati, F., Magonette, G., Marazzi, F. (2006). *Technology of semi-active devices and applications in vibration mitigation*. John Wiley and Sons Ltd.

[40] Yang, G., Spencer Jr, B. F., Jung, H. J., Carlson, J. D. (2004). Dynamic modeling of large-scale magnetorheological damper systems for civil engineering applications. *J. Eng. Mech.-ASCE*. 130, 1107–14.

[41] Domaneschi, M., Martinelli, L. (2013). Deepening the ASCE bridge benchmark: transversal response under seismic loading. Accepted in *Journal of Bridge Engineering ASCE*. DOI: 10.1061/(ASCE) 13 BE.1943-5592.0000532.

[42] Domaneschi, M., Martinelli, L. (2014). Robustness of passive and semi-active control schemes on a cable stayed bridge under extreme loading conditions. *7th International Conference on Bridge Maintenance, Safety and Management (IABMAS 2014)*. Shanghai, China, July 7-11.

[43] Domaneschi, M., Martinelli, L., E. Po (2014). Robustness issues and hybridization of a Tuned Mass Damper system on a suspension bridge model under variable wind buffeting. *7th International Conference on Bridge Maintenance, Safety and Management (IABMAS 2014)*. Shanghai, China, July 7-11.

[44] Domaneschi, M., Martinelli, L. (2013). Optimal Passive and Semi-active Control of a Wind Excited Suspension Bridge. *Structure and Infrastructure Engineering*. 9(3):242-259.

[45] Domaneschi, M., Martinelli, L. (2014). Refined Optimal Passive Control of Buffeting-induced Wind Loading of a Suspension Bridge. *Wind and Structures*. 18, 1-20.

In: Hysteresis
Editor: José Carlos Dias

ISBN: 978-1-63321-336-4

Chapter 8

PERCOLATION ANALYSIS OF TENDENCIES OF RELATIVE PERMEABILITY HYSTERESIS BEHAVIOR

V. V. Kadet and A. M. Galechyan*
Department of Petroleum and earth hydromechanics, Gubkin Russian State University of Oil and Gas, Moscow, Russian Federation

ABSTRACT

The phenomenon of relative permeability hysteresis is observed during the process of developing the oil field by methods where the flow direction changes and the displacement of oil by water changes into the displacement of water by oil and vice versa. This Chapter represents the model of relative permeability hysteresis for drainage and imbibition based on percolation theory. The phenomenon of active oil components adsorption on the rock-forming minerals is chosen for the first mechanism of hysteresis origin. In the process of drainage this causes surface hydrophobization of initially hydrophilic rock which leads to phase relative permeability change.

To describe this phenomenon, percolation model for the various media with micro heterogeneous wettability is used. The second mechanism is plastic properties change, described by percolation model as well. Obtained numerical solution is represented as relative permeability curves and is qualitatively confirmed by the experimental data.

The behavior of relative permeability hysteresis is observed and analyzed for various differential radius distribution curves, capillary network coordination numbers, saturation models, hydrophobization degree and plastic properties. It allows establishing general tendencies of relative permeability hysteresis behavior. Further introduced methodology can be put into practice for relative permeability calculation considering hysteresis in any porous media to reduce the time spent.

Keywords: relative permeability, drainage, imbibition, hysteresis, percolation theory

* Corresponding author: A. M. Galechyan. Department of Petroleum and Earth Hydromechanics, Gubkin Russian State University of Oil and Gas, 65 Leninsky Prospekt, Moscow, 119991, Russian Federation. Phone: +79152310568, e-mail: matvey0508@yandex.ru.

Introduction

During the process of developing the oil field using such methods as cycling waterflooding and changing the flow direction the displacement of oil by water is changed into the displacement of water by oil and vice versa. Such change in flow character depends on relative permeability as function of water saturation. This phenomenon is called relative permeability hysteresis in drainage and imbibition.

In different scientific articles drainage and imbibition is explained differently. In this Chapter, in accordance with the introduced experimental methodology [1], drainage is given as displacement of wetting phase by non-wetting phase under the pressure gradient. As for imbibition it is presented as displacement of non-wetting phase by wetting phase under the pressure gradient. Naturally for hydrophilic sample drainage is the displacement of water by oil, as imbibition is the displacement of oil by water. In case of hydrophobic rock the displacement of water by oil is imbibition and the displacement of oil by water is drainage.

Processes of fluid flow in porous media can be mathematically described with percolation theory which was introduced by Broadbent and Hammersley [2]. The first percolation approach for relative permeability calculation was introduced by Heiba et al. [3]. This approach was used by many authors including Dixit et al. [4] (who used this method to calculate relative permeability considering hysteresis for different coordination numbers and fractions of oil-wet pores in the network) and Phirani et al. [5] (who compared this method with correlation functions for relative permeability in case of hydrate formation). But there was no ultimate analytical expression for relative permeability in these works. Kadet and Selyakov obtained this solution based on *r*-chain hierarchy [6]. It was used by some authors [7].

The represented percolation model of relative permeability hysteresis employs analytical expressions for media with micro heterogeneous wettability and for fluids with different plastic properties obtained by Selyakov and Kadet [8] based on the approach introduced at [6]. It allows to describe hysteresis phenomenon and to explain the mechanism of its origin.

The innovation of this paper is using percolation approach introduced at [8] which allows calculation of relative permeability hysteresis for any type of reservoir rocks. It is important to notice that analytical expressions for relative permeability give understanding of nature and physics of relative permeability hysteresis which is not finally investigated till nowadays.

Experimental Research of Relative Permeability Hysteresis Review

Presence of relative permeability hysteresis was noted in quite a few experimental articles [1, 9-13]. The most obvious results are introduced at [1], where the following experimental methodology was used. Ten core samples (their numbers: 113, 114, 9, 15, 22, 72, 4, 6, 138, 139) covering five reservoir rock types (RRT) were selected based on whole core analysis.

The samples 113, 114 were formed by RRT 1; samples 9, 15 – RRT 2; samples 22, 72 – RRT 3; samples 4, 6 – RRT 4; samples 138, 139 – RRT 5. The samples 113, 114 were dolomite-rich and the other samples – calcite-rich.

The samples 113, 114 had differential radius distribution 1 (Distr. 1); samples 9, 15, 22, 72, 4, 6 – Distr. 2; samples 138, 139 – Distr. 3. Every sample was totally imbibed by wetting fluid (water).

In case of water-wet sample initially saturated with water it was displaced by oil under the pressure gradient up to the maximal oil saturation for drainage. Then oil was displaced by water under the pressure gradient up to the maximum water saturation for imbibition.

The received experimental data are represented in the article mentioned above.

MATHEMATICAL MODELS OF DRAINAGE AND IMBIBITIONS

Model of Porous Media

The regular network of capillaries (primitive cubic, body-centered cubic or face-centered cubic) is taken for the model of porous media. Capillaries have the model density of radius distribution – lognormal function, which is qualitatively close to real differential radius distribution curve:

$$f(r)=\left(\sqrt{2\pi}\sigma_d r\right)^{-1}\exp\left(-\frac{(\ln(r)-\mu)^2}{2\sigma_d^2}\right),$$

where; σ_d = 0.25, μ = 2 for model distribution shown on Figure 3-5 (which was used for illustration); $\sigma_d = 0.5$, $\mu = 0.9$ for Distr. 1; $\sigma_d = 0.25$, $\mu = -0.1$ for Distr. 2; $\sigma_d = 0.3$, $\mu = -1$ for Distr. 3 (Figure 1).

Water saturation S can be considered to be proportional to the number of water saturated capillaries if the sizes of the pores do not differ significantly (Model I) or proportional to the volume of water saturated capillaries (Model II). According to Model I water saturation is the area of zone of capillaries occupied by water in the $f(r)$ plot.

The following phenomena are chosen for the main mechanisms of hysteresis origin. In the process of drainage active oil components are adsorbed on the rockforming minerals [14, 15]. It causes two possible effects, which are considered in this paper. The first is surface hydrophobisation and the second – changing of plastic properties in both fluids. The second effect also occurs because of mixing oil and water during drainage, when some oil components get into water.

The two percolation models of drainage and imbibition based on these mechanisms – hydrophobisation model and plastic model – are considered further.

Hydrophobisation Model

Drainage Model

Water-wet sample oil first of all flows into large pores because of small hydrodynamic resistance while small pores water is kept by capillary forces that prevent oil flow through them.

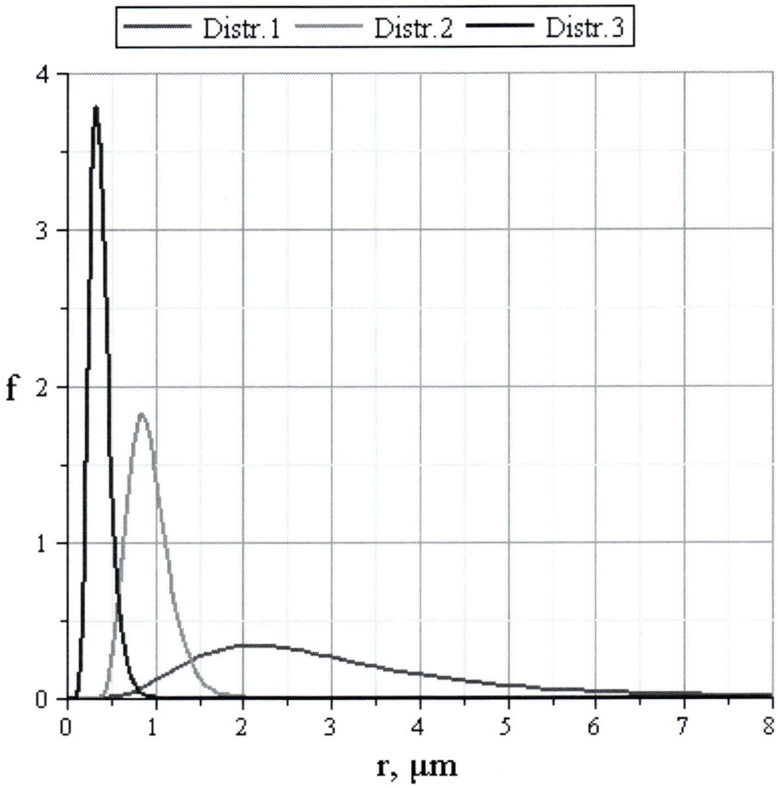

Figure 1. Differential radius distribution curves.

Thereby oil is contained in capillaries with radius more than r_k and water – in capillaries with radius less than r_k, where r_k is the minimal radius of capillary from which the displacement of wetting phase (water) occurs having set pressure difference in liquids Δp [16]

$$r_k = \frac{2\chi\cos\theta}{\Delta p},$$

where; χ is interfacial tension factor, and θ is the wetting angle.

According to this, density of radius distribution function for capillaries occupied by oil is equivalent to:

$$f_o(r) = \begin{cases} 0, & r < r_k, \\ f(r), & r \geq r_k. \end{cases} \tag{1}$$

Then analytical expression for oil relative permeability $k_o(r_k)$ [8] can be written as

$$k_o(r_k) = K_0^{-1} \int_{r_k}^{r_c} \frac{\left[\int_{r_1}^{r_c} f(r)dr\right]^{\nu} f(r_1)}{\frac{8}{\pi}\int_{r_1}^{\infty} \frac{f(r)}{r^4}dr \left[\int_{r_1}^{\infty} f(r)dr\right]^{-1}} dr_1. \tag{2}$$

here K_0 is absolute permeability of sample, v is correlation radius index and r_c is determined by the relation

$$\int_{r_c}^{\infty} f(r)dr = P_c^b = \frac{D}{z(D-1)}, \tag{3}$$

where; $P_c^{\,b}$ is percolation threshold, D is dimensionality of network (D=3), z is coordination number (z=6 for primitive cubic network; z=8 for body-centered cubic network; z=12 for face-centered cubic network).

The equation for permeability K_0 has the form of:

$$K_0 = \int_0^{r_c} \frac{\left[\int_{r_1}^{r_c} f(r)dr\right]^{v} f(r_1)}{\frac{8}{\pi}\int_{r_1}^{\infty} \frac{f(r)}{r^4}dr\left[\int_{r_1}^{\infty} f(r)dr\right]^{-1}} dr_1. \tag{4}$$

Density of radius distribution function for capillaries occupied by water is the following:

$$f_w(r) = \begin{cases} f(r), & r \le r_k, \\ 0, & r > r_k. \end{cases} \tag{5}$$

Accordingly, the analytical expression for water relative permeability $k_w(r_k)$ [8] is equal to:

$$k_w(r_k) = K_0^{-1} \int_0^{r_c'} \frac{\left[\int_{r_1}^{r_c'} f(r)dr\right]^{v} f(r_1)}{\frac{8}{\pi}\int_{r_1}^{\infty} \frac{f(r)}{r^4}dr\left[\int_{r_1}^{\infty} f(r)dr\right]^{-1}} dr_1. \tag{6}$$

here r_c' is determined by the relation:

$$\int_{r_c'}^{r_k} f(r)dr = P_c^b. \tag{7}$$

Connection of the dependences (2), (6) with water saturation S using Model I is determined by the following relation:

$$S = \int_0^{r_k} f(r)dr. \tag{8}$$

and for Model II:

$$S = \int_0^{r_k} f(r) r^2 dr \left[\int_0^{\infty} f(r) r^2 dr \right]^{-1}. \tag{9}$$

Thereby formulas (1) – (9) determine the parametric relative permeability dependence on water saturation for above mentioned Models I and II.

Imbibition Model

Surface properties of some part of capillaries are changed due to oil flow through initially hydrophilic porous media. To describe this phenomenon percolation model for media with micro heterogeneous wettability is used [8]. Two parameters κ and α, are used here: κ – part of capillaries with unchanged surface properties (type 1), $(1 - \kappa)$ – part of capillaries with changed surface properties (type 2), $\alpha = \cos\theta_2/\cos\theta_1$, where θ_1 and θ_2 are wetting angles in capillaries of the 1st and the 2nd type respectively. According to this, surface properties may be changed in two different ways (Figure 2): either hydrophilic properties are decreased ($0<\alpha<1$) or capillary surface becomes hydrophobic at all ($\alpha<0$).

For $0<\alpha<1$ water displaces oil firstly flowing into small capillaries. Capillaries of the 1st type with radius less than r_{k1} and capillaries of the 2nd type with radius less than r_{k2} contain water (Figure 3). Radiuses r_{k1} and r_{k2} are determined by equation [16]

$$r_{ki} = \frac{2\chi_i \cos\theta_i}{p_k}, \qquad i = 1, 2, \tag{10}$$

where; χ_i is interfacial tension factor in i-type capillaries, p_k is capillary pressure. To decrease the number of parameters the condition $\chi_1=\chi_2=\chi$ is taken because this parameter appears only in product $\chi_i \cos\theta_i$. Then (10) follows the condition $r_{k2}=\alpha r_{k1}$.

Density of radius distribution function for capillaries occupied by oil is the following:

$$f_o(r) = \begin{cases} 0, & r < r_{k2}, \\ (1-\kappa) f(r), & r_{k2} \le r \le r_{k1}, \\ f(r), & r > r_{k1}. \end{cases}$$

Analytical expression for oil relative permeability $k_o(r_{k1})$ is written in form of

$$k_o(r_{k1}) = 0, \quad r_{k1} > r_{c1},$$

$$k_o(r_{k1}) = \frac{\pi(1-\kappa)^2}{8K_0} \int_{\alpha r_{k1}}^{r_{co}} \left[\int_{r'}^{r_{co}} f(r)dr \right]^{\nu} f(r') \left((1-\kappa)\int_{r'}^{r_{k1}} f(r)dr + \int_{r_{k1}}^{\infty} f(r)dr \right) \times$$

$$\times \left((1-\kappa)\int_{r'}^{r_{k1}} \frac{f(r)}{r^4} dr + \int_{r_{k1}}^{\infty} \frac{f(r)}{r^4} dr \right)^{-1} dr', \qquad r_c < r_{k1} \le r_{c1},$$

$$k_o(r_{k1}) = \frac{\pi}{8K_0} \left\{ (1-\kappa) \int_{\alpha r_{k1}}^{r_{k1}} \left[(1-\kappa)\int_{r'}^{r_{k1}} f(r)dr + \int_{r_{k1}}^{r_c} f(r)dr \right]^{\nu} f(r') \left((1-\kappa)\int_{r'}^{r_{k1}} f(r)dr + \right. \right.$$

$$\left. + \int_{r_{k1}}^{\infty} f(r)dr \right) \left((1-\kappa)\int_{r'}^{r_{k1}} \frac{f(r)}{r^4} dr + \int_{r_{k1}}^{\infty} \frac{f(r)}{r^4} dr \right)^{-1} dr' + \int_{r_{k1}}^{r_c} \left[\int_{r'}^{r_c} f(r)dr \right]^{\nu} f(r') \times$$

$$\left. \times \int_{r'}^{\infty} f(r)dr \left(\int_{r'}^{\infty} \frac{f(r)}{r^4} dr \right)^{-1} dr' \right\}, \qquad 0 < r_{k1} \le r_c,$$

where; r_{c1} is radius of the 1st type capillaries starting from which the infinite cluster (IC) is formed by capillaries of both types occupied by oil; r_{co} is the r_c analog as a function of r_{k1}. Radius r_c is calculated from (3).

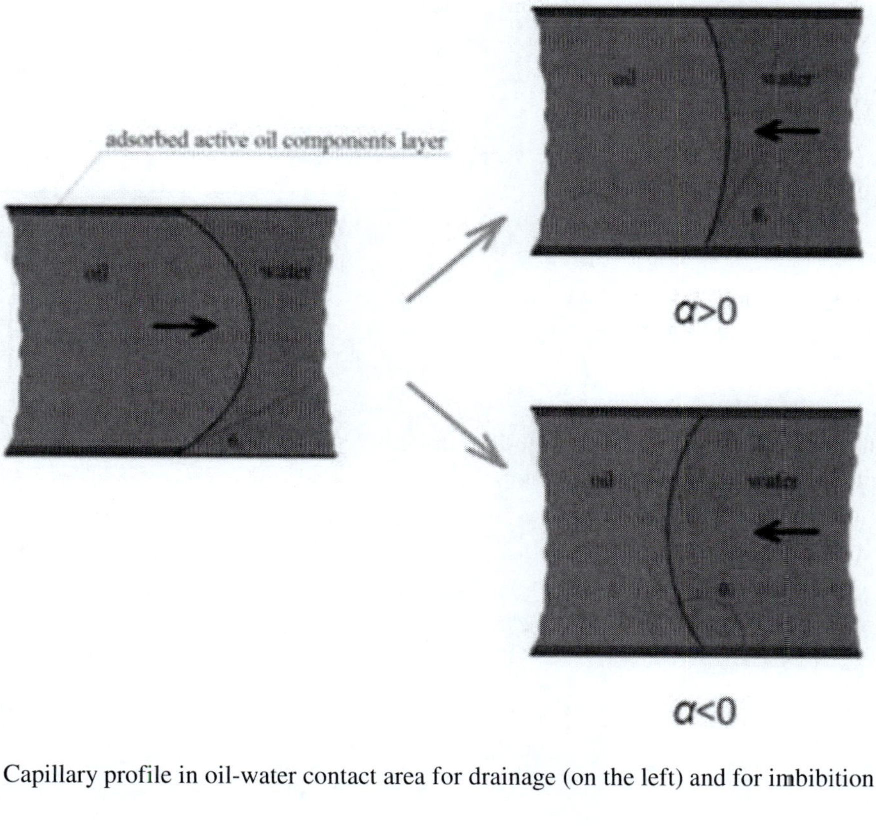

Figure 2. Capillary profile in oil-water contact area for drainage (on the left) and for imbibition (on the right).

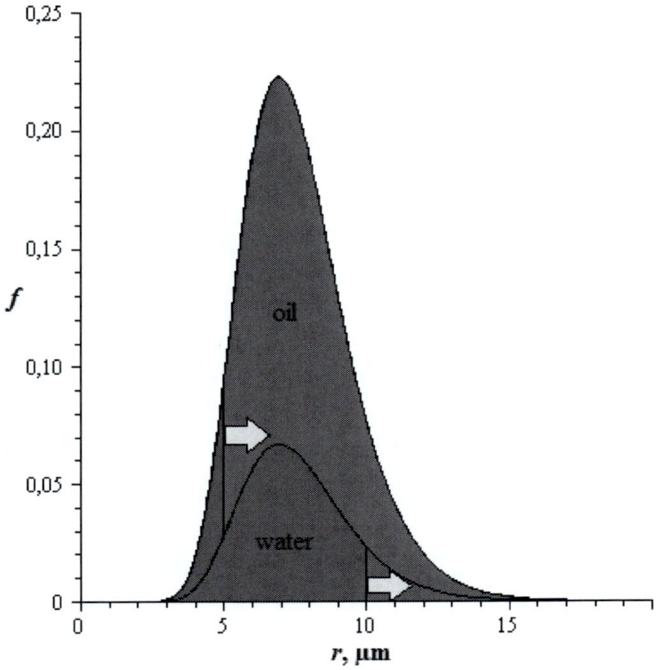

Figure 3. Density of radius distribution curves for capillaries occupied by water and for capillaries occupied by oil for imbibition in some instant of time in case of α>0.

Quantity r_{c1} is determined by relation

$$(1-\kappa)\int_{\alpha r_{c1}}^{r_{c1}} f(r)dr + \int_{r_{c1}}^{\infty} f(r)dr = P_c^b$$

r_{co} is determined by expression

$$(1-\kappa)\int_{r_{co}}^{r_{k1}} f(r)dr + \int_{r_{k1}}^{\infty} f(r)dr = P_c^b.$$

According to this, r_{co} changes within the limits $\alpha r_{c1} \le r_{co} \le r_c$ and $r_{co}=r_c$ when $r_{k1} \le r_c$. Density of radius distribution function for capillaries occupied by water is the following:

$$f_w(r) = \begin{cases} f(r), & r < r_{k2}, \\ \kappa f(r), & r_{k2} \le r \le r_{k1}, \\ 0, & r > r_{k1}. \end{cases}$$

Analytical expression for water relative permeability $k_w(r_{k1})$ is written in the following form:

$$k_w(r_{k1})=\frac{\pi}{8K_0}\int_0^{r_{cw}}\left[\int_{r'}^{r_{cw}}f(r)dr\right]^{\nu}f(r')\left(\int_{r'}^{\alpha r_{k1}}f(r)dr+\kappa\int_{\alpha r_{k1}}^{r_{k1}}f(r)dr\right)\left(\int_{r'}^{\alpha r_{k1}}\frac{f(r)}{r^4}dr+\right.$$

$$\left.+\kappa\int_{\alpha r_{k1}}^{r_{k1}}\frac{f(r)}{r^4}dr\right)^{-1}dr',\qquad r_{k1}>r_{c1w},\quad r_{cw}<\alpha r_{k1},$$

$$k_w(r_{k1})=\frac{\pi}{8K_0}\left\{\int_0^{\alpha r_{k1}}\left[\int_{r'}^{\alpha r_{k1}}f(r)dr\right]^{\nu}f(r')\left(\int_{r'}^{\alpha r_{k1}}f(r)dr+\kappa\int_{\alpha r_{k1}}^{r_{k1}}f(r)dr\right)\left(\int_{r'}^{\alpha r_{k1}}\frac{f(r)}{r^4}dr+\right.\right.$$

$$\left.\left.+\kappa\int_{\alpha r_{k1}}^{r_{k1}}\frac{f(r)}{r^4}dr\right)^{-1}dr'+\kappa^2\int_{\alpha r_{k1}}^{r_{cw}}\left[\int_{r'}^{r_{cw}}f(r)dr\right]^{\nu}f(r')\int_{r'}^{r_{k1}}f(r)dr\left(\int_{r'}^{r_{k1}}\frac{f(r)}{r^4}dr\right)^{-1}dr'\right\},$$

$$r_{k1}>r_{c1w},\quad r_{cw}\geq\alpha r_{k1},$$

where; $r_{cw}\,(r_{k1})$ is the same as $r_{co}\,(r_{k1})$ in case of forming ICW (infinite cluster of capillaries occupied by water) while $r_{co}\,(r_{k1})$ is obtained in forming ICO (infinite cluster of capillaries occupied by oil). According to this, function $r_{cw}\,(r_{k1})$ changes from 0 for some minimal sense $r_{k1}=r_{c1w}$ to sense r_c for $r_{k1}=\infty$. The condition for determination of function $r_{cw}\,(r_{k1})$ has two variants of notation.

$$\int_{r_{cw}}^{\alpha r_{k1}}f(r)dr+\kappa\int_{\alpha r_{k1}}^{r_{k1}}f(r)dr=P_c^b,\qquad r_{cw}<\alpha r_{k1},$$

$$\kappa\int_{r_{cw}}^{r_{k1}}f(r)dr=P_c^b,\qquad r_{cw}\geq\alpha r_{k1},$$

Radius r_{c1w} is determined by relation

$$\int_0^{\alpha r_{c1w}}f(r)dr+\kappa\int_{\alpha r_{c1w}}^{r_{c1w}}f(r)dr=P_c^b.$$

Now the case $\alpha<0$ will be considered. According to this, water first of all displaces oil from the 1st type capillaries starting from small ones (Figure 4).

After all the 1st type capillaries are occupied by water the displacement of oil from the 2nd type capillaries will begin starting from large ones (Figure 5). Different variants of two-phase flow are possible depending on value κ.

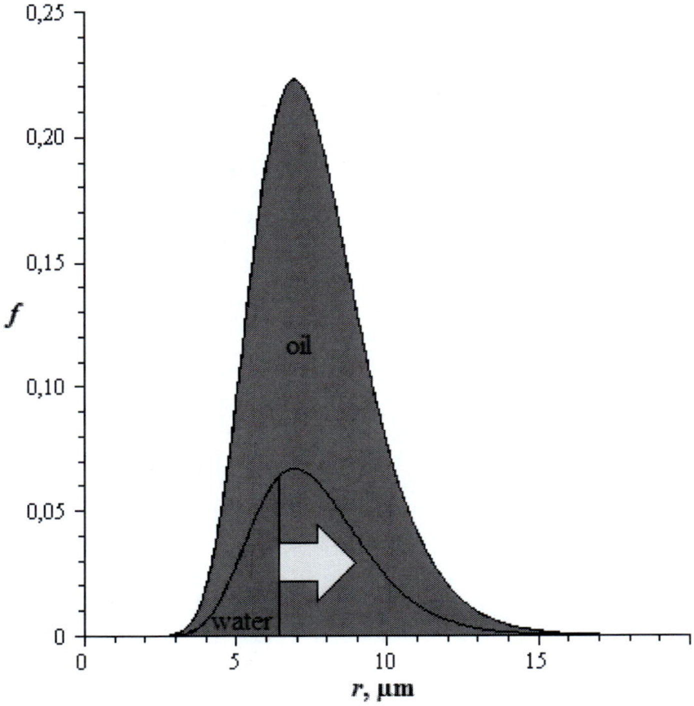

Figure 4. Density of radius distribution curves for capillaries occupied by water and for capillaries occupied by oil for imbibition in some instant of time for p_k<0 in case of α<0.

In capillary pressure increase from -∞ to 0 the 1st type capillaries part in ICO decreases. Density of radius distribution function for them will have the form as:

$$f_o(r)=\begin{cases}(1-\kappa)f(r), & r<r_{k1},\\ f(r), & r\geq r_{k1}.\end{cases}$$

If $r_{k1}<r_c$ then $r_{co}(r_{k1})=r_c$ and for k_o we have

$$k_o(r_{k1})=\frac{\pi}{8K_0}\left\{\int_0^{r_{k1}}\left[(1-\kappa)\int_{r'}^{r_{k1}}f(r)dr+\int_{r_{k1}}^{r_c}f(r)dr\right]^{\nu}f(r')\left(\int_{r'}^{\infty}f(r)dr-\kappa\int_{r'}^{r_{k1}}f(r)dr\right)\times\right.$$

$$\times\left[(1-\kappa)\int_{r'}^{r_{k1}}\frac{f(r)}{r^4}dr+\int_{r_{k1}}^{\infty}\frac{f(r)}{r^4}dr\right]^{-1}dr'+\int_{r_{k1}}^{r_c}\left[\int_{r'}^{r_c}f(r)dr\right]^{\nu}f(r')\int_{r'}^{\infty}f(r)dr\times$$

$$\left.\times\left(\int_{r'}^{\infty}\frac{f(r)}{r^4}dr\right)^{-1}dr'\right\},\qquad r_{k1}<r_c$$

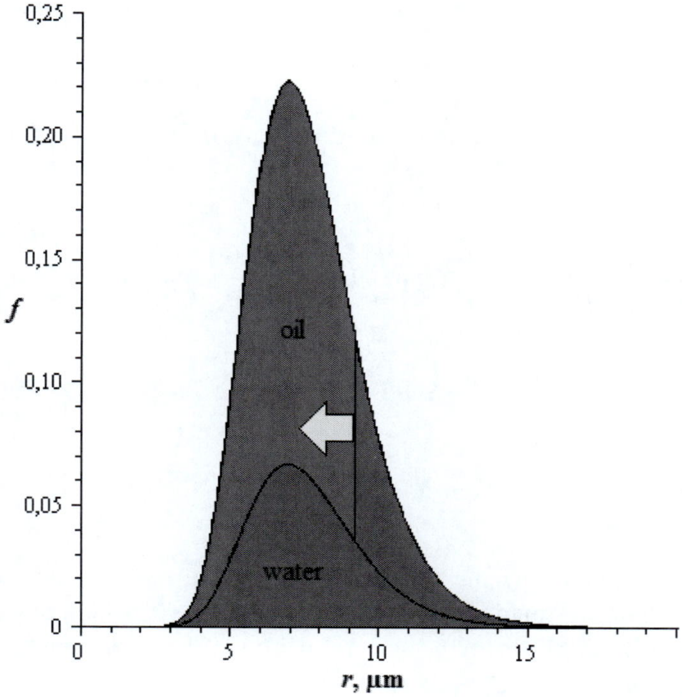

Figure 5. Density of radius distribution curves for capillaries occupied by water and for capillaries occupied by oil for imbibition in some instant of time for $p_k>0$ in case of $\alpha<0$.

If $r_{k1}>r_c$ then dependence $r_{co}(r_{k1})$ will have the form

$$(1-\kappa)\int_{r_{co}}^{r_{k1}} f(r)dr + \int_{r_{k1}}^{\infty} f(r)dr = P_c^b.$$

If $(1-\kappa) <P_c^b$ then capillary pressure increase ($p_k<0$) results in ICO disappearance for some $p_k=p_{co}<0$ because the number of the 2nd type capillaries for forming ICO is not enough. If $(1-\kappa)>P_c^b$ then ICO will not disappear until $p_k=0$ ($r_{k1}=\infty$) while oil flow stops only for $p_k>0$, i.e. when $p_{co}=p_{c2o}$ and $r_{c1o}=\infty$. As a result for $r_c<r_{k1}<r_{c1o}$ k_o is the following

$$k_o(r_{k1}) = \frac{\pi(1-\kappa)^2}{8K_0}\int_0^{r_{co}}\left[\int_{r'}^{r_{co}} f(r)dr\right]^{\nu} f(r')\left(\int_{r'}^{\infty} f(r)dr - \kappa\int_{r'}^{r_{k1}} f(r)dr\right)\left[(1-\kappa)\int_{r'}^{r_{k1}}\frac{f(r)}{r^4}dr + \right.$$
$$\left. + \int_{r_{k1}}^{\infty}\frac{f(r)}{r^4}dr\right]^{-1} dr', \qquad r_c < r_{k1} < r_{c1o}.$$

If $(1-\kappa)>P_c^b$ then in case of $p_k>0$ only the 2nd type capillaries with $r<r_{k2}$ are occupied by oil. In p_k increase the value $k_o(r_{k2})$ decreases until it becomes zero for $r_{k2}=r_{c2o}$ where r_{c2o} is calculated from the formula:

$$(1-\kappa)\int_0^{r_{c2o}} f(r)dr = P_c^b.$$

In the interval $0<p_k<p_{c2o}$ ($r_{k2}>r_{c2o}$), the value $k_o(r_{k2})$ is determined by the following relation:

$$k_o(r_{k2})=\frac{\pi(1-\kappa)^2}{8K_0}\int_0^{r_{co}}\left[\int_{r'}^{r_{co}} f(r)dr\right]^{\nu} f(r')\int_{r'}^{r_{k2}} f(r)dr\left(\int_{r'}^{r_{k2}}\frac{f(r)}{r^4}dr\right)^{-1}dr', \qquad r_{k2}>r_{c2o}.$$

Function $r_{co}(r_{k2})$ is found from the expression

$$(1-\kappa)\int_{r_{co}}^{r_{k2}} f(r)dr = P_c^b.$$

Dependences for calculating the value k_w are written using the same scheme. If $\kappa<P_c^b$ then water flow cannot start until water occupies all the 1st type capillaries which number is not enough to form ICW. In further p_k increase there will be a moment when ICW will be formed for the first time by the aggregate of the 1st and the 2nd type capillaries.

According to this, r_{c2} is calculated from the following relation

$$\kappa+(1-\kappa)\int_{r_{c2}}^{\infty} f(r)dr = P_c^b.$$

Function $r_{cw}(r_{k2})$ is determined by expression

$$\kappa\int_{r_{cw}}^{r_{k2}} f(r)dr+\int_{r_{k2}}^{\infty} f(r)dr = P_c^b. \tag{11}$$

It has values from 0 to r_c, while r_{k2} changes from r_{c2} to 0. Water relative permeability k_w (r_{k2}) is calculated from formulas

$$k_w(r_{k2})=0, \qquad r_{k2}>r_{c2},$$

$$k_w(r_{k2})=\frac{\pi\kappa^2}{8K_0}\int_0^{r_{cw}}\left[\int_{r'}^{r_{cw}} f(r)dr\right]^{\nu} f(r')\left(\kappa\int_{r'}^{r_{k2}} f(r)dr+\int_{r_{k2}}^{\infty} f(r)dr\right)\left(\kappa\int_{r'}^{r_{k2}}\frac{f(r)}{r^4}dr+\right.$$

$$\left.+\int_{r_{k2}}^{\infty}\frac{f(r)}{r^4}dr\right)^{-1}dr', \qquad r_c<r_{k2}\le r_{c2}, \tag{12}$$

$$k_w(r_{k2}) = \frac{\pi}{8K_0}\left\{\kappa\int_0^{r_{k2}}\left[\kappa\int_{r'}^{r_{k2}} f(r)dr + \int_{r_{k2}}^{r_c} f(r)dr\right]^{\nu} f(r')\left(\kappa\int_{r'}^{r_{k2}} f(r)dr + \int_{r_{k2}}^{\infty} f(r)dr\right)\times\right.$$

$$\times\left(\kappa\int_{r'}^{r_{k2}}\frac{f(r)}{r^4}dr + \int_{r_{k2}}^{\infty}\frac{f(r)}{r^4}dr\right)dr' + \int_{r_{k2}}^{r_c}\left[\int_{r'}^{r_c} f(r)dr\right]^{\nu} f(r')\int_{r'}^{\infty} f(r)dr\times$$

$$\left.\times\left(\int_{r'}^{\infty}\frac{f(r)}{r^4}dr\right)^{-1} dr'\right\}, \qquad 0 < r_{k2} \le r_c.$$

(13)

Let $\kappa > P_c^{\,b}$ now. Then ICW formation is possible during water occupation of the 1st type capillaries. For $-\infty < p_k < 0$

$$f_w(r) = \begin{cases} \kappa f(r), & r < r_{k1}, \\ 0, & r \ge r_{k1}. \end{cases}$$

Analytical expression for $k_w(r_{k2})$ will be written in the following form:

$$k_w(r_{k1}) = 0, \qquad r_{k1} < r_{c1w},$$

$$k_w(r_{k1}) = \frac{\pi\kappa^2}{8K_0}\int_0^{r_{cw}}\left[\int_{r'}^{r_{cw}} f(r)dr\right]^{\nu} f(r')\int_{r'}^{r_{k1}} f(r)dr\left(\int_{r'}^{r_{k1}}\frac{f(r)}{r^4}dr\right)^{-1} dr', \qquad r_{k1} \ge r_{c1w}.$$

After capillary pressure becomes above zero the 2nd type capillaries occupied by water will be included in ICW and all the 1st type capillaries will be already filled by water. The technique of further calculations of $k_w(r_{k2})$ is totally the same as that of for the case $\kappa < P_c^{\,b}$. This technique of calculations is still described by formulas (12), (13) where $r_{c2} = \infty$ and r_{cw} is determined by expression (11).

In terms of obtained formulas, calculations of relative permeabilities $k_o(S)$ and $k_w(S)$ were carried out for different parameters of micro heterogeneity α and κ. According to Model I for $0 < \alpha < 1$ water saturation S was determined by the following relation

$$S = \int_0^{\alpha r_{k1}} f(r)dr + \kappa\int_{\alpha r_{k1}}^{r_{k1}} f(r)dr,$$

for $\alpha < 0$:

$$S = \kappa\int_0^{r_{k1}} f(r)dr, \qquad S \le \kappa,$$

$$S = \kappa + (1-\kappa)\int_{r_{k2}}^{\infty} f(r)dr, \qquad S > \kappa.$$

In case of Model II for $0<\alpha<1$ we have

$$S = \left(\int_0^{\alpha r_{k1}} f(r)r^2 dr + \kappa \int_{\alpha r_{k1}}^{r_{k1}} f(r)r^2 dr\right)\left(\int_0^{\infty} f(r)r^2 dr\right)^{-1},$$

for $\alpha<0$:

$$S = \left(\kappa \int_0^{r_{k1}} f(r)r^2 dr\right)\left(\int_0^{\infty} f(r)r^2 dr\right)^{-1}, \qquad S \le \kappa,$$

$$S = \left(\kappa + (1-\kappa)\int_{r_{k2}}^{\infty} f(r)r^2 dr\right)\left(\int_0^{\infty} f(r)r^2 dr\right)^{-1}, \qquad S > \kappa.$$

The represented technique of calculations therefore enables one to calculate relative permeability as a function of water saturation for above mentioned Models I and II of porous media with micro heterogeneous wettability.

Plastic Model

According to this model we assume water to be Newtonian fluid during drainage and pseudoplastic fluid during imbibition. If oil has pseudoplastic properties during drainage it becomes Newtonian during imbibition and if it is Newtonian fluid during drainage it becomes pseudoplastic or Bingham plastic. This approach is based on percolation model for fluids with different plastic properties [8].

The general equation for power-low fluids (pseudoplastic and dilatant) is the following:

$$\tau = \mu_1 \dot{\gamma}^{n},$$

where; τ – shear stress, μ_1 – analogue of viscosity for power-low fluids, $\dot{\gamma}$ – shear rate, n – experimental index ($n < 1$ – pseudoplastic fluids, $n > 1$ – dilatant fluids).

Density of radius distribution functions for capillaries occupied by oil or water are equivalent to (1) and (5). In this case the equations for relative permeability are the following:

$$k_o(r_k) = \frac{\pi}{8} K_0^{-1} \frac{2^{3-1/n}}{3+1/n} \int_{r_k}^{r_c} \left[\frac{\int_{r_1}^{\infty} f(r)dr}{\int_{r_1}^{\infty} \frac{f(r)}{r^{3n+1}} dr}\right]^{1/n} \left[\int_{r_1}^{r_c} f(r)dr\right]^{\nu} f(r_1)dr_1, \tag{14}$$

$$k_w(r_k)=\frac{\pi}{8}K_0^{-1}\frac{2^{3-1/n}}{3+1/n}\int_0^{r_c'}\left[\frac{\int_{r_1}^{\infty}f(r)dr}{\int_{r_1}^{\infty}\frac{f(r)}{r^{3n+1}}dr}\right]^{1/n}\left[\int_{r_1}^{r_c'}f(r)dr\right]^{\nu}f(r_1)dr_1. \tag{15}$$

The general equation for Bingham plastic fluid is the following:

$$\tau=\tau_0+\mu\dot{\gamma},$$

where; μ – analogue of viscosity for Bingham plastic, τ_0 – yield stress.

Density of radius distribution functions for capillaries occupied by oil or water are still equivalent to (1) and (5). So the equations for relative permeability are the following:

$$k_o(r_k)=\frac{\pi}{8}K_0^{-1}\int_{r_k}^{r_c'}\left[\int_{r_1}^{r_c}f(r)dr\right]^{\nu}\frac{\int_{r_1}^{\infty}f(r)\left(1-\frac{4}{3}\frac{r_m'}{r}\right)dr}{\int_{r_1}^{\infty}\frac{f(r)}{r^4}dr}f(r_1)\eta\left(1-\frac{4}{3}\frac{r_m'}{r_1}\right)dr_1, \tag{16}$$

$$k_w(r_k)=\frac{\pi}{8}K_0^{-1}\int_0^{r_c'}\left[\int_{r_1}^{r_c'}f(r)dr\right]^{\nu}\frac{\int_{r_1}^{\infty}f(r)\left(1-\frac{4}{3}\frac{r_m'}{r}\right)dr}{\int_{r_1}^{\infty}\frac{f(r)}{r^4}dr}f(r_1)\eta\left(1-\frac{4}{3}\frac{r_m'}{r_1}\right)dr_1, \tag{17}$$

where; $\eta(x)$ – Heaviside function, r_m' – minimal radius of capillaries, where the flow occurs, $r_m'=2\tau_0/\nabla p$.

Linking the dependences (14) – (17) with water saturation is determined by (8) for Model I and (9) for Model II.

NUMERICAL CALCULATION RESULTS

It is proven that with z increase, irreducible oil and water saturations also decrease. The best accordance with experimental data [1] is achieved for z=6. It means that primitive cubic network of capillaries is the best model of porous media structure of the analyzed samples.

The hydrophobisation model gives the following results. For α>0 and any κ oil relative permeability curves for imbibition are lower than for drainage using Model I for water saturation. For Model II they are above. Water relative permeability curves for imbibition are lower than for drainage for both models. For invariable value κ with α increase smoothing of relative permeability curves for imbibition and their approach to curves for drainage are observed.

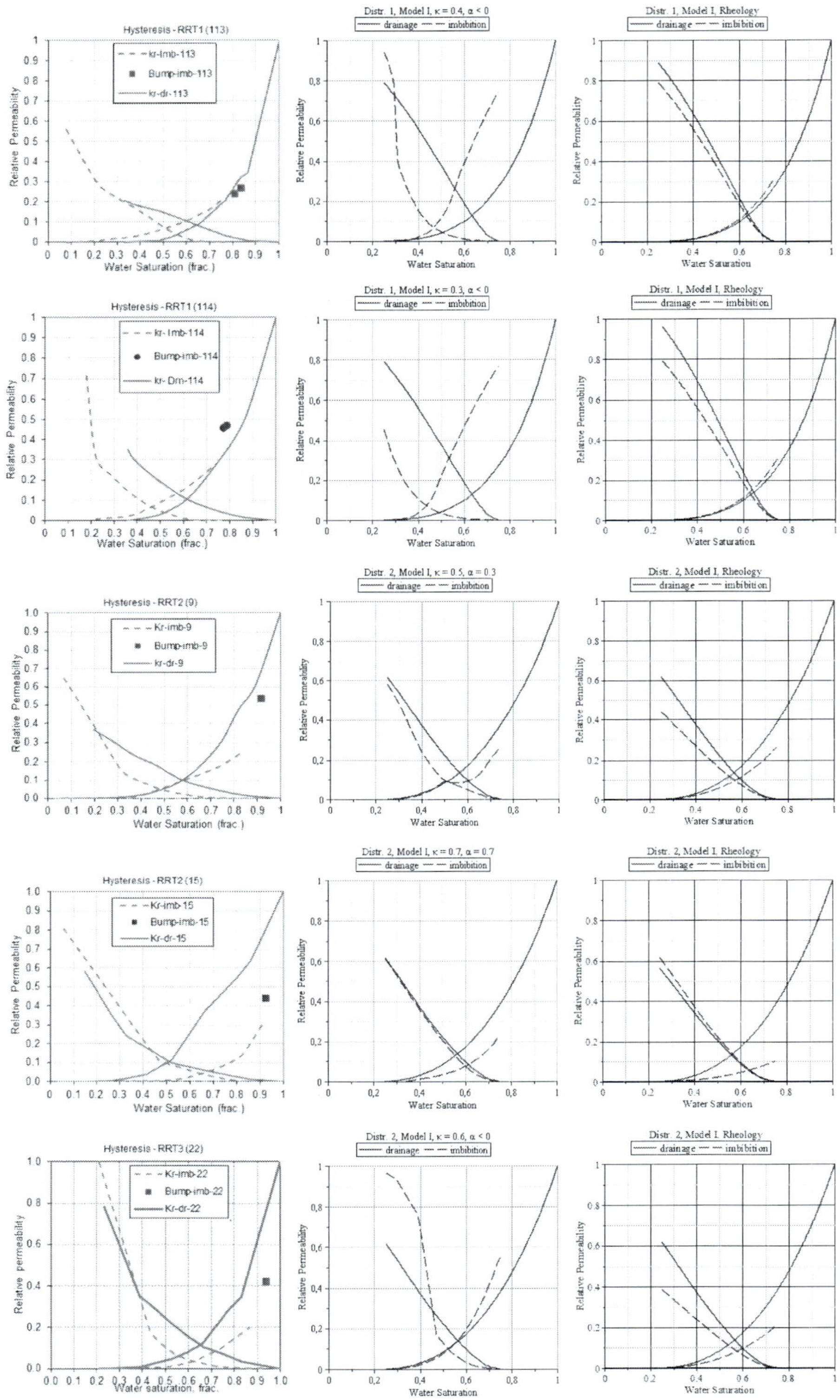

Figure 6. Comparison of experimental (on the left) and calculated with hydrophobisation model (in the center) and with plastic model (on the right) relative permeability curves (samples 113, 114, 9, 15, 22).

This can be explained by the following: the closer α is to the unit, the lesser the difference in surface properties of the 1st and the 2nd type capillaries and accordingly the lesser the difference between drainage and imbibition. In extreme case α=1 (that means the lack of the difference in surface properties of the 1st and the 2nd type capillaries) relative permeability curves for drainage and imbibition totally match. Maximal deflection of relative permeability curves for imbibition from curves for drainage is observed for lower relative permeability values if α is constant while κ increases (in other words part of capillaries with changed surface properties decreases). Obviously for κ=1 and κ=0 relative permeability curves for drainage and imbibition match, i.e. if surface properties of all capillaries are unchanged (κ=1) or if surface properties of all capillaries are changed (κ=0) being hydrophilic (because α>0) then the difference between drainage and imbibition disappears.

For α<0 with κ increase the shift of relative permeability curves for imbibition to water saturation increase is observed for Model I. For κ about 1-P_c^b this shift is maximal. During further κ increase, relative permeability curves for imbibition and drainage approach each other and in extreme case (κ=1) these curves match. For Model II with κ increase the shift of water relative permeability curves for imbibition to water saturation decrease occurs. And oil relative permeability increases for water saturation less than 0.2 – 0.3 and decreases for water saturation more than 0.2 – 0.3.

Concerning the plastic model we can note that relative permeability behavior for power-low fluids depends on $f(r)$ function. For Distr. 1 with n increase oil and water relative permeability decrease, while for Distr. 2 and Distr. 3 with n increase relative permeability increase. In case of Bingham plastic fluids the tendency is always the same: with r_m' (or τ_0) increase oil and water relative permeability decrease.

The values of n and r_m' (where necessary) for ten samples under study are represented on table 1.

It is important to note that relative permeability curves for Model II are located in area of low water saturation values. It means that for Model II irreducible water saturation is nearly 0.05 – 0.15 and irreducible oil saturation is about 0.35 – 0.65 for different differential radius distributions. Model I is more accordant with the experiment. Differential radius distributions Distr. 1, Distr. 2 and Distr. 3 give qualitatively the same results different quantitatively. This difference is very small for Model I but it is more appreciable for Model II.

Table 1. Types of fluids and values of n and r_m' for different samples according to the plastic model

Sample #	Drainage						Imbibition					
	Oil			Water			Oil			Water		
	Type of fluid	n	r_m', μm	Type of fluid	n	r_m', μm	Type of fluid	n	r_m', μm	Type of fluid	n	r_m', μm
113	Pseudoplastic	0,8		Newtonian	1		Newtonian	1		Pseudoplastic	0,7	
114	Pseudoplastic	0,7		Newtonian	1		Newtonian	1		Pseudoplastic	0,7	
9	Newtonian	1		Newtonian	1		Pseudoplastic	0,7		Pseudoplastic	0,7	
15	Pseudoplastic	0,9		Newtonian	1		Newtonian	1		Pseudoplastic	0,4	
22	Newtonian	1		Newtonian	1		Bingham plastic		0,3	Pseudoplastic	0,6	
72	Newtonian	1		Newtonian	1		Pseudoplastic	0,9		Pseudoplastic	0,6	
4	Newtonian	1		Newtonian	1		Pseudoplastic	0,7		Pseudoplastic	0,7	
6	Newtonian	1		Newtonian	1		Pseudoplastic	0,7		Pseudoplastic	0,7	
138	Newtonian	1		Newtonian	1		Pseudoplastic	0,8		Pseudoplastic	0,7	
139	Newtonian	1		Newtonian	1		Pseudoplastic	0,8		Pseudoplastic	0,7	

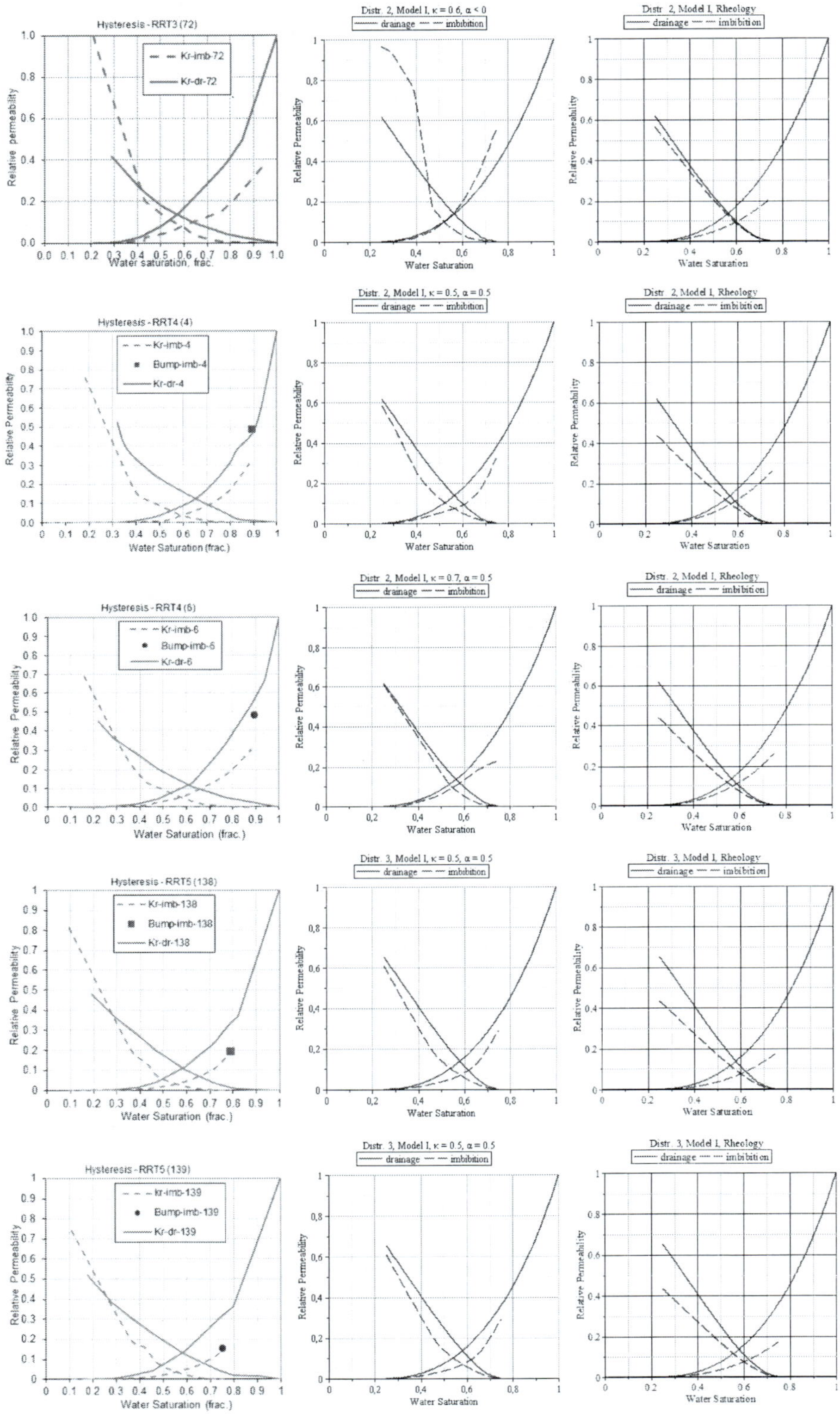

Figure 7. Comparison of experimental (on the left) and calculated with hydrophobisation model (in the center) and with plastic model (on the right) relative permeability curves (samples 72, 4, 6, 138, 139).

The experimental curves and calculation results are represented on Figure 6, 7 (for all the shown calculated curves z=6 and Model I is used).

CONCLUSION

The represented model above is rather universal and enables one to consider different mechanisms of porous media surface properties change. The best representation with referred experiment is achieved by using hydrophobization of porous media part during water displacement by oil as the first mechanism and oil and water plastic properties change as the second mechanism. Values of parameters κ and α or n and r_m' for concrete process are determined by comparison of calculated relative permeability curves and experimental data.

The results of this work based on experimental data [1] allow us to conclude that the average values of κ and α are the following: for dolomite-rich samples (113, 114) κ=0.35 and α<0; for calcite-rich samples (9, 15, 4, 6, 138, and 139) κ=0.5 and α=0.6. Samples 22 and 72 are not taken into consideration because of not satisfying accordance between experimental and calculated relative permeability curves. These results are for the hydrophobisation model, while for plastic model the results are the following. For dolomite-rich samples during drainage oil is pseudoplastic fluid with n=0.75 and water is Newtonian fluid (n=1); during imbibition oil becomes Newtonian (n=1) and water becomes pseudoplastic (n=0.7). For calcite-rich samples during drainage oil and water are Newtonian fluids (n=1); during imbibition oil becomes pseudoplastic (n=0.8) and water becomes pseudoplastic too (n=0.64).

Further represented methodology can be used for relative permeability calculation in any porous media having only mineral composition data and differential radius distribution curve $f(r)$. Experimental core research for the purpose of relative permeability curves determination requires considerable amount of time as compared to experimental $f(r)$ measurements. Having carried out relative permeability hysteresis behavior classification of different mineral composition samples, it is possible to calculate parameters κ and α, like it was done previously for sandpacks [17] and for dolomite- and calcite rich samples in this paper, or n and r_m', which were determined for dolomite- and calcite rich samples as well. Thus on the basis of the presented methodology one can calculate relative permeability with hysteresis using these values and measurements $f(r)$.

In future, we look forward to build a generalized model which would include both considered mechanisms. The results obtained in this paper allow us to suggest that such model will be in better accordance with the experiment.

APPENDIX: NOTATION

$f(r)$	density of capillary radius distribution function
r	capillary radius
σ_d	shape index of lognormal function
μ	scale index of lognormal function
r_k	minimal radius of capillary where the displacement of wetting phase occurs by given pressure gradient

Δp	fluid pressure disturbance
X	interracial tension factor
Θ	wetting angle
$f_o(r)$	density of radius distribution function for capillaries occupied by oil
k_o	oil relative permeability
K_0	absolute permeability
v	correlation radius index
r_c	critical radius, from which ICO appears
P_c^b	percolation threshold
D	dimensionality of network
z	coordination number
$f_w(r)$	density of radius distribution function for capillaries occupied by water
k_w	water relative permeability
r_c'	critical radius, from which ICW appears
S	water saturation
κ	micro heterogeneous parameter, responding for part of capillaries with unchanged surface properties
α	micro heterogeneous parameter, responding for surface hydrophobisation (ratio of wetting angle cosines)
θ_i	wetting angle in capillaries of type i
r_{ki}	maximal radius of type i, where wetting phase is contained
χ_i	interracial tension factor in capillaries of type i
p_k	capillary pressure
r_{c1}	capillary radius of type I, from which ICO of capillaries of both types appears in case of $0<\alpha<1$
r_{co}	analogue of r_c, being function of r_{k1}
r_{cw}	analogue of r_c', being function of r_{k1}
r_{c1w}	minimal value of r_{k1}, wherein $r_{cw}=0$
p_{co}	capillary pressure, wherein ICO disappears in case of further pressure increase for $\alpha<0$ and $(1-\kappa)<P_c^b$
p_{c2o}	capillary pressure, wherein ICO disappears in case of further pressure increase for $\alpha<0$ и $(1-\kappa)>P_c^b$
r_{c1o}	maximal value of r_{k1}, wherein ICO exists
r_{c2o}	minimal value of r_{k2}, wherein ICO exists for $\alpha<0$
r_{c2}	type 2 capillary radius from which while decrease ICW of both types capillaries appears in case of $\alpha<0$ and $\kappa<P_c^b$
τ	shear stress
μ_1	analogue of viscosity for power-low fluids
γ	shear rate
n	experimental index ($n<1$ – pseudoplastic fluids, $n>1$ – dilatant fluids)
μ	analogue of viscosity for Bingham plastic
τ_0	yield stress
$\eta(x)$	Heaviside function
r_m'	minimal radius of capillaries, where the flow of Bingham plastic fluid occurs

REFERENCES

[1] Dernaika, M. R., Basioni, M. A., Dawoud, A., Kalam, M. Z., Skjæveland, S. M., (2012). Variations in bounding and scanning relative permeability curves with different carbonate rock types. *Paper SPE 162265 presented at the 2012 Abu Dhabi International Petroleum Conference and Exhibition*, Abu Dhabi, UAE, Nov. 11–14.

[2] Broadbent, S. R., Hammersley, J. M., (1957). Percolation processes. *Proc. Cambr. Phil. Soc.* 53, 629–645.

[3] Heiba, A. A., Sahimi, M., Scriven, L. E., Davis, H. T., (1982). Percolation theory of two-phase relative permeability. *Paper SPE 11015 presented at the 1982 SPE 57th Annual Fall Technical Conference and Exhibition*, New Orleans, Sept. 26–29.

[4] Dixit, A. B., McDougall, S. R., Sorbie, K. S., (1998). Analysis of relative permeability hysteresis trends in mixed-wet porous media using network models. *Paper SPE 39656 presented at the 1998 SPE/DOE Improved Oil Recovery Symposium,* Tulsa, Oklahoma, Apr. 19–22.

[5] Phirani, J., Pitchumani, R., Mohanty, K. K., (2009). Transport properties of hydrate bearing formations from pore-scale modeling. *Paper SPE 124882 presented at the 2009 SPE Annual Technical Conference and Exhibition,* New Orleans, Louisiana, Oct. 4–7.

[6] Kadet, V. V., Selyakov, V. I., (1987). Percolation model of two-phase flow in porous media. *Fluid Dynamics* 22, 75-82.

[7] Salomao, M. C., (1997). Analysis of flow in spatially correlated systems by applying the percolation theory. *Paper SPE 39039 presented at the 1997 Latin American and Caribbean Petroleum Engineering Conference,* Rio de Janeiro, Brazil, Aug. 30 – Sept. 3.

[8] Selyakov, V. I., Kadet, V. V., (1996). *Percolation models for transport in porous media.* Kluwer Academic Publishers, Dordrecht/ Boston/ London.

[9] Braun, E. M., Holland, R. F., (1995). Relative permeability hysteresis: laboratory measurements and a conceptual model. *SPE Reservoir Engineering* 10, 222-228.

[10] Chang, L. C., Chen, H. H., Shan, H. Y., Tsai, J. P., (2009). Effect of connectivity and wettability on the relative permeability of NAPLs. *Environmental Geology* 56, 1437-1447.

[11] Parvazdavani, M., Masihi, M., Ghazanfari, M. H., Sherafati, M., Mashayekhi, L., (2012). Investigation of the effect of water based nano-particles addition on hysteresis of oil-water relative permeability curves. *Paper SPE 157005 presented at the 2012 SPE International Oilfield Nanotechnology Conference,* Noordwijk, The Netherlands, June 12–14.

[12] Spiteri, E. J., Juanes, R., Blunt, M. J., Orr Jr., F. M., (2005). Relative-permeability hysteresis: trapping models and application to geological CO2 sequestration. *Paper SPE 96448 presented at the 2005 SPE Annual Technical Conference and Exhibition*, Dallas, Texas, Oct. 9–12.

[13] Wei, J. Z., Lile, O. B., (1993). *Influence of wettability and saturation sequence on relative permeability hysteresis in unconsolidated porous media*. [Electron. resource]. http://www.onepetro.org/mslib/app/Preview.do?paperNumber=00 025282andsociety Code=SPE.

[14] Drozdov, A. N., Drozdov, N. A., (2012). Laboratory researches of the heavy oil displacement from the Russkoye field's core models at the SWAG injection and development of technological schemes of pump-ejecting systems for the water-gas mixtures delivering. *Paper SPE 157819 presented at the 2012 SPE Heavy Oil Conference Canada*, Calgary, Alberta, Canada, June 12–14.

[15] Yusupova, T. N., Romanova, U. G., Petrova, L. M., Romanov, G. V., Ovchinnikov, V. V., Muslimov, R. Kh., Mukhametshin, R. Z., (1997). Hydrophobization of reservoir rock in bed conditions. *Paper PS 97-125 presented at the 1997 Annual Technical Meeting,* Calgary, Alberta, June 8–11.

[16] Landau, L. D., Lifshitz, E. M., (1987). *Fluid Mechanics.* Vol. 6 (2nd ed.). Butterworth–Heinemann.

[17] Kadet, V. V., Galechyan, A. M., (2013). Percolation model of relative permeability hysteresis. *Journal of Applied Mechanics and Technical Physics* 54, 423-432.

In: Hysteresis
Editor: José Carlos Dias

ISBN: 978-1-63321-336-4

Chapter 9

THRESHOLD/FLAT-BAND VOLTAGE HYSTERESIS OF NON-VOLATILE SEMICONDUCTOR MEMORY DEVICES

Zs. J. Horváth*

Óbuda University, Kandó Kálmán Faculty of Electrical Engineering,
Institute of Microelectronics and Technology,
Budapest, Hungary
Hungarian Academy of Sciences, Research Centre for Natural Sciences,
Institute for Technical Physics and Materials Science,
Budapest, Hungary

ABSTRACT

The types and structure of different non-volatile semiconductor memory devices and the principle of their operation are summarised. Analysis of memory hysteresis curves is ppresented. The effect of layer thicknesses and current levels on the memory hysteresis of MNOS structures is studied by computer simulation. Experimental results obtained in MNOS structures containing Ge or Si nanocrystals are interpreted on the basis of the results of computer simulation. It is shown that the presence of nanocrystals enhances the charge injection properties of MNOS strucures for thin oxide layers only (up to 3 nm). There is a narrow range of oxide thickness or depth of nanocrystals from the oxide/nitride interface, when the charging properties change very fast. Retention exhibits a very sharp dependence on the oxide thickness or depth of nanocrystals as well.

Keywords: hysteresis, MNOS, nanocrystals, non-volatile memory, silicon nitride, tunneling

* horvath.zsolt@kvk.uni-obuda.hu.

INTRODUCTION

Hysteresis is a frequent phenomenon in nature. It is related to a delayed response of materials to the change of external parameters. It may be useful for some applications (e.g., in hard magnetic materials), or harmful in other applications (e.g., in soft magnetic materials). Complex systems can also exhibit hysteresis behaviour. Such systems are the non-volatile memory devices as well.

The information storage in non-volatile semiconductor memory devices is based on the injection of electrical charge into the layered dielectric structure under the gate of Field Effect Transistors (FETs). The charge is stored in or between the dielectric layers under the gate, and changes the threshold voltage of the FET or the flat-band voltage of the capacitor based on the same system [1]. Charge is injected by high electrical voltage pulses applied to the gate. If the charging pulse amplitude is increased and then decreased gradually step by step, the change of the threshold voltage and of the flat-band voltage exhibit nice hysteresis [2]. Such experimental threshold voltage and flat-band voltage hysteresis curves are presented in Figure 1 obtained on a FET or a capacitor prepared on the same semiconductor structure [3, 4].

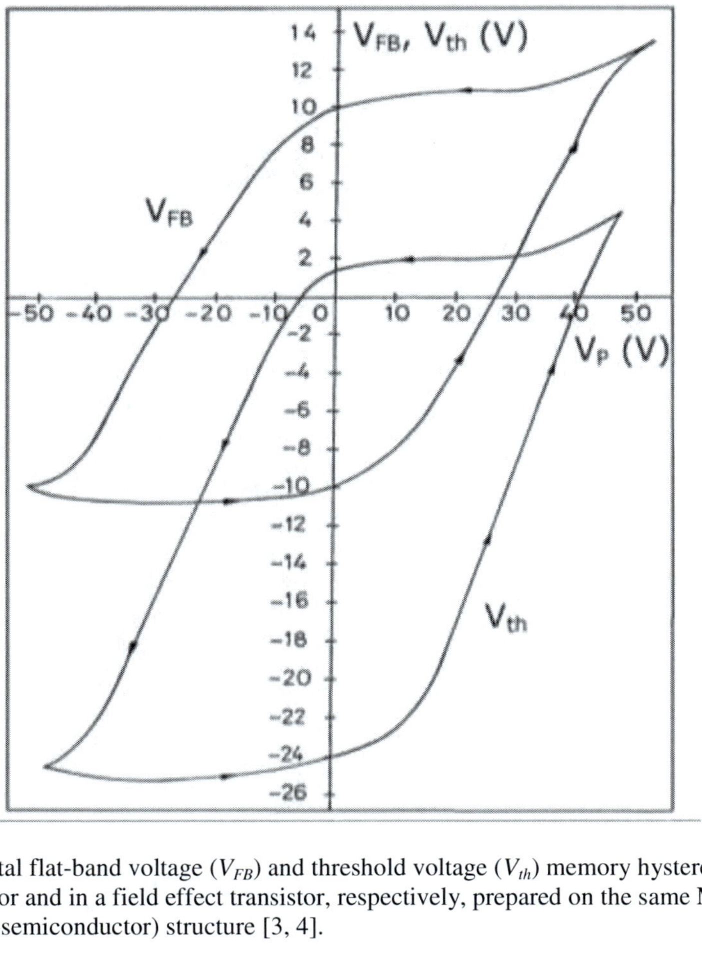

Figure 1. Experimental flat-band voltage (V_{FB}) and threshold voltage (V_{th}) memory hysteresis curves obtained in a capacitor and in a field effect transistor, respectively, prepared on the same MNOS (metal-nitride-oxide-semiconductor) structure [3, 4].

After this short Introduction (first Section) in the second Section of this Chapter the types and structure of different non-volatile semiconductor memory devices and the principle of their operation are presented. In the third Section the shape of memory hysteresis curves is analysed. In the fourth Section the tunneling probability of charge carriers into MNOS structures is discussed. In the fifth Section the effect of layer thicknesses and current levels on the memory hysteresis of MNOS structures is studied by computer simulation. In the sixth Section experimental results obtained in MNOS structures containing Ge or Si nanocrystals are interpreted on the basis of the results of computer simulation. In the last Section the main results are summarized.

NON-VOLATILE SEMICONDUCTOR MEMORY DEVICES

Memories that hold the information without supplying power are called non-volatile. A part of basic elements of non-volatile memories are based on Metal-Oxide-Semiconductor (MOS) FETs prepared on silicon. The structure of an MOS FET is presented in Figure 2. Two regions are formed with opposite conductivity, than that of the substrate, for the drain and source, which are isolated by two p-n junctions between them. If an appropriate voltage is applied to the gate, which covers the area between source and drain, a conductive layer will be formed between the source and drain with opposite conductivity than that of the substrate, called inversion layer. The inversion layer surves as a channel for electrical current between the source and drain [5]. The lowest voltage applied to the gate necessary for the formation of the inversion layer is called the threshold voltage (V_{th}) of the transistor. The threshold voltage depends on electrical charge being present in the insulator layer under the gate. So, if charge can be injected and stored in the insulator for a long time, the threshold voltage can be changed. As a consequence, transfer characteristics (the current via the channel vs. gate voltage) shifts as shown in Figure 3. This is the principle of information storage in non-volatile semiconductor structures [1]. If charge of opposite sign is injected and stored, the threshold voltage and the transport characteristics shift back. So, using a reading voltage V_r between the two threshold voltage values V_{th1} and V_{th2} (see Figure 3), the transistor is either open or closed depending on the stored charge, representing the two different logic states.

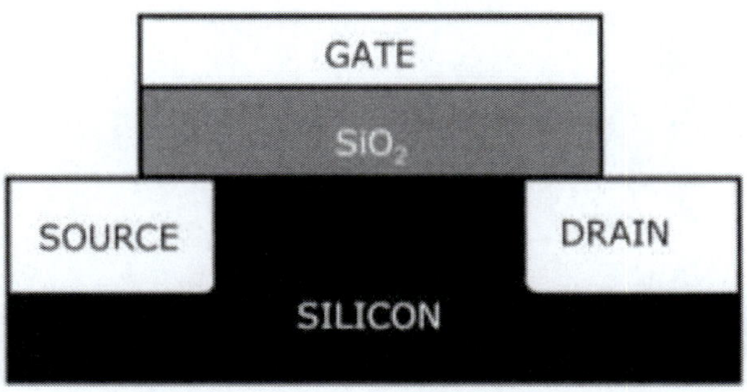

Figure 2. Schematic structure of an MOS FET. Type of doping (n vs. p) of source and drain regions are opposite, than that of silicon substrate. Gate covers the area between source and drain and it is isolated by a thin silicon dioxide layer.

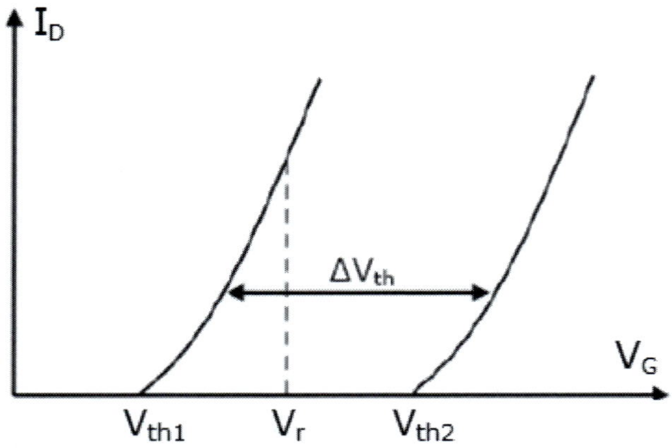

Figure 3. The memory effect: shift of the transfer characteristics (drain current I_D vs. gate voltage V_G) and of the treshold voltage (V_{th}) of memory transistor due to the different stored charge. V_r is the voltage reading the (logic) memory state.

The charge present in the insulator affects not only the threshold voltage of the transistor, but the flat-band voltage of the capacitor, prepared on the same structure. So, not only the transfer characteristics of the transistor, but the capacitance-voltage (C-V) characteristics of the capacitor shifts either, as it can be seen in Figure 4 [6]. The shift of C-V characteristics can be monitored by the change of flat-band voltage V_{FB}. Flat-band voltage can be obtained from the C-V characteristics: it is the voltage that should be applied to obtain a well-defined capacitance value C_{FB} [5] (see Figure 4).

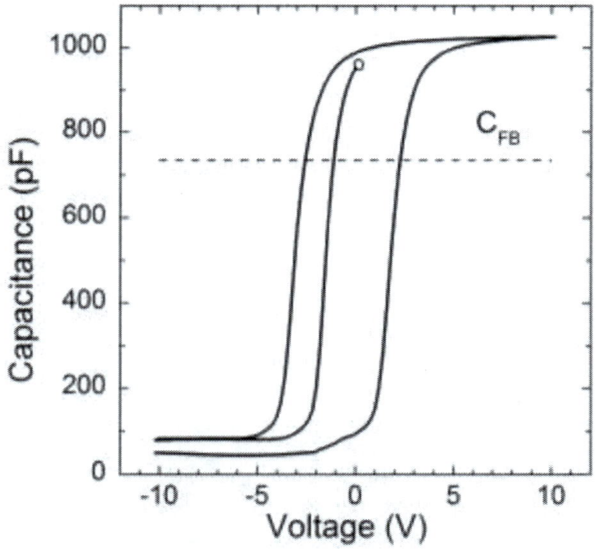

Figure 4. Capacitance-voltage characteristics of an MNOS structure with Si nanocrystals embedded at the oxide/nitride interface, obtained for the bias range of ±10 V with step of 0.2 V [6]. Dots indicate the beginning (first obtained point) of measurement. Dashed line indicate the value of flat-band capacitance C_{FB}. The characteristics are shifted after positive and negative bias due to opposite charge injection.

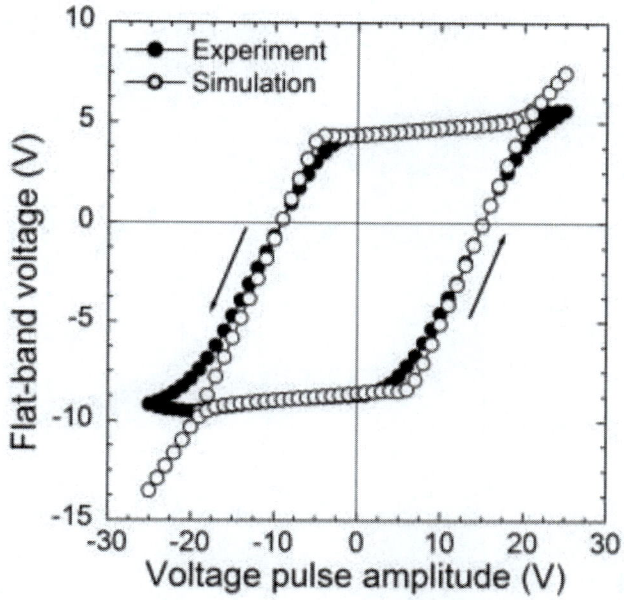

Figure 5. Experimental (solid symbols) and fitted theoretical (open symbols) flat-band voltage memory hysteresis curves for an MNS (metal-niride-semiconductor) structure containing Si nanocrystals [7].

Charge is injected from the silicon substrate by a high electrical voltage pulse applied to the gate. The actual mechanism of charge injection is tunneling (details see below). If the charging pulse amplitude V_P is increased and then decreased gradually step by step, the change of the threshold voltage of transistor or of the flat-band voltage of capacitor exhibit a hysteresis as shown in Figure 1 for the same MNOS (metal-nitride-oxide-semiconductor) structure [3, 4]. Another example of experimental flat-band voltage hysteresis is presented in Figure 5 obtained on an MNS (metal-nitride-semiconductor) structure containing Si nanocrystals. Figure 5 presents a calculated hysteresis curve as well, which was fitted to the experimental hysteresis by computer simulation [7]. Measurement of memory hysteresis is a useful method for study the charge injection behaviour of the non-volatile semiconductor structures and elements.

Another way to study the charge injection behaviour of non-volatile memory structures is the memory window measurement. Memory window means the change of the threshold voltage of a transistor or the flat-band voltage of a capacitor, if alternating positive and negative voltage pulses of the same amplitude are applied to the gate several times. The memory window is usally studied as a function of voltage pulse amplitude Some results of memory window measurement are presented in Figure 6 [1, 2].

In addition to charge injection behaviour charge storage behaviour are very important for non-volatile memories as well. It can be characterized by the rate of charge loss or by the retention time. The change of threshold or flat-band voltage is usually monitored after switching off the charging pulse during a few hours. Such experimental results are presented in Figure 7 [6]. The rate of charge loss can be obtained from the slope of the curves, while the retention time is obtained from the extrapolation of dependences as the time, when the memory window srinks to zero.

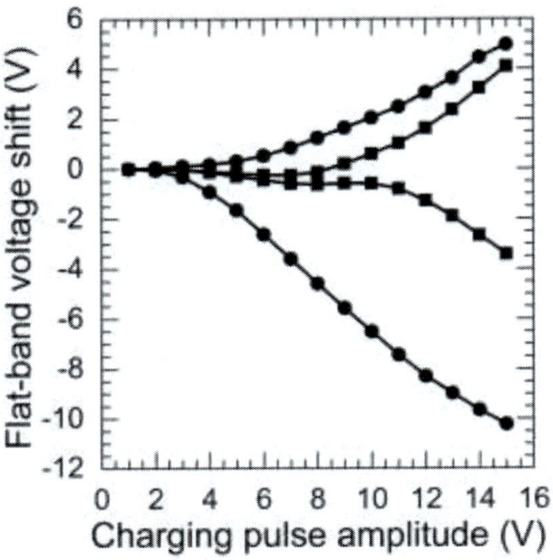

Figure 6. Experimental memory window for an MNOS structure containing Si nanocrystals at a depth of about 3 nm from the oxide/nitride interface in the nitride layer (circles) and for a similar MNOS structure without Si nanocrystals (squares) as a function of charging pulse amplitude [1, 2]. Upper curves are for positive, lower curves for negative charging pulses.

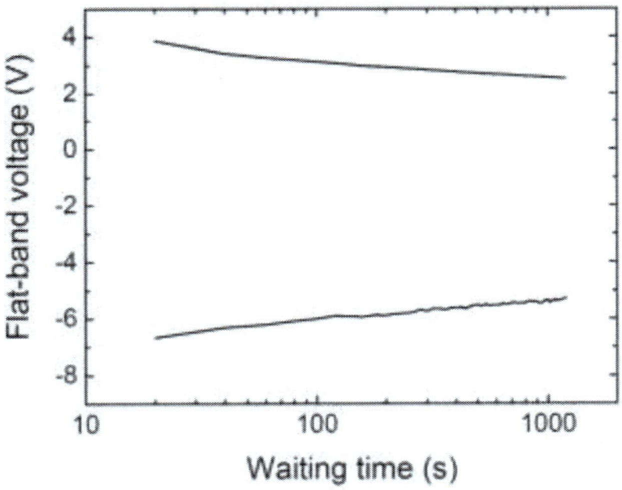

Figure 7. The change of flat-band voltage obtained on an MNOS structure containing Si nanocrystals after switching of the charging voltage pulse with amplitude of 20 V and pulse width of 400 ms: upper curve after positive, lower curve after negative charging pulse [6].

As mentioned above, the information is hold by charge captured and stored in the insulator layer under the gate of a FET. But the thermal silicon dioxide under the gate [5] is a very good insulator, it cannot capture and store electrical charge in such high amount, which is necessary for memory purposes. To solve the problem an isolated floating gate having no outer electrical contacts is built in the oxide layer. Such a floating gate memory FET is shown in Figure 8. Charge carriers can be injected to the floating gate via the bottom thin silicon dioxide layer, which is called the tunnel layer. The upper gate is called the control gate and the oxide layer between the two gates is called the control layer. Nowadays both gates are prepared of polycrystalline silicon, which does not react with silicon dioxide, as, e.g., aluminum does.

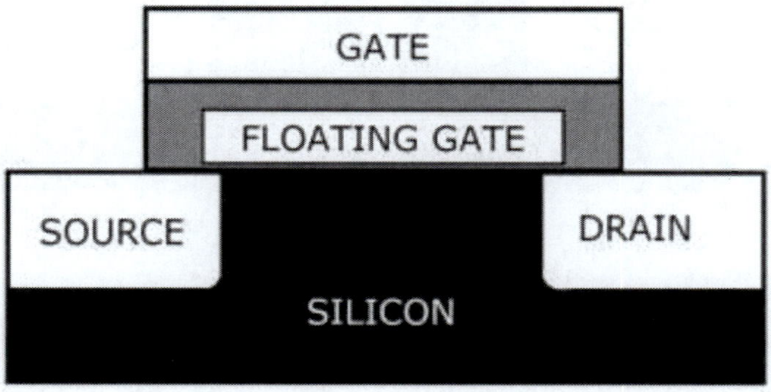

Figure 8. Schematic structure of a floating gate FET. The SiO_2 layer under the floating gate is a few nm thick only to make possible charge injection by tunneling.

Nevertheless, due to technology development reliability problems arised in floating gate memories. Srinking of lateral dimensions resulted in the decrease of logic voltage levels, which involved the reduction of thickness of both the tunnel and control layers [8]. So, the role of defects in the tunnel oxide increased. If a defect or weak point is present in the tunnel oxide, the stored charge can escape via it and the information will be lost. But, if the floating gate is replaced by isolated semiconductor nanocrystals, the charge will escape from a few of them only, which are close to the defect, and the information will be held [1, 9, 10]. The schematic structure of a nanocrystal memory FET is presented in Figure 9. Usually Si or Ge nanocrystals are embedded in silicon dioxide [1, 11-16], but wide research is carried out with other insulators as well [1, 17-19], mainly with HfO_2.

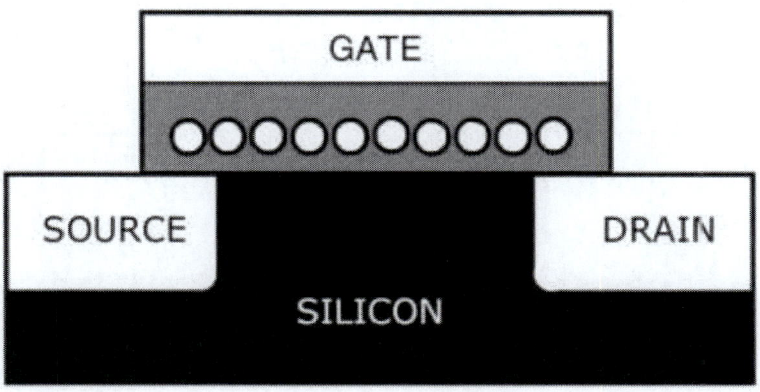

Figure 9. Schematic structure of a nanocrystal memory FET. The floating gate is replaced by isolated semiconductor nanocrystals.

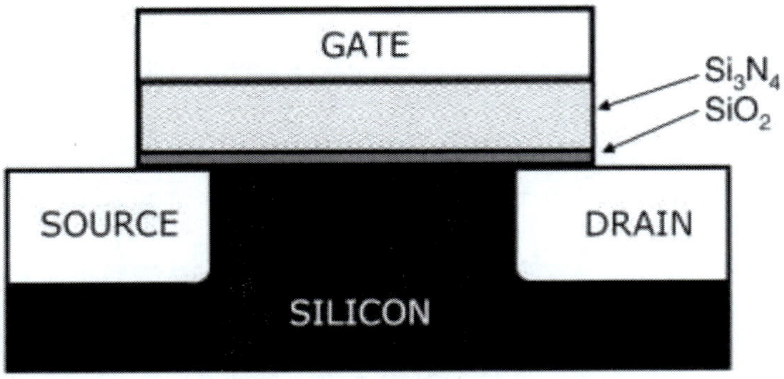

Figure 10. Schematic structure of an MNOS memory FET. The injected charge is captured and stored in the nitride layer close to the oxide/nitride interface by traps.

Most of the other insulators are not so dense as thermal silicon dioxide. They contain defects that act as traps for charge carriers. These insulators can be used as charge storage media in metal-insulator-semiconductor (MIS) FETs under the gate. Charge carriers injected into the insulator layer can be captured and stored in the taps. Such materials are, e.g., Al_2O_3 and Si_3N_4. The charge storage behaviour of the latter has been demonstrated in Figure 5. However, the stored charge escapes relatively fast form the traps in silicon nitride to the substrate [20]. (The required information storage time for non-volatile memories is 10 years.) For improve the retention time, a thin silicon dioxide layer is grown on the silicon surface before nitride deposition. The structure of such an MNOS transistor is shown in Figure 10.

During the charging process carriers are injected from the substrate via the tunnel oxide and most of them is captured by traps in the nitride layer. But a part of charge carriers can travel via the whole nitride layer and reach the gate. This means that a part of the injected charge is lost for the memory effect. To decrease this loss another oxide layer is grown on the top of the nitride layer. (SiO_2 has wider forbidden gap, than Si_3N_4.) The structure of such SONOS (silicon-oxide-nitride-oxide-silicon) memory transistors can be seen in Figure 11.

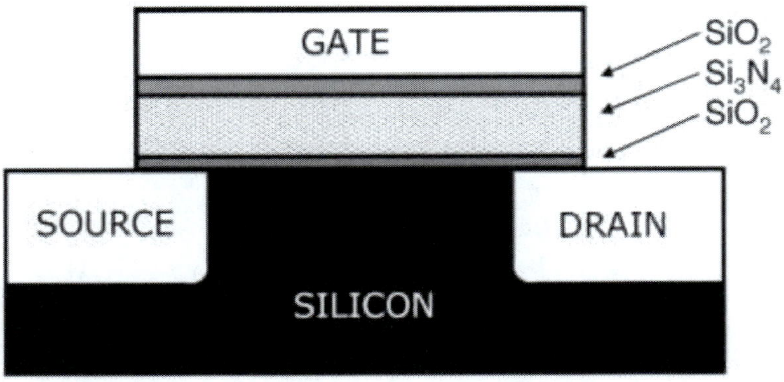

Figure 11. Schematic structure of an SONOS memory FET. The upper oxide layer is grown to decrease charge loss to the gate.

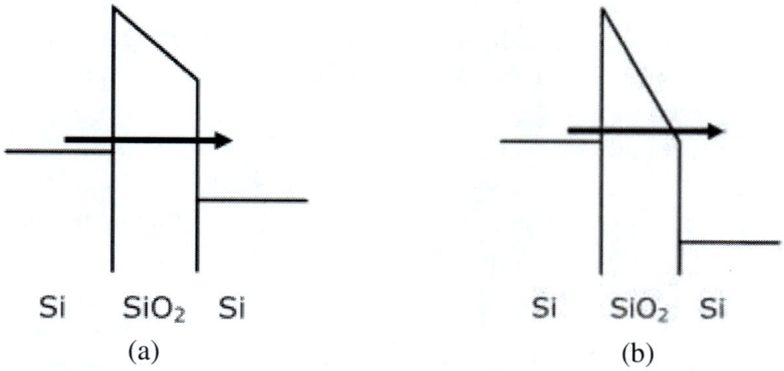

Figure 12. Schematic conduction band of a memory structure with embedded floating gate or Si nanocrystals. Arrows indicate electron injection by direct tunneling at low electric fields (a) or Fowler-Nordheim tunneling at high electric fields (b).

In floating gate or nanocrystal devices the actual tunneling mechanism depends on the tunnel oxide thickness and electric field in the tunnel layer. At thin oxide layers and low electric fields the direct tunneling via a trapesiodal energy barrier to the floating gate or nanocrystals is the injection mechanism (see Figure 12.a), while at high electric field and thick oxide layers the Fowler-Nordheim tunneling via a triangular energy barrier to the oxide conduction band (see Figure 12.b) [21]. In MNOS and SONOS structures an additional tunneling mechanism, the modified Fowler-Nordheim tunneling takes place at low electric fields (Figure 13.a). At medium electric fields direct tunneling to the nitride conductance band via the oxide layer (Figure 13.b), and at high electric fields Fowler-Nordheim tunneling to the oxide conduction band take place (Figure 13.c) [21, 22].

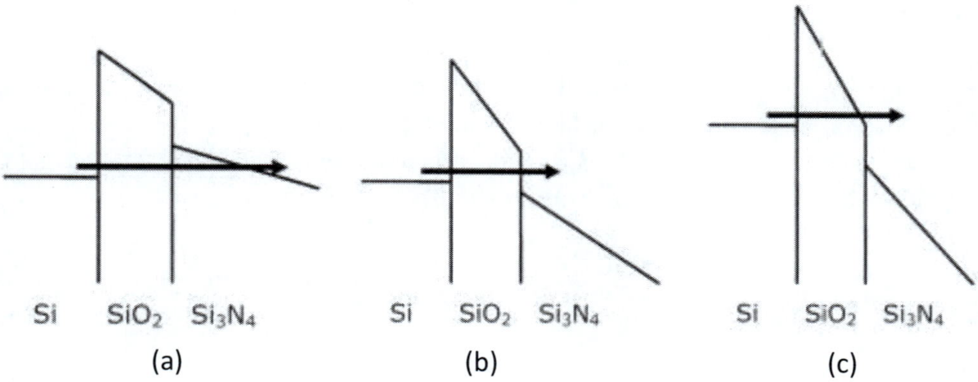

Figure 13. Schematic conduction band of an MNOS structure. Arrows indicate electron injection by modified Fowler-Nordheim tunneling to the nitride conduction band at low electric fields (a), by direct tunneling to the nitride conduction band at medium electric fields (b) or by Fowler-Nordheim tunneling to the oxide conduction band at high electric fields (c) [21, 22].

Floating gate and SONOS memory transistors are the most common building blocks of recent semiconductor non-volatile memories. Nanocrystal non-volatile memories are in production as well [23]. Two types of non-volatile semiconductor memories are used nowadays, namely the Electrically Erasable Programmable Read Only Memories (EEPROMs) and the flash memories [8]. The latter memories are the newest and most dinamically developing ones applied, e.g., in mobile phones, pen drives, music players, digital cameras, pocket computers, hybrid hard disks, etc. The main difference between EEPROMS and flash memories is in the organization of memory arrays. In EEPROMs memory cells are erased one by one, while in flash memories by blocks using a "flash", i.e., a common erasing voltage pulse [1].

In addition to semiconductor non-volatile memories phase-change, ferroelectric and magnetoresistive memories as well as memristors are intensively studied for the development of the future non-volatile memories [24, 25].

ANALYSIS OF FLAT-BAND VOLTAGE MEMORY HYSTERESIS

The first experimental and simulated flat-band voltage memory hysteresis curves were obtained on MNOS structures [26]. Memory hysteresis measurements provide useful information about the charging behaviour of the structures and their simulation is a tool for the interpretation of experimental results. Nevertheless, to the best of our knowledge, only a few works presented results of memory hysteresis measurements and our is the only other group that has studied memory hysteresis behaviour in more detail [2-4, 7, 21, 27-32]. Charge injection properties are usually studied either by simple C-V or by memory window measurements.

The relation between the charging voltage pulse amplitude and the electric fields in the control (E_1) and tunnel (E_2) layers at the moment of cut-off the charging voltage pulse with amplitude V_P can be expressed as [7, 29]

$$V_P = E_1 d_1 + E_2 d_2 + q\varphi_s + q\varphi_k \tag{1}$$

where q is the elementary charge, ϕ_s is the band bending at the semiconductor surface, ϕ_k is the work function difference between the semiconductor and metal, and d_1 and d_2 are the thickness of the control and tunnel layers for structures with embedded floating gate or nanocrystals, while for MNOS and SONOS structures d_1 and d_2 reads:

$$d_1 = d_N - d_C \tag{2a}$$

$$d_2 = d_{OX} + \frac{\varepsilon_2}{\varepsilon_1} d_C \tag{2b}$$

where d_N and d_{OX} are the nitride and oxide thickness and d_C is the depth of charge centroid form the oxide/nitride interface. The polarity of charging pulse corresponds to the polarity applied to the gate.

Assuming no fixed charge at interfaces, on floating gate and in nanocrystals and dielectric layers, the charge (per unit area) injected and stored in the structure on the floating gate, in nanocrystals or in traps reads:

$$\sigma_{ST} = \varepsilon_2 E_2 - \varepsilon_1 E_1 \tag{3}$$

where ε_1 and ε_2 are the permittivities of the control and tunnel layers, respectively.

The flat-band voltage can be calculated via the stored charge as

$$V_{FB} = -\frac{\sigma_{ST}}{\varepsilon_1} d_1 + q\varphi_k \tag{4}$$

When a charging voltage pulse with appropriate amlitude is applied tc the gate, the flat-band voltage is changing during the pulse duration due to charge injection and storage. The captured charge changes the electric fields in the tunnel and control layers as well: the electric field and the current decrease via the tunnel layer, while they increase via the control layer during the charging pulse. The change of the stored charge can be obtained by the integration of the difference of currents via the control and tunnel layers over the time:

$$\delta\sigma_{ST} = -\int_0^{\tau} (J_2 - J_1) dt \tag{5}$$

where J_1 and J_2 are current densities via the control and tunnel layers, respectively, and τ is the width of charging voltage pulse. If both current densities are very low, there is no practical change of the stored charge during the charging pulse. This is the case for the horisontal segments of the hysteresis curve. If the tunneling current is higher, than the current via the control layer (the integral is negative), the flat-band voltage increases. In this case for positive charging pulses the flat-band voltage becomes positive, as shown in Figures 1, and 5. In the case of injection from the gate (positive integral), the hysteresis loop is opposite, i.e., for positive charging pulses the flat-band voltage becomes negative, as it is demonstrated in Figure 14 for the lowest oxide current level by simulation [4]. Such dependences are obtained by experiments as well.

At high electric fields, when the tunneling process occurs, the band bending at semiconductor surface is a very weak function of the electric field [7]. So, assuming that the charging pulse amplitude step is small enough, taking the differences between two successive points of the hysteresis curve, on the basis of Eqs. (1)-(4) the slope of the memory hysteresis curve $\delta V_{FB}/\delta V_P$ can be expressed as

$$\frac{\delta V_{FB}}{\delta V_P} = \frac{g_2 - \frac{\varepsilon_2}{\varepsilon_1} g_1 \frac{\delta J_2}{\delta J_1}}{g_2 + \frac{d_2}{d_1} g_1 \frac{\delta J_2}{\delta J_1}} \tag{6}$$

where g_1 and g_2 are the differential conductivities of the control and tunnel layers at the actual current levels J_1 and J_2:

$$g_1 = \frac{dJ_1}{dE_1} \tag{7a}$$

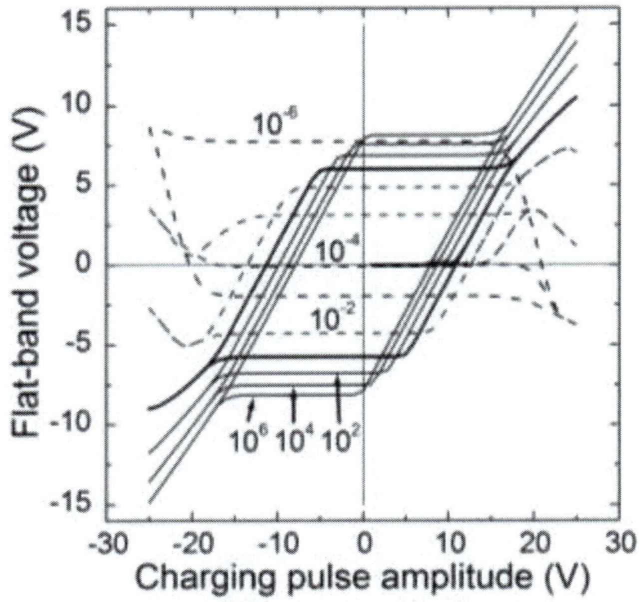

Figure 14. Simulated memory hysteresis curves for a constant nitride current level but different oxide current levels in MNOS structures without nanocrystals. Thick solid curve is for the reference oxide current level (the actual parameters see in Section „Simulation of MNOS memory hysteresis..."), thin solid curves are for increased oxide current levels multiplied by factors 10^2, 10^4 and 10^6, dashed curves for decreased oxide current levels multiplied by factors 10^{-2}, 10^{-4} and 10^{-6}. Note the opposite hysteresis loop for the lowest oxide current level indicating charge injection from the gate [4].

and

$$g_2 = \frac{dJ_2}{dE_2} \tag{7b}$$

and δJ_1 and δJ_2 are differences of current densities via the control and tunnel layers at the moment of cut-off of two successive charging voltage pulses.

As the control layer is thicker, than the tunnel layer, if

$$g_2 >> \frac{\varepsilon_2}{\varepsilon_1} g_1 \frac{\delta J_2}{\delta J_1} \tag{8}$$

i.e., the second terms in the numerator and denominator of Eq. (6) can be neglected, the slope of the memory hysteresis curve is unity. If these terms cannot be neglected, the slope will be lower.

So, the relation of differential (AC) conductivity and of the current differences via the tunnel and control layers determine the slope of the V_{FB} vs. V_P relation. It has been obtained for abrupt segments of the hysteresis curve by our simulations [2, 4, 29, 30] (details see below) that both the tunneling current and the differential conductivity of the tunneling oxide layer are much higher in MNOS structures, than the current via the control nitride layer and its differential conductivity. Nevertheless, the slope of the abrupt parts of the calculated hysteresis curves was very close to unity in most cases (see Figures 5 and 14). The slope was lower only for high currents via the control nitride layer in comparision to the tunneling current, when the height of hysteresis curve is relatively low, as it can be seen in Figure 14.

In the case of experimental flat-band voltage hysteresis curves obtained in MNS and MNOS structures the slope of the abrupt part is in the range of 0.7-0.9. This indicates higher charge loss due to higher current via the control nitride layer. However, increasing depth of charge centroid d_C with increasing amount of stored charge, which is indicated in MNOS structures by experiments [33-35], can also be the origin of this phenomenon. (A constant position of charge centroid is assumed in the calculation above.) Indeed, it can be easily obtained that increasing depth of charge centroid decreases the slope of the hysteresis curve, as presented below. Further on, the escape of injected charge after swithching off the charging pulse until and during the measurement of flat-band voltage results in charge loss either. Moreover, the injection of opposite charge from metal side can reduce the effect of charge injection from the substrate as well.

At the end of abrupt segments of hysteresis curves the slope becomes lower. In these parts of the curves the currents via the tunnel and control layers become comparable, and the slope of the curve is determined by Eq. (6). If steady-state is reached during the charging pulse, i.e., the two currents equal at the moment of cut-off the pulse, curves up and down will be the same, as it can be seen in Figure 14. In this case the differences of the currents equal as well, and Eq. (6) reduces as

$$\frac{\delta V_{FB}}{\delta V_P} = \frac{g_2 - \dfrac{\varepsilon_2}{\varepsilon_1} g_1}{g_2 + \dfrac{d_2}{d_1} g_1} \tag{9}$$

In this part of the hysteresis curve the slope is determined by the differential conductivities, thickness and dielectric constants of the layers. It can be clearly seen in Figure 14 that the slope of that part of hysteresis curve, when steady-state occurs, monotonously decreases, as the oxide current is decreased. The V_{FB} vs. V_P dependence even changes its sign from positive to negative, as the differential conductivity of nitride layer increases (see hysteresis curves with decreased oxide current levels).

Coming back with decreasing absolute values of charging pulse amplitude, the horizontal segment is reached, where both currents are reduced to so low levels when they do not contribute practically to the stored charge during the pulse duration. It is seen in Figure 14 that the flat-band voltage for the horisontal segments of hysteresis curves is very close to the value at that steady-state is reached. For lower oxide currents steady state is reached at higher charging pulse amplitudes, but at lower flat-band voltages (absolute values). Thus, for lower oxide currents the horisontal segments correspond to lower flat-band voltages, i.e., the height of hysteresis curve is lower.

In principle, the horizontal segments of hysteresis curve can also be connected to the „filling up" of traps or nanocrystals. If they cannot capture more charge, e.g., since Coulomb blockade, the whole amount of injected charge travels via the control layer and reaches the gate. In this case no shift of flat-band voltage occurs as well. So, the height of hysteresis curve is connected to the charge storage capability of the structure due to charge loss to gate.

On the basis of Eqs. (1)-(4) another expression can be obtained for the slope of the memory hysteresis curve [32]:

$$\frac{\delta V_{FB}}{\delta V_P} = 1 - (\frac{\varepsilon_2}{\varepsilon_1} d_1 + d_2) \frac{\delta E_2}{\delta V_P} \tag{10}$$

Eq. (10) indicates that if the slope of the abrupt segment of the hysteresis curve equals unity, the electric field in the control layer at the moment of the cut-off the charging pulse is constant along the whole abrupt segment. Its value can be considered, as a threshold electric field that is necessary for the charge injection for the actual charging pulse width. The threshold electric field can be extracted from the width of the hysteresis curve [32].

If it is considered that the charge centroid shifts into the nitride with increasing amount of stored charge, Eq. (10) becomes:

$$\frac{\delta V_{FB}}{\delta V_P} = 1 - (\frac{\varepsilon_2}{\varepsilon_1} d_1 + d_2) \frac{\delta E_2}{\delta V_P} - d_1 E_2 \frac{\delta}{\delta V_P} \left(\frac{d_2}{d_1} \right) \tag{11}$$

As d_1 decreases and d_2 increases with increasing charging pulse amplitude (see Eq. (2)), the shift of the charge centroid decreases the slope of the abrupt part of the hysteresis curve.

In the above calculation no fixed charge was taken into account. If fixed charge is present near the substrate/tunnel layer interface (within the tunneling depth), the hysteresis curve shifts along the x-axis [32]. Therefore the asymmetry of the hysteresis curve along x-axis (i.e., the difference between positive and negative charging pulses necessary for setting zero flat-band voltage) holds information about the charge fixed at or close to the substrate/tunnel layer interface [31, 32].

So, it can be concluded on the basis of Eqs. (6) and (11) that the slope of the abrupt part of hysteresis curve is connected to the efficiency of charge injection. If there is no loss of the injected charge via the control layer to the metal layer or after switching off the charging pulse, there is no charge injection from the metal side, and the centroid of captured charge does not depend on the amount of stored charge, the slope of the abrupt part of the hysteresis curve is unity. In opposite case it is lower.

The width of the hysteresis curves is strongly related to charge injection properties. On the basis of Eqs. (10) and (11) it can be concluded that if the slope of the hysteresis curve is close to unity, the electric field in the tunnel layer is almost constant in the moment of swithching off the charging pulse, along the whole abrupt part of the hysteresis curve. This value can be considered, as a threshold electric field, which is necessary for charge injection at the given charging pulse duration.

The height of hysteresis curve is limited by the charge storage capability of the structure due to charge loss to the gate either by high current level via the conrol layer or by „filling up" energy levels for charge capture and storage.

The asymmetry of the hysteresis curve along x-axis holds information about the charge fixed near the substrate/tunnel layer interface.

TUNNELING PROBABILITIES IN MNOS AND SONOS STRUCTURES

The charge injection into non-volatile semiconductor structures is determined by the tunneling probability via the potential barrier shown in Figures 12 and 13 for the different structures and electric field strengths. In our simulations the tunneling probability as a function of electric field was calculated for MNOS or SONOS structures by the WKB approximation using the approach presented by Lundström and Svensson [36]. Both MNOS structures without and with embedded Si or Ge nanocrystals were considered. The band structures used in calculations are presented in Figure 15 [36]. The actual barrier height values and band discontinuities presented in Figure 15 were also taken from Ref. 36. For both the electron and hole effective masses $m^*=0.42m_0$ was used [36], where m_0 is the free electron mass. No image force lowering and band bending at the silicon surface were considered.

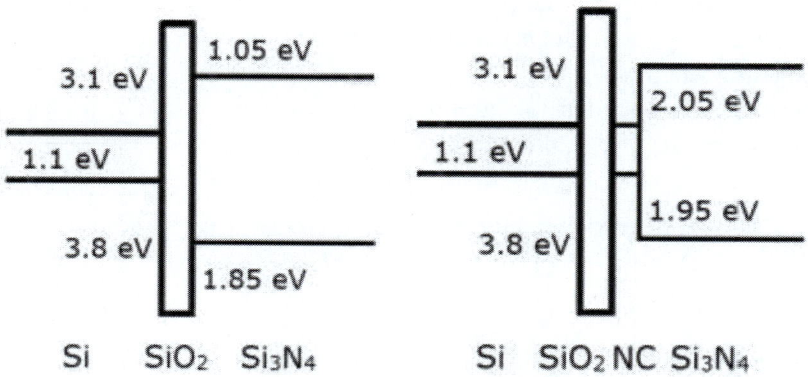

(a) (b)

Figure 15. Band diagram of MNOS structures without (a) and with embedded Si nanocrystals (NCs) used for the calculations. The actual values of band gaps and discontinuities are taken from Ref. 36.

Figure 16 presents the two extremes of probability of electron tunneling to MNOS structures with or without embedded semiconductor nanocrystals as a function of the oxide thickness and the electric field in the oxide layer [2, 21]. One of these extrems is the tunneling probability to the conduction band of the nitride layer for MNOS structures without nanocrystals. This is the electron tunneling probability for MNOS structures with nanocrystals either, if nanocrystals are located deep in the nitride layer. If nanocrystals are located deeper, than the tunneling length (see Figure 13.a), they do not influence the tunneling process directly. The other extreme is the tunneling probability to the nanocrystals located at the SiO_2/Si_3N_4 interface. It is just the tunneling probability via the oxide layer. The tunneling probability via nanocrystals located at the interface and via or to nanocrystals or traps [37] located deeper, but within the tunneling depth, is between these two extremes.

The tunneling probability of electrons to the nitride conductance band is higher for higher electric fields. It exhibits maximum as a function of oxide thickness, which depends on the actual electric field. The higher the electric field the thinner the oxide layer for the maximum probability. The increase of tunneling probability in the presence of a thin oxide layer in comparison with MNS structures without oxide layer, is due to a high potential drop on the oxide layer caused by its lower dielectric constant: the area of potential barrier above the electron energy is smaller, as it can be deduced from Figure 13.a [22]. At higher oxide thicknesses and/or electric fields the charge injection mechanism changes to direct tunneling to the nitride conductance band via the oxide layer, as shown in Figure 13.b. In this case the tunneling probability decreases with the further increase of oxide thickness. At high oxide thickness the tunneling probability saturates at a certain level. At this point the potential drop on the oxide layer just equals the barrier height at the SiO_2/Si_3N_4 interface, and the current mechanism changes to the Fowler-Nordheim tunneling of electrons to the oxide conduction band, as presented in Figure 13.c [38, 39].

The tunneling probability via the oxide layer (direct tunneling to nanocrystals located at the SiO_2/Si_3N_4 interface) doesn't depend strongly on the electric field, but exhibits fast decrease with increasing the oxide thickness. For a given oxide thickness the two extremes merge at a certian electric field, when the modified Fowler-Nordheim tunneling to the nitride conductance band via the oxide and nitride layer (Figure 13.a) changes to direct tunneling via the oxide layer (Figure 13.b). So, the presence of nanocrystals enhances the charge injection at thin oxide layers only, and if nanocrystals are located close to the SiO_2/Si_3N_4 interface (within the tunneling length).

Similar results were obtained for hole tunneling, but the maximum probabilities were obtained for thinner oxides (1.5-2.5 nm) [38, 39].

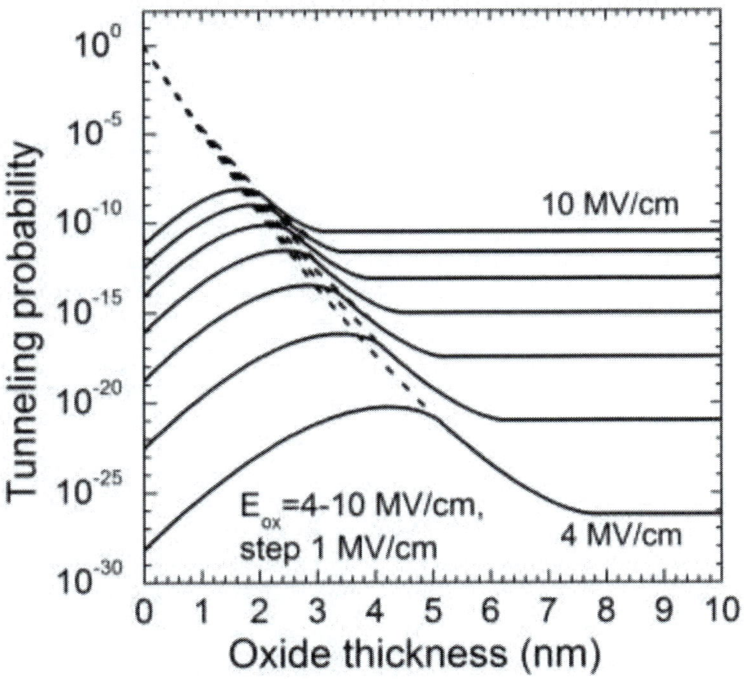

Figure 16. Electron tunneling probability to MNOS or SONOS structures without nanocrystals: tunneling to the nitride conduction band (solid lines), and to MNOS or SONOS structures containing nanocrystals located at the oxide/nitride interface: direct tunneling to nanocrystals (dashed lines) as a function of oxide thickness and electric field in the oxide layer [2, 21].

SIMULATION OF MNOS MEMORY HYSTERESIS – EFFECT OF LAYER THICKNESSES AND CURRENT LEVELS

The memory hysteresis curves were simulated on the basis of the calculated tunneling probabilities for MNOS structures both with and without embedded semiconductor nanocrystals. The effect of oxide thickness was studied in both cases. In MNOS structures without nanocrystals the effect of nitride thickness and the depth of charge centroid was investigated as well as the effect of current levels. The main goal of the simulations with MNOS structures containing nanocrystals was the interpretation of our earlier experimental results.

The memory hysteresis curves of MNOS structures were simulated calculating the flat-band voltage shift due to the charge injected and stored in traps or in nanocrystals embedded in the nitride layer using Eqs. (1)-(5) by iteration [7, 29]. The net charge can be calculated by the integration of the difference of the current flowing into the structure via the tunnel oxide layer and that of flowing out of the structure via the control nitride layer, as given in Eq. (5).

The current density via the tunnel oxide layer was calculated via the tunneling probability [32]:

$$J_2 = C_{FN} E_2^2 P \qquad (12)$$

where C_{FN} is a constant and P is the tunneling probability via the potential barrier. A C_{FN} value of 2.2×10^{-6} A/V^2 was used for both the electron and hole injection. The expression for current through the nitride layer was obtained by experiment [7, 20]:

$$J_1 = J_{PF} + J_{EX} \qquad (13)$$

where J_{PF} is the Poole-Frenkel current:

$$J_{PF} = C_{PF1} \cdot E_1 \cdot \exp\left(C_{PF2} \cdot \sqrt{E_1}\right) \qquad (14)$$

Here C_{PF1} and C_{PF2} are parameters depending on the insulator properties. J_{EX} is an exponential excess current density obtained at low voltages [7, 20]:

$$J_{EX} = C_{EX1} \cdot \exp\left(C_{EX2} \cdot E_1\right) \qquad (15)$$

where C_{EX1} and C_{EX2} are parameters evaluated from our experiments [7, 20]. A C_{PF1} value of 8×10^{-26} A/Vcm was used for both charging pulse polarities.

The details of integration is described elsewhere [7, 29]. The hysteresis curves were begun to be calculated from zero pulse amplitude and zero flat-band voltage. For the simulation of hysteresis curves the voltage pulse amplitude was increased or decreased by 1 V step by step. The hysteresis curves were calculated for a charging pulse amplitude range of ±25 V in most cases. The pulse width used for simulations was 10 ms.

The effect of oxide thickness was studied in MNOS structures without nanocrystals in the range of 0.1-10 nm with a total thickness of nitride layer of 30 nm [29]. A constant depth of charge centroid of 6 nm was considered. The height and width of the hysteresis curves simulated for MNOS structures without nanocrystals are presented in Figure 17 as a function of the oxide thickness. Increasing the oxide thickness from 0.1 nm to 2 nm, the calculated hysteresis curve exhibits increasing height for a little lower charging voltage pulse amplitude. This indicates increased tunneling current for an oxide thickness of 2 nm in comparison with lower oxide thickness values, while the nitride current does not increase. It is in agreement with the calculated tunneling probabilities presented in Figure 16. The highest and thinnest hysteresis curve was simulated for an oxide thickness of 2 nm, as it is seen in Figure 17. This one is the reference hysteresis curve in Figure 14. Increasing the oxide thickness further, the height of the hysteresis curve, i.e., the amount of the possible stored charge, decreases fast, while the pulse amplitude necessary for the charging process increases abruptly up to 4 nm (see Figure 17). This effect indicates decreasing oxide current in agreement with tunneling probabilities. From 4 nm up to 10 nm the highest possible amount of charge does not change further, while the pulse amplitude increases slowly. It is connected to the change of charge injection mechanism from direct to Fowler-Nordheim tunneling. If Fowler-Nordheim tunneling occurs, the oxide current does not depend on oxide thickness for a constant electric field (see Figures 13.c and 16), but the voltage pulse amplitude increases with increasing oxide thickness.

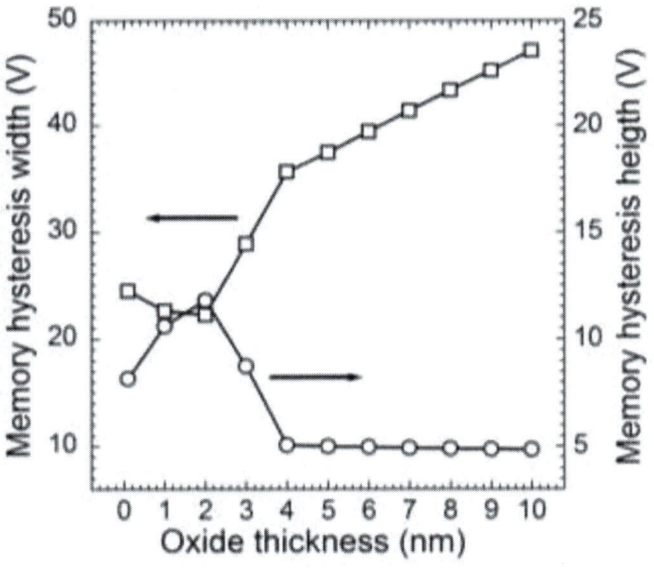

Figure 17. The width (squares) and height (circles) of memory hysteresis curves simulated for MNOS structures without nanocrystals as a function of oxide thickness.

The effect of nitride thickness was studied in the range of 10-40 nm for MNOS structures without nanocrystals with an oxide thickness of 2 nm. For depth of charge centroid two different approaches were studied. In one case the charge cetroid was located at a fixed distance of 6 nm from the oxide/nitride interface, in the other case it was located at the 1/5 part of the total nitride thickness [29]. In the first case the shape of the simulated hysteresis curves were similar, as the reference curve in Figure 14. Changing the nitride thickness, both the height and width of hysteresis curves changed proportional to the nitride thickness. But in the case of a fixed depth of charge centroid, the height of the hysteresis changed faster than its width. This case is probably closer to reality. So, in real cases the memory window shrinks faster with decreasing nitride thickness, than the charging pulse amplitude decreases. Therefore, a thin nitride layer decreases the efficiency of the injected charge.

The effect of the depth of charge centroid was also studied [29]. (No dependence of its depth on the amount of stored charge was considered.) It was obtained that for the same voltage pulse amplitude the height of hysteresis decreased linearly with inreasing depth of charge centroid, as it follows from Eqs. (2) and (4).

The effect of current levels was studied by a simple multiplication of currents given in Eqs. (12)-(15) with a multiplication factor [4, 30]. To study the effect of oxide current level C_{FN} value was varied from 2.2×10^{-12} A/V^2 to 2.2×10^{9} A/V^2, i.e., the reference oxide current for C_{FN} value of 2.2×10^{-6} A/V^2 was multiplied by factors from 10^{-6} to 10^{15}. To study the effect of nitride current level C_{PF1} value was changed from of 8×10^{-30} A/Vcm to 8×10^{-23} A/Vcm, i.e., the reference nitride current was multiplied by factors from 10^{-4} to 10^{3}. It was observed during simulations that the excess current J_{EX} had not influence the hysteresis curves, so its value was taken zero during the study of the influence of nitride current.

The effect of oxide current level on the simulated hysteresis curves is presented in Figure 14. The width of the hysteresis curve increased, while its height decreased with decreasing oxide current. For multiplication factor of 10^{-4} the hysteresis curve became small. For multiplication factor of 10^{-6} the hysteresis curve changed even its direction, positive charging pulse yielded negative flat-band voltage and vice versa. In this case the oxide current is lower than the nitride current during the charging voltage pulse, as discussed above. For increased oxide currents the hysteresis width decreased slowly, while the height of curves increased. The higher the oxide current the narrower the hysteresis and higher its height.

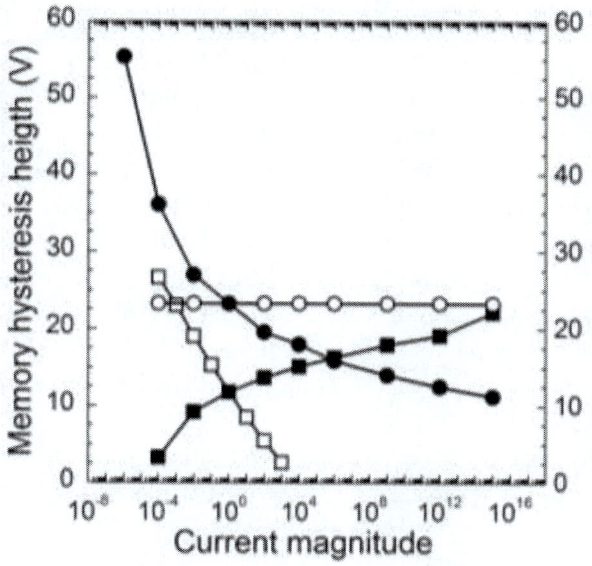

Figure 18. The height (sqares) and width (circles) of simulated memory hysteresis curves as a function of current magnitude (multiplication factor) of oxide (solid symbols) or nitride (open symbols) current [4].

Figure 18 summarizes the effect of oxide and nitride current levels on memory hysteresis height and width data obtained by the simulation [4]. The memory hysteresis width depends strongly on the oxide current level, in particular at low current levels. Increased voltage pulse amplitudes are required for charge injection, if the oxide current level is reduced. (The width of hysteresis for multiplication factors of 10^{-4} and 10^{-6} have been obtained for zero nitride current [4].) The abrupt dependence of hysteresis width on oxide current level at low oxide currents is probably connected to the change of the dominating current mechanism from modified Fowler-Nordheim or direct tunneling to the Fowler-Nordheim tunneling, when the tunneling probability is a relative weak function of oxide electric field [21, 38, 39].

The nitride current level affects the hysteresis height only due to charge loss via the nitride layer. The width of hysteresis does not depend on the level of nitride current, its value is the same as in the reference case. The height of the hysteresis decreases linearly with the logarithm of nitride current.

Interpretation of Experimental Results Obtained in MNOS Structures with Embedded Si or Ge Nanocrystals

Replacement of floating gate with semiconductor nanocrystals yielded improved behaviour of non-volatile semiconductor memories. It had been demonstrated by Rao and co-workers for SONOS structures either [40] that formation of semiconductor nanocrystals in silicon nitride based structures can enhance their performance. Nevertheless, only very few works had been devoted to silicon nitride based memory structures with embedded nanocrystals, even it seemed obvious to merge the advantages of nanocrystal and silicon nitride based memory devices.

For improving the charging and/or retention behaviour of MNOS structures we studied the effect of Si or Ge nanocrystals on their memory behaviour by memory window and retention measurements. Indeed, it was obtained that the presence of Si or Ge nanocrytals enhanced the charge injection properties of MNOS structures [1, 6, 15, 16]. The memory window was wider by about 5-20% in structures containing nanocrystals at the SiO_2/Si_3N_4 interface, than in reference structures prepared in a similar way, but without nanocrystals. But, there was no explicit improvment of the retention behaviour, in some structures it was better, in the other structures it was worse, than in the reference structures [1, 6, 15, 16].

Si nanocrystals were embedded into MNS structures as well [20], but in these structures both the charge injection and retention behaviour became worse due to the presence of nanocrystals. As a matter of fact, nanocrystals in MNS structures were located very deep in the nitride layer, at about 15 nm.

For further possible improvment of memory properties an additional thin nitride layer was grown between the oxide layer and the sheet of nanocrystals, i.e., nanocrystals (NCs) were shifted into the nitride layer forming SiO_2/Si_3N_4/Si NC/Si_3N_4 or SiO_2/Si_3N_4/Ge NC/Si_3N_4 sandwich structures between the Si substrate and the gate [1, 2]. But the results were contradictory. For structures with Si nanocrystals the charging behaviour improved much, as presented in Figure 6, while their retention properties degraded considerably. In the case of structures with Ge nanocrystals there was a low enhancement of charging behaviour only, while for a structure the retention time improved up to near 4000 years [1, 2]. Further on, the significant increase of memory window width for structures with Si nanocrystals seemed contradictory with the calculated tunneling probabilities: the deeper the location of nanocrystals the lower their effect on charge injection. To understand the origin of these apparent contradictions, an intensive study of charge injection and retention properties of MNOS structures with embedded nanocrystals was performed by simulation [2].

The simulations were carried out for three different assumptions: 1) MNOS structures without nanocrystals, tunneling to the nitride conduction or valence band, 2) MNOS structures with nanocrystals, tunneling to the nitride conduction or valence band via nanocrystals, and 3) MNOS structures with nanocrystals, direct tunneling to nanocrystals. The effects of oxide thickness and the depth of location of nanocrystals were studied by the simulation of memory window, memory hysteresis and retention behaviour [2].

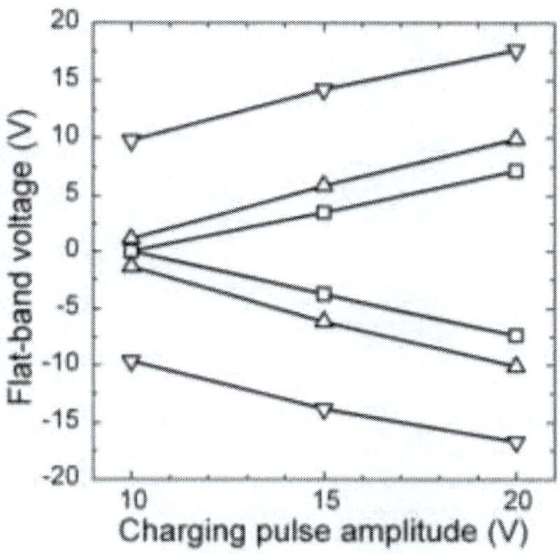

Figure 19. Simulated memory window for MNOS structures containig Si or Ge nanocrystals at a depth of 1 nm from the SiO_2/Si_3N_4 interface. Oxide and nitride layer thicknesses are 1 nm and 30 nm, respectively. Si nanocrystal size is 3 nm, the depth of charge centroid is 6 nm, the pulse width is 10 ms: squares - reference structure without nanocrystals, triangles down - direct tunneling to nanocrystals, triangles up - tunneling via nanocrystals to the nitride conduction or valence bands [2].

The typical results of memory window simulations is presented in Figure 19 [2]. The memory window for tunneling to nitride layer via nanocrystals is somewhat wider, than that for MNOS structures without nanocrystals, as it was obtained by experiments in MNOS structures containing Si or Ge nanocrystals at the oxide/nitride interface and Ge nanocrystals at a depth of 3 nm in the nitride layer. But, the memory window was about two times wider for direct tunneling to nanocrystals, than that for MNOS structures without nanocrystals. Such experimental results were obtained in structures with Si nanocrystals embedded at a depth of 3 nm in the nitride layer.

However, memory hysteresis simulations yielded rather unexpected results [2]. The memory window for tunneling to nitride layer via nanocrystals was somewhat thinner, than that for MNOS structures without nanocrystals, as expected (see Figure 20). But, the memory hysteresis was very thin for tunneling to nanocrystals, as it can also be seen in Figure 20. In some cases its width was even zero, i.e., the hysteresis curve collapsed to a straight line with a slope of unity, as shown in Figure 21 [2]. This phenomenon is due to very high tunneling probability and so, to very high current level via the oxide layer. The system is very close to equilibrium. But, no sign of the collapse of memory behaviour was obtained by the simulation of memory window, as presented in Figure 19 for the same structure with same layer thicknesses, size and location of nanocrystals.

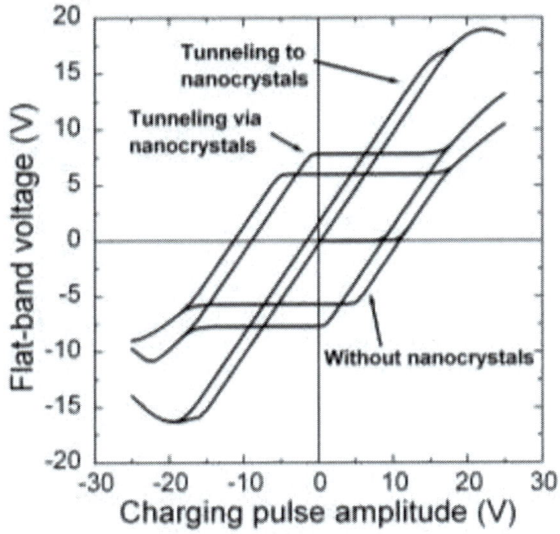

Figure 20. Simulated memory hysteresis curves of MNOS structures containing Si or Ge nanocrystals at the SiO_2/Si_3N_4 interface. The oxide and nitride thicknesses are 2 nm and 30 nm, respectively, the depth of charge centroid is 6 nm, nanocrystal size is 3 nm, the pulse width is 10 ms [2].

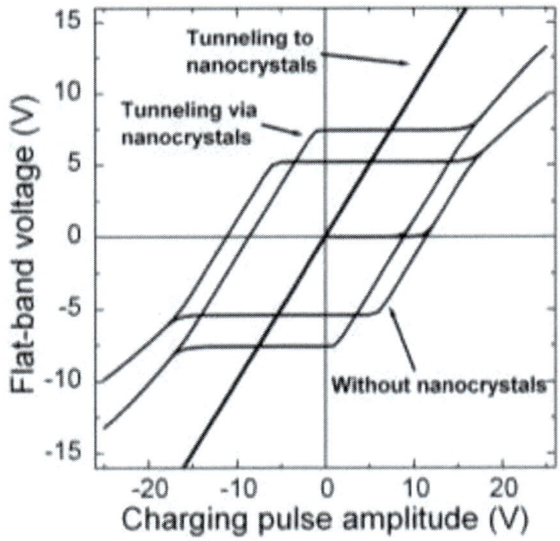

Figure 21. Simulated memory hysteresis curves exhibiting the collapse of memory behaviour at high tunneling probabilities (oxide currents). Nanocrystals are located at a depth of 1 nm from the oxide/nitride interface. The oxide and nitride thicknesses are 1 nm and 30 nm, respectively, the depth of charge centroid is 6 nm, nanocrystal size is 3 nm, the pulse width is 10 ms [2].

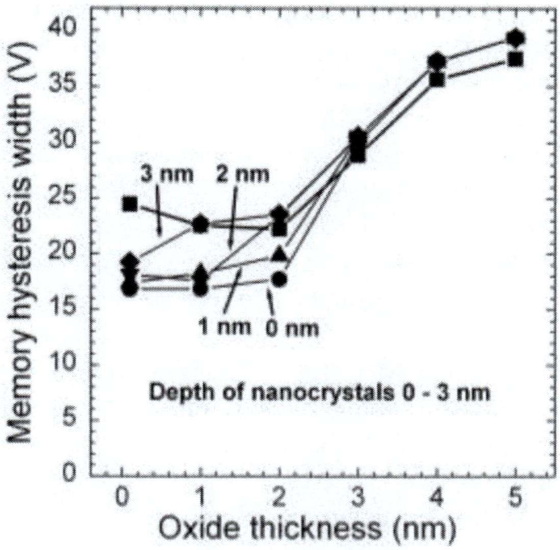

Figure 22. The width of memory hysteresis simulated for structures without (sqares) and with nanocrystals at different depth from the SiO_2/Si_3N_4 interface: 0 nm (circles), 1 nm (triangles up), 2 nm (triangles down) and 3 nm (diamonds). The size of nanocrystals is 3 nm, the pulse width is 10 ms, tunneling via nanocrystals [2].

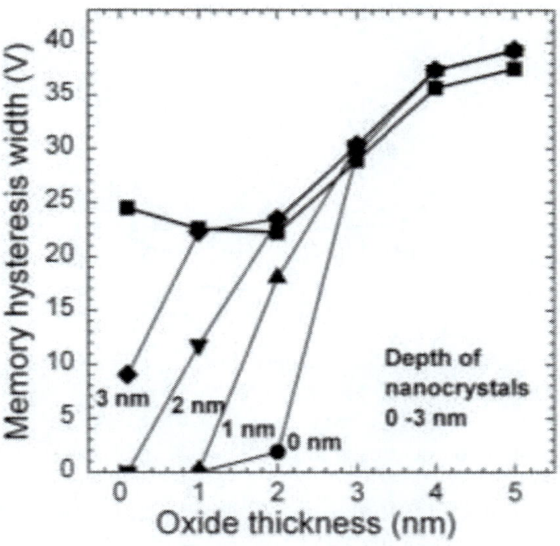

Figure 23. The width of memory hysteresis simulated for structures without (sqares) and with nanocrystals at different depth from the SiO_2/Si_3N_4 interface: 0 nm (circles), 1 nm (triangles up), 2 nm (triangles down) and 3 nm (diamonds). The size of nanocrystals is 3 nm, the pulse width is 10 ms, direct tunneling to nanocrystals [2].

The width of memory hysteresis for tunneling to the nitride conduction or valence band via nanocrystals is compared with that for MNOS structures without nanocrystals in Figure 22 as a function of oxide thickness and depth of nanocrystal location. It can be seen that nanocrystals enhances charge injection for oxide thickness up to 3 nm only and in the case, if they are close to the oxide/nitride interface. The increased width of the hysteresis at and above an oxide thickness of 3 nm is due to the potential drop on nanocrystals resulting lower electric field in the oxide layer for the same charging pulse amplitude.

The width of memory hysteresis for direct tunneling to nanocrystals is presented in Figure 23 as a function of oxide thickness and depth of nanocrystal location. These dependences show the collapse of memory behaviour. There is a very sharp change of the width of memory hysteresis with a narrow range of oxide thickness for each value of nanocrystal depth. Here not the actual thickness values are important, but the tendencies. The parameters of real structure (barrier heights, effective masses, etc.) are probably different than those used for the simulation.

Some results of simulation of retention behaviour are presented in Figure 24. The retention time for electron escape varies about 10 orders of magnitude from several hundred microseconds to several months, while the oxide thickness changes from 2 nm to 4 nm. There is a sharp effect of the location of nanocrystals in the nitride layer either. So, there is a very strong dependence of retention behaviour on the insulator thickness as well.

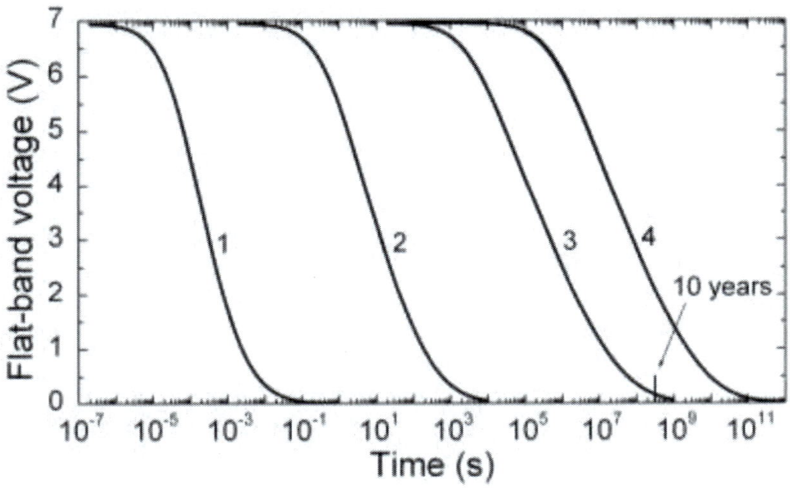

Figure 24. The change of flat-band voltage due to electron escape from Si nanocrystals. The initial charge is -10^{13} q/cm^2 (-1.6×10^{-6} C/cm^2), total nitride thickness 30 nm. 1, 2, and 3 - nanocrystals located at the oxide/nitride interface, oxide thickness 2 nm, 3 nm, and 4 nm, respectively, 4 - nanocrystals located at a depth of 3 nm from the oxide/nitride interface, oxide thickness 2 nm [2].

On the basis of the results of simulation it can be concluded that in the case, when nanocrystals are close to the oxide/nitride interface, there is no charge storage in nanocrystals, the injection properties are enhanced by tunneling via them. The injected charge captured by nanocrystals escapes very fast after switching off the charging pulse and cannot be observed by measurements. This was the case in our first experiments, where some improvment of the injection properties was obtained by nanocrystals located at the oxide/nitride interface [1, 6, 15, 16]. In the case, when nanocrystals are located deep in the nitride (deeper, than the tunneling depth), they do not affect the charging behaviour directly, but the memory behaviour are degraded by the voltage drop on the nanocrystals. Such effect has been obtained in MNS structures with Si nanocrystals embedded deep in the nitride layer [20]. In a narrow intermediate case, when the tunneling probability is high, both types of tunneling occur, the injected charge is stored by nanocrystals and in traps as well, and very good charging behaviour are obtained, but the retention time is short due to the charge escape from the nanocrystals. Such a case was otained for our SiO_2/Si_3N_4/Si NC/Si_3N_4 structures.

In the case of SiO_2/Si_3N_4/Ge NC/Si_3N_4 structures the direct tunneling did not play important role in the charging process, as no considerable enhancement was obtained. The good retention behaviour obtained in one of these structures can be due to the injection of charge carriers via Ge nanocrystals to traps located relatively deep in the nitride layer.

CONCLUSION

The operation of floating gate, nanocrystal and silicon nitride based non-volatile memory devices is based on charge injection and storage on floating gate, in nanocrystals or in traps in the nitride layer, respectively.

It can be concluded on the basis of analysis of flat-band voltage hysteresis presented that the slope of the abrupt part of hysteresis curve is connected to the efficiency of charge injection. If there is no loss of the injected charge via the control layer to the metal layer or after switching off the charging pulse, there is no charge injection from the metal side, and the centroid of captured charge does not depend on the amount of stored charge, the slope of the abrupt part of the hysteresis curve is unity. In opposite case it is lower.

The width of the hysteresis curves is strongly related to charge injection properties. If the slope of the hysteresis curve is close to unity, the electric field in the tunnel layer is almost constant in the moment of swithching off the charging pulse, along the whole abrupt part of the hysteresis curve. This value can be considered, as a threshold electric field, which is necessary for charge injection at the given charging pulse duration.

The height of hysteresis curve is limited by the charge storage capability of the structure due to charge loss to gate either by high current level via the conrol layer or by „filling up" energy levels for charge capture and storage.

The asymmetry of the hysteresis curve along *x*-axis holds information about the charge fixed near the substrate/tunnel layer interface.

Calculation of tunneling probabilities into MNOS structures and simulation of their memory hysteresis behaviour indicated that the optimal oxide thickness is about 2 nm. Both the height and width of memory hysteresis srink proportionally with decreasing nitride thickness. So, choosing the nitride thickness of the structure one should make a compromise between the low charging pulse amplitude and the memory window width.

The height of memory hysteresis decreases proportionally with increasing depth of charge centroid either. Therefore, for optimal charging behaviour the depth of injected charge centroid should be minimized. However, it is rather complicated to influence the depth of charge centroid by technology, on one hand. On the other hand, the retention time is better for deeper location of the injected charge. So, if even one can influence the location of charge centroid, he has to make a compromise between the memory window width and retention time.

On the basis of calculation of tunneling probabilities and simulation of memory hysteresis and retention behaviour of MNOS structures it can be concluded that the presence of nanocrystals enhances the charge injection only for structures with thin tunnel oxide layer (below 3 nm), and if the nanocrystals are located close to the oxide/nitride interface (within the tunneling depth). In the case of very high tunneling probability (high tunneling currents) the system approaches equilibrium and the memory behaviour collapses. There is a narrow range of oxide thickness or depth of nanocrystals, when the charging properties change very fast. Retention exhibits a very sharp dependence on the oxide thickness or depth of nanocrystals as well.

If nanocrystals are close to the oxide/nitride interface, there is no charge storage in nanocrystals, the injection properties are enhanced by tunneling via them. In the case, when nanocrystals are located deep in the nitride (deeper, than the tunneling depth), they do not affect the charging behaviour directly, but the memory behaviour are degraded by the voltage drop on the nanocrystals. In a narrow intermediate case, when the tunneling probability is high, the injected charge is stored in nanocrystals and in traps as well, and very good charging behaviour are obtained, but the retention time is short due to the charge escape from the nanocrystals. Our experimental results can be explained on the basis of these conclusions.

Acknowledgements

The author wish to thank P. Basa, T. Jászi, Gy. Molnár, K. Z. Molnár, M. Németh-Sallay, A. E. Pap, G. Stubnya, P. Szöllősi and P. Tüttő for their valuable contribution to the results.

References

[1] Horváth, Zs.J., Basa, P., (2008). Chapter 7. Nanocrystal memory structures. in: *Nanocrystals and Quantum Dots of Group IV Semiconductors*, (Eds. T. V. Torchynska, Yu. V. Vorobiev), American Scientific Publishers, 2010, Stevenson Ranch, California, USA, ISBN: 1-58883-154-X, 225-251, and references therein.

[2] Horváth, Zs.J., Basa, P., Molnár, K.Z., Molnár, Gy., Jászi, T., Pap, A.E., (2013). Effect of location of Si or Ge nanocrystals on the memory behaviour of MNOS structures. *Physica E* 51, 104-110.

[3] Horváth, Zs.J., Németh-Sallay, M., Tüttő, P., Stubnya, G., (1982). The influence of film thicknesses on the switching behaviour of the MNOS memory devices. *Proc. Conf. Microelectronics'82*, Siófok, Hungary, 131-132.

[4] Horváth, Zs.J., Molnár, K.Z., (2013). Effect of current levels on memory behaviour of MNOS structures - limits of computer simulation. *Proc. IEEE 9th Int. Conf. Computational Cybernetics, ICCC 2013*, Tihany, Hungary, 303-307.

[5] Sze, S.M., (1981). Physics of Semiconductor Devices. 2nd Ed., Wiley, New York.

[6] Horváth, Zs.J., Basa, P., Jászi, T., Pap, A.E., Dobos, L., Pécz, B., Tóth, L., Szöllősi, P., Nagy, K., (2008). Electrical and memory properties of Si_3N_4 MIS structures with embedded Si nanocrystals. *J. Nanosci. Nanotechnol.* 8, 812-817, 2008.

[7] Horváth, Zs.J., Hardy, V., (2008). Simulation of memory behaviour of non-volatile structures. *J. Nanosci. Nanotechnol.* 8, 834-840.

[8] Parat, K.K., (2004). Flash Memory Technology - Recent Advances and Future Outlook. in: *Physics of Semiconductor Devices*, (Eds. K. N. Bath and A. DasGupta), Narosha Publishing House, New Delhi, 433-438.

[9] S. Tiwari, S., Rana, F., Hanafi, H., Hartstein, A., Crabbé, E.F., Chan, K., (1996). A silicon nanocrystals based memory. *Appl. Phys. Lett.* 68, 1377-1379.

[10] Horváth, Zs.J., (2006). Semiconductor nanocrystals in dielectrics: Optoelectronic and memory applications of related silicon based MIS devices. *Current Appl. Phys.* 6, 145-148.

[11] Lombardo, S., De Salvo, B., Gerardi, C., Baron, T., (2004). Silicon nanocrystal memories. *Microel. Eng.* 72, 388-394.

[12] Duguay, S., Slaoui, A., Grob, J.J., Kanounb, M., Burignat, S., Souifi, A., (2005). Structural properties of Ge-implanted SiO_2 layers and related MOS memory effects. *Mater. Sci. Eng. B* 124-125, 488-493.

[13] A. Kanjilal, A. Hansen, J.L., Gaiduk, P., Nylandsted Larsen, A., Normand, P., Dimitrakis, P., Tsoukalas, D., Cherkashin, N,. Claverie, A., (2005). Size and aerial density distributions of Ge nanocrystals in a SiO_2 layer produced by molecular beam epitaxy and rapid thermal processing. *Appl. Phys. A* 81, 363-366.

[14] Kapetanakis, E., Normand, P., Holliger, P., (2008). Electrical properties of metal-oxide-semiconductor structures with low-energy Ge-implanted and annealed thin gate oxides. *J. Appl. Phys.* 103 064515-13.

[15] Basa, P., (2008). Semiconductor nanocrystals in dielectrics for memory purposes. *PhD Thesis*, Budapest University of Technology and Economics, and references therein; http://dept.phy.bme.hu/phd/dissertations/basa_dissertation.pdf

[16] Horváth, Zs.J., Basa, P., (2009). Nanocrystal non-volatile memory devices. *Mater. Sci. Forum* 609, 1-9.

[17] Ng, T.H., Chim, W.K,. Choi, W.K., Ho, V., Teo, L.W., Du, A.Y., Tung, C.H., (2004). Minimization of germanium penetration, nanocrystal formation, charge storage, and retention in a trilayer memory structure with silicon nitride/hafnium dioxide stack as the tunnel dielectric. *Appl. Phys. Lett.* 84, 4385-4387.

[18] Lu, J., Lin, Ch.-H., Kuo, Y., (2008). Nanocrystalline Zinc-Oxide-Embedded Zirconium-Doped Hafnium Oxide for Nonvolatile Memories. J. Electrochem. Soc. 155, H386-H389.

[19] Dimitrakis, P., Mouti, A., Bonafos, C., Schamm, S., Ben Assayag, G., Ioannou-Sougleridis, V., Schmidt, B., Becker, J., Normand, P., (2009). Ultra-low-energy ion-beam-synthesis of Ge nanocrystals in thin ALD Al_2O_3 layers for memory applications. *Microel. Eng.* 86, 1838-1841.

[20] Basa, P., Horváth, Zs.J., Jászi, T., Pap, A.E., Dobos, L., Pécz, B., Tóth, L., Szöllősi, P., (2007). Electrical and memory properties of silicon nitride structures with embedded Si nanocrystals. *Physica E* 38, 71-75.

[21] Horváth, Zs.J., Basa, P., Jászi, T., Molnár, K.Z., Pap, A.E., Molnár, Gy., (2013). Charging behaviour of silicon nitride based non-volatile memory structures with embedded semiconductor nanocrystals. *Appl. Surf. Sci.* 269, 23-28.

[22] Horváth, Zs.J., Stubnya, G., Tüttő, P., Németh-Sallay, M., Balázs, J., (2000). Charge injection and storage properties of MNOS structures. in: *Physics of Semiconductor Devices*, Allied Publisher Ltd., New Delhi, India, (Eds. V. Kumar and S. K. Agarwal), 383-386.

[23] http://media.freescale.com/phoenix.zhtml?c=196520&p=irol-newsArticle&ID=831808&highlight=

[24] Horváth, Zs.J., Basa, P., (2009). New trends in non-volatile semiconductor memories. *Studies in Computational Intelligence* 243, 323–333.

[25] A. Geresdi, M. Csontos, A. Gubicza, A. Halbritter, G. Mihály, (2013). GHz Operation of Nanometer-Scale Metallic Memristors: Highly Transparent Conductance Channels in Ag_2S Devices. http://arxiv.org/abs/1310.4964

[26] D. Frohman-Bentchkowsky, M. Lenzlinger, (1969). Charge Transport and Storage in Metal-Nitride-Oxide-Silicon (MNOS) Structures. *J. Appl. Phys.* 40, 3307-3319.

[27] Stubnya, G., I. C. Szép, G. Hoffman, Horváth, Zs., Tüttő, P., (1978). Distribution and role of N-H and Si-H bonds in MNOS structures. *Rev. Phys. Appl.* 13, 679-682.

[28] Horváth, Zs.J., (1980). Memory hysteresis measurements on silicon oxynitride films. *Solid-State Electron.* 23, 1053-1054.

[29] Molnár, K.Z., Horváth, Zs.J., (2012). Simulation of MNOS memory hysteresis - Effect of layer thicknesses. *Proc. 7th IEEE Int. Symp. Applied Computational Intelligence and Informatics, SACI 2012*, Timisoara, Romania, 365-369.

[30] Horváth, Zs.J., Molnár, K.Z., Turmezei, P., Kovács, B., (2013). Effect of current levels on the memory hysteresis behaviour of MNOS and SONOS sturctures. *Proc. 19th Int. Conf. Applied Physics of Condensed Matter APCOM 2013*, Štrbské Pleso, Slovakia, 145-148; http://apcom.fyzika.uniza.sk/proceedings/pdf/145_Horvath.pdf.

[31] Horváth, Zs.J., (1978). Evaluation of the hysteresis of MNOS memory elements. *Proc. 3rd Int. Sci. Coll. T.U. Ilmenau*, Ilmenau, GDR, Heft 7, 81-84 (in Russian).

[32] Horváth, Zs.J., (2013) Evaluation of flat-band voltage memory hysteresis of non-volatile structures. Proc. 8th Solid State Surfaces and Interfaces, SSSI 2013, November 25-28, 2013, Smolenice, Slovakia, p.58.

[33] Lehovec, K., Crain, D.W., (1976). Charge centroid and trapping model for MNOS structures. *J. Appl. Phys.* 47, 2763-2764.

[34] Endo, N., (1978). Charge distributions in silicon nitride of MNOS devices. *Solid-State Electron.* 21 1153-1156.

[35] Tüttő, P., Stubnya, G., Horváth, Zs.J., (1982). Depth inhomogeneity of CVD Si_3N_4 layers. *IEE Proc. Pt.I.: Solid State Electron Dev.* 129, 103-104.

[36] Lundström, K.I., Svensson, C.M., (1972). Properties of MNOS structures. IEEE Trans. Elec. Dev., ED-19, 826-836.

[37] Maes, H.E., Overstraeten, R.J., (1976). Low-field transient behaviour of MNOS devices. *J. Appl. Phys.* 47, 664-666.

[38] Horváth, Zs.J., Molnár, K.Z., Molnár, Gy., Basa, P., Jászi, T., Pap, A.E., Lovassy, R., Turmezei, P., (2012). Charging behaviour of MNOS structures with embedded Ge nanocrystals. *Phys. Stat. Sol. (C)* 9, 1370-1373.

[39] Horváth, Zs.J., Basa, P., Molnár, K.Z., Jászi, T., Pap, A.E., Molnár, Gy., Dobos, L., Tóth, L., Pécz, B., (2013). Charging behaviour of MNOS memory structures with embedded Si or Ge nanocrystals. *Nanosci. Nanotechnol. Lett.* 5, 513-517.

[40] Rao, R.A., Steimle, R.F., Sadd, M., Swift, C.T., Hradsky, B., Straub, S., Merchant, T., Stoker, M., Anderson, S.G.H., Rossow, M., Yater, J., Acred, B., Harber, K., Prinz, E.J., White, B.E. Jr., Muralidhar, R., (2004). Silicon nanocrystal based memory devices for NVM and DRAM applications. *Solid-State Electron.* 48, 1463-1473.

In: Hysteresis
Editor: José Carlos Dias
ISBN: 978-1-63321-336-4

Chapter 10

BEHAVIOR PATTERNS, ORIGIN OF PROBLEMS AND SOLUTIONS REGARDING HYSTERESIS PHENOMENA IN COMPLEX BATTERY SYSTEMS

***Mohammad Rezwan Khan*[*,1], *Jorge V. Barreras*[1], *Ana I. Stan*[1], *M. Swierczynski*[1], *Søren J. Andreasen*[1] and *Søren K. Kær*[1]**
[1]Department of Energy Technology, Aalborg University, Denmark

ABSTRACT

One of the common phenomena for most of the battery cell chemistries is hysteresis. Since an open circuit voltage (OCV) path is not identical for the charge and discharge of the battery cell at different states of charge (SoC) level, the battery cell show the hysteresis effect. Usually, the OCV i.e. voltage with zero current after previous charge is higher than the OCV after discharge at the same SoC level. It embodies the hysteresis of the battery cell. The OCV is principally subjected to previous operating condition and cannot be taken as self-regulating from the operating history. Therefore, an accurate knowledge of the hysteresis of OCV is vital for various applications and battery models. This is because currently Battery Management Systems (BMS) use the well-defined OCV-SoC representative curve for SoC estimation and power prediction. Particularly lithium-ion batteries with iron-phosphate cathode material show a complex OCV behavior including a hysteresis that is dependent on the previous operation i.e. the previous state - charging or discharging. The extent of drawbacks related to hysteresis varies on specific cell chemistry. Batteries exhibiting the hysteresis are susceptible to catastrophes due to failures to estimate the correct SoC or energy content of the battery consequently limiting operation life of the battery pack. Therefore, novel and more cutting-edge hysteresis management strategies are required to be able to prevent securely the energy storage system from ever facing critical circumstances, such as thermal run away, internal short circuit etc. In case of the exchanging between the charging and discharging curves (due to exposure of charge and discharge pulse) makes the SoC estimation problematic since switching from the OCV leads to sudden changes in the

---

* Corresponding Author: Email: rezwankhn@gmail.com ;mrk@et.aau.dk; Tel.: +45- 99 40 33 74; Fax: +45- 98 15 14 11.

output of SoC that physically is not consistent for a battery cell. In this chapter, the characteristics of the OCV in function of relaxation times for different Li-ion chemistries are presented. Moreover, a novel hysteresis compensation structure is introduced based on Kalman filtering and statistical error functions. These combinations are able to meet the high requirements of estimating accurate SoC using open circuit voltage. In addition, the issues of safety and reliability are going to be increased for the integration of the newly established compensation scheme for the hysteresis phenomenon. A hysteresis compensation framework is elaborated in the subsequent sections of this chapter.

Keywords: battery management system; hysteresis compensation; open circuit voltage; state of charge; lithium, ion batteries

INTRODUCTION

For consumer applications such as laptops, smartphones, hybrid electric vehicles (HEVs) and electric vehicles (EVs), $LiCoO_2$ based cathode material is currently the preferred choice but the future trend is shifting towards mixed oxide cathodes ($LiNi_xCo_yAl_zO_2$ (NCA) and $LiNi_xMn_yCo_zO_2$ (NMC)) or $LiFePO_4$ owing to sufficient power density and energy density [1]. Lithium-ion batteries demonstrate a multifaceted behavior concerning their OCV particularly iron-phosphate cathode material lithium-ion batteries show an intricate OCV behavior including a hysteresis depending on the preceding operation. The OCV is predominantly subjective to the previous operating state and cannot be taken as self-regulating from the operating history. A precise understanding of the OCV is therefore indispensable for various applications especially in the case of the application that is operated using battery models e.g. finding SoC using battery models. Generally state of the art Li-ion batteries are typically made of a carbon/graphite based anode; the diversity of cathode materials is comparatively high. The OCV characteristic of $LiFePO_4$ and Nickel Manganese Cobalt Oxide (NMC) battery system is analyzed in-depth qualitatively and quantitatively. A distinguishing focus is on the parameters that influence the occurring OCV, namely temperature, SoC and previous history. OCV and hysteresis effects are investigated in order to analyze the long-term stability of the hysteresis. Because of the safety and cost related issues associated with the olivine-type cathode materials especially $LiFePO_4$ is getting more and more attention.

The electrical properties in a Li-ion battery are improved considerably through the developments in the production methods and new materials in the past few years. The chemical composition and the structure of its anode and cathode dictate the physical and electrochemical behavior of a Li-ion battery [2]. Sleight et al. [3] reported on hysteresis and memory effects in lithium batteries. A voltage hysteresis effect depending on the sign of the previous current is observed. They explained this behavior through a first-order phase transition when lithium (Li) is inserted in Li_xMnO_2. Moreover an influence on previous history of the cell operation called "new terrain" effect is observed. The "new terrain" effect is traced back to the idea of Li_xMnO_2 as a disordered two-phase system, with the memory of the previous cycles stored through the positions of the phase boundaries. In the two phase systems hysteresis is caused by the dissipation energies when the phase transitions take place in the crystal structure of the electrode. Barker et al. [4] reported a voltage hysteresis for a

Li_xCoO_2 system. As per the Li-metal (LiM) system, Li_xCoO_2 shows two-phase transition during the intercalation and de-intercalation of Li-ions in the cathode structure. The hysteresis effect for Li_xCoO_2 is noticed as relatively small and explained by the thermodynamic properties during the phase transition processes. The structural changes and two-phase transition characteristics in $LiCoO_2$ systems are deliberated by Carlier et al. [5]. They observed a two-phase transition, which is represented by the voltage plateaus in the OCV-SoC characteristics of a system. This is in harmony with a constant value of lithium chemical potential measured in the intervening time. One of the first who reported on the structure and behavior of phosphor-olivine structured cathode materials mainly lithium iron phosphate ($LiFePO_4$) was Padhi et al. [6]. They contribute on the explanation for the insertion process of Li-ions in a $FePO_4$ structure. Also the flat open-circuit characteristic of $LiFePO_4$ was observed. They reduced these results to structural changes during chemical delithiation of $LiFePO_4$ or lithiation of $FePO_4$ is noticed. The change was summarized as a first-order transition in the phosphor-olivine crystal structure. The phase structure is supplemented with the investigation by Dodd et al. [7]. They studied the transformation of the two-phase mixture, $xLiFePO_4$ and $(1-x)$ $FePO_4$, to a single-phase Li_xFePO_4 for several compositions x and found some hysteresis caused by this phase transition process. The batteries, made of $LiFePO_4$ electrodes charge/discharge characteristics in a Lithium battery, are investigated in detail by Srinivasan and Newman [8]. They established a discharge model for the electrodes, the shrinking-core approach, which displays the association of the two phases mentioned earlier in an iron-phosphate system and the movement of the phase boundary. The shrinking-core approach with its hard phase boundaries, and thus pre-history influenced potential boundaries is a promising approach. It is found to be a proper approach for explanation of OCV and hysteresis effects on the electrode materials showing a two-phase transition. The same authors enhanced their model approach in [9] finding the existence of path-dependence or impact of previous (cycling) history on the behavior of a $LiFePO_4$ battery. Zhang and White developed a similar modeling approach for the $LiCoO_2$ electrode [10]. Thus the shrinking-core model approach seems to be an appropriate approach for description for all electrodes showing a first-order phase transition. Zheng et al. [11, 12] noticed the hysteresis effects related to the lithium insertion in the anode materials for the lithium-ion batteries namely carbons. During their studies they established a strong relationship between the hydrogen content of the carbon materials and the characteristic of the occurring voltage hysteresis. The hysteresis effect is explained by the interaction between Li-ion and hydrogen-containing carbon structure during the intercalation: When inserting lithium in the carbon, some lithium atoms bind on the hydrogen-terminated edges of the carbon fragments. In order to intercalate, lithium must overcome an energy barrier before it can be bonded near hydrogen. The energy barrier seems to regulate the size of the hysteresis. Therefore, the lower hydrogen content means an inferior energy barrier consequently less hysteresis. Likewise Yazami et al. [13] reported on $LiCoO_2$ and $LiFePO_4$ cathodes effects also designated phase or structure transition on the graphite anodes. The graphite exhibits a plateau in its voltage characteristics depending on the degree of lithiation or de-lithiation. The transition process between these two stages is categorized by a hysteresis. These could be instigated by the entropy changes for the duration of the intercalation processes. There is not yet a very clear and substantive explanation of the hysteresis phenomena in lithium-ion batteries. There are many indications that hysteresis effects are caused by phase-transition processes, but an ultimate proof for this hypothesis is non-existent. Furthermore an interaction between effects occurring on the anode and the cathode of the Li-ion systems cannot be excluded [14, 15].

In the literature there are also some suggestions as to how the hysteresis effects can be counteracted or accounted for. Among them, [16] used extended Kalman Filter (EKF) to counteract hysteresis related effects while using battery models. In [15] fitted hysteresis polynomials are used to encounter a similar hysteresis problem. In recent work, [17] proposed error function based hysteresis compensation in tandem with a battery model.

Figure 1. Battery Test System including ajoining thermal chamber and impedance spectrometer.

Experimental Setup

The following subsections describe the applied test procedures.

Test Equipment

All the tests are carried out using battery test benches made by Maccor® Series 4000 Automated Test System. The used facility has an 8 channel maximum charge/discharge current rate of 30A and a voltage/current accuracy of ±0.1 % related to the set-point

(Figure 1). During long-term equilibrium measurements and self-discharge tests cell voltages are logged.

OCV Characteristics Measurement

To measure the characteristics of OCV of a battery cell, the fully charged cells are discharged with 0.5 C-rate in 5 % or 10 % ΔSoC steps until end-of-discharge voltage i.e. battery is full empty is reached. The pause after the discharge steps has the maximum duration of three hours. In charge direction the measurement is repeated reversely by charging the cells with 0.5C-rate until end of charge voltage is reached i.e. the battery is full.

ORIGIN OF HYSTERESIS

The measurements have identified the OCV as something which is considerably deviating from the equilibrium voltage in a thermodynamic definition. The measured OCV is a complex superposition of relaxation, diffusion and self-discharge processes on the one hand and a hysteresis caused by different material phases on the other hand. It is hard to separate the single processes which occur when current is cut-off. During the first period of time the OCV is dominated by faster relaxation and diffusion processes and shows a clear dependency on cell temperature and SoC. Afterwards, the OCV is dominated by the slower solid-state diffusion processes and self-discharge. A comparison of the cell types shows that hysteresis of OCV seems to be a question of definition and considered time period. The hysteresis phenomenon nearly disappears for many systems if batteries under test are relaxed into infinite time. This disappearance level depends on the composition of the battery. The impact of these different phases has to be further investigated since it influences the paths between the OCV boundary curves, which occur during technically relevant micro and partial cycle operation. It seems to be likely that the hysteresis phenomena are mainly caused by the cathode and its characteristics [14, 15, 18]. The higher the capacity throughput or the current rate per time, the lower the hysteresis. A higher current rate per time yields a higher local heating inside the active mass accelerating relaxation shortly after current is cut off. These temperature effects could superpose current impacts on hysteresis. Foremost it is critical to design test procedures which ensure a temperature equilibrium inside the cell to uncouple the current from the temperature impacts. Nevertheless it is likely that the relaxation and diffusion is strongly dependent on the characteristics of the anode and cathode material including variations in the stoichiometry of the mixed oxides or the coating of the iron phosphate based cathode. Moreover the composition of cell and the manufacturing process might yield different OCV phenomena [14, 15, 18].

Relaxation of Overvoltages

To confirm the presence of hysteresis after long test periods, tests are needed. Vanishing effects after long rest periods are caused only by diffusion processes. Additionally the effect of the temperature is investigated. Measurements are carried out at different operating points, which are adjusted by employing a current rate of 0.5C-rate. After adjustment of SoC levels, the cell voltage is recorded. The measurements are accomplished both after preceding discharge and preceding charge.

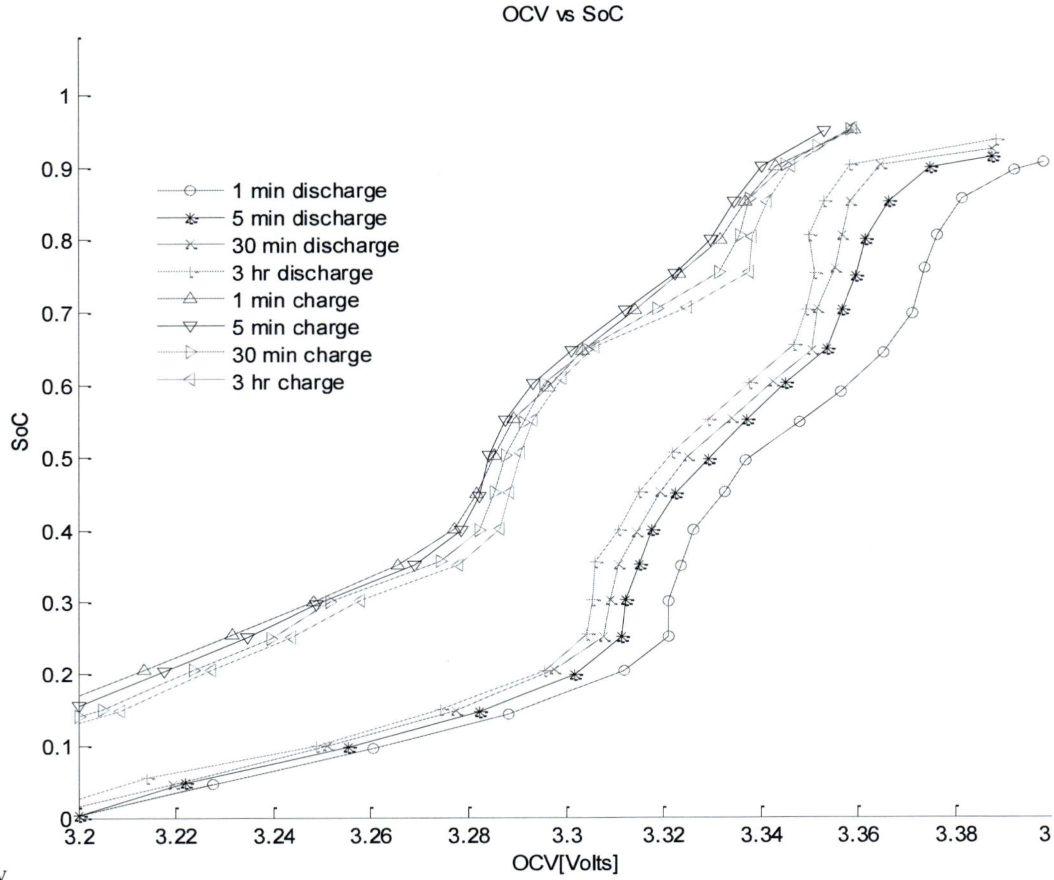

v

Figure 2. The effect of pause interval on OCV of a $LiFePO_4$ battery.

Figure 2 illustrates the effect of the pause of charge or discharge for a $LiFePO_4$ battery. One of the notable effects is that for same SoC level the charging and discharging OCV represents different level of displacement among the curves.

Relaxation of Over Voltages Equilibrium Voltage

Lithium-ion batteries show a hysteresis phenomena, which cannot be explained by the diffusion and relaxation processes alone. In the further context "hysteresis" of an OCV describes a thermodynamically steady difference between OCV after previous charge and OCV after previous discharge. At this point all the relaxation or diffusion processes caused by the states of the non-equilibrium inside the battery have decayed. It is often difficult to differentiate between a real hysteresis effect and a virtual hysteresis. A virtual hysteresis describes a difference between OCV still occurring after a rest period of a few hours, when slow and very slow solid-state diffusion processes are still ongoing [14, 15, 18].The following Figure 3 represents the terminal voltage change in terms of different SoC levels.

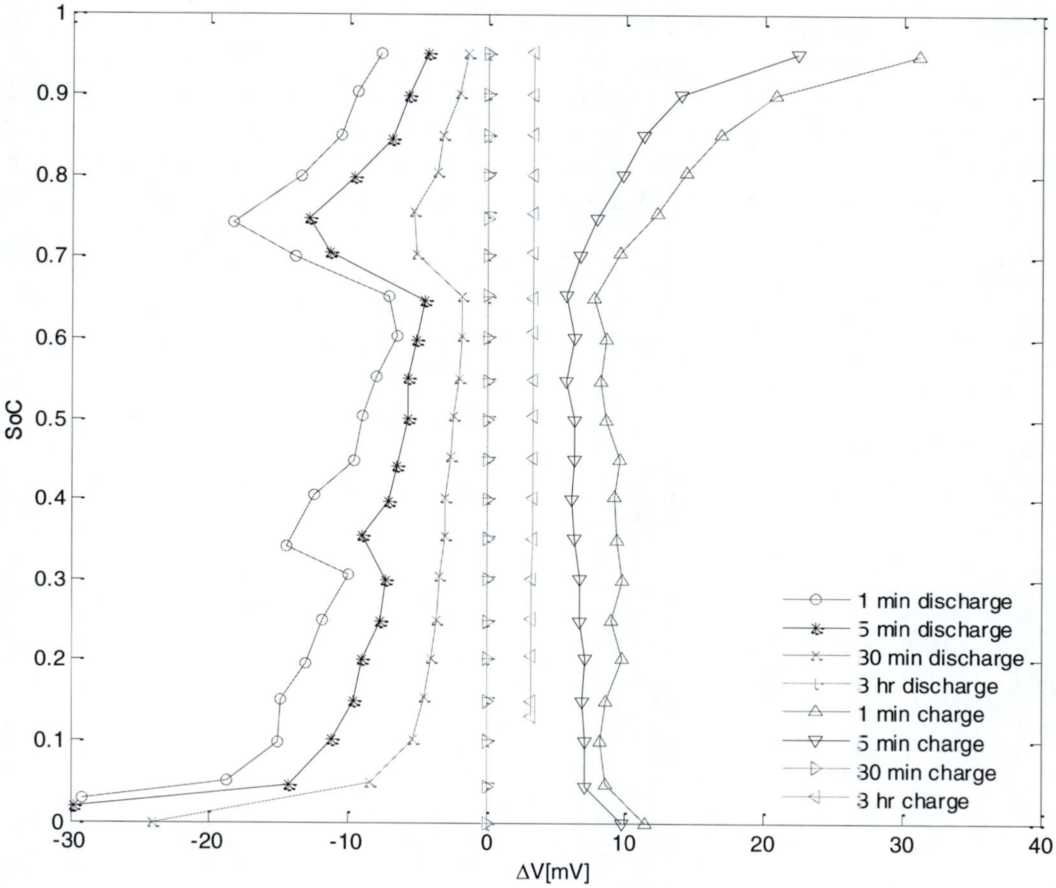

Figure 3. The effect of change of voltage in respect of different SoC levels of a $LiFePO_4$ battery. Negative voltage change appears on discharge and positive change appears on charge.

Impact of Temperature and SOC on the Hysteresis Phenomenon

It is obvious that SoC influences the voltage relaxation. At higher SoC the decay of ΔOCV appears much faster than at medium and lower SoC (Figure 3). This effect might be caused by the behavior of either cathode or anode, one of the electrodes seem to limit the equalization process for high SoC.

Hysteresis Compensation

There is no opportunity of the model to remedy the hysteresis problem by itself. So an introduction of a correction mechanism is required.

Since the battery modeling is out of the scope of the chapter, the details of model buildup is skipped. The results that are shown in the chapter uses the model as introduced in [17]. In case of swapping between charging and discharging curves make the state of charge estimation tricky as there are two different voltage level for equal SoC level. When switching between these two graphs (Charge to discharge curve switching or vice versa in Figure 4). Therefore, an abrupt jump may occur between voltages on the same SoC level that is not quite characteristic of real life battery behavior. This problem is noticeable on the portion of the hysteresis graph that they lie further apart. The OCV recovery is showed in Figure 4, plotted over SoC, where the differences between the measured OCVs after certain rest durations and the OCV curves after 3h rest, which are assumed to be almost stable [15].

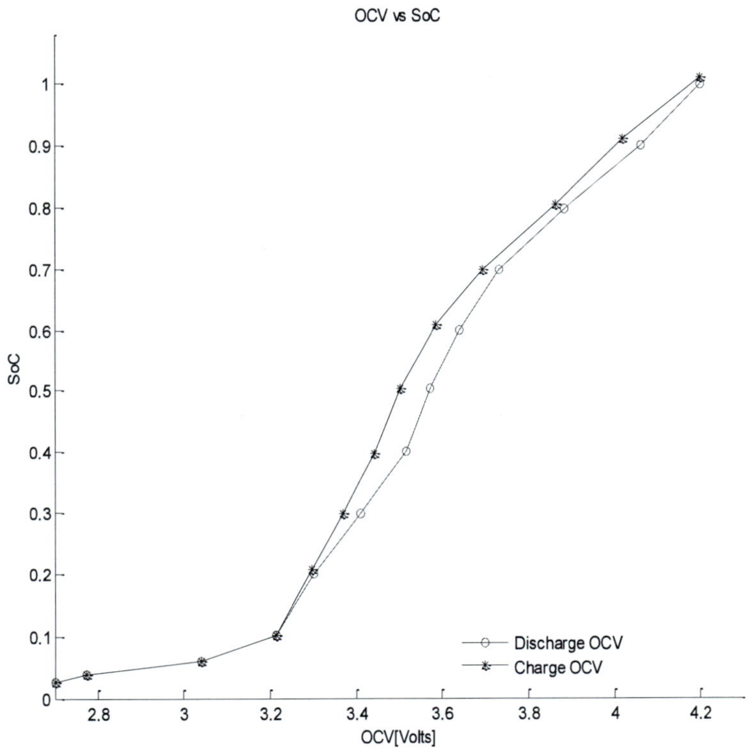

Figure 4. OCV vs. SoC graph of NMC battery.

Introduction of Control Parameter

To regulate this transition one of the probabilistic control parameter introduced is named *Confidence Level, CL. Confidence level* that signifies the probability of a battery being on charge state (state 1) or discharge state (State 0) or different hybrid states flanked by charge and discharge (0<state<1).

Charge state means in particular SoC level taking the voltage from the charge voltage graph solely. By contrast, in discharge state the voltage is taken from discharge voltage graph. In case of intermediate position between them it takes in-between positions among charge and discharge curve as determined by the confidence level parameter whereas error function regulates the transition path.

Mathematical Representation

The charge state in this chapter is determined based on the manifestation of the successive charge pulse amount. This can be controlled by the user adjusting the amount of the consecutive charge pulse through the user defined parameter *Transition time for charge state*. The discharge state can be found using the duplication of same method by user defined parameter *Transition time for discharge state. Transition time for charge or discharge state* is a chosen parameter to accommodate the smooth transition between discharge to charge curve.

This control parameter is to be reset initially as 0 in case of the first battery pulse is found discharge pulse or idle state (zero current), in case if first pulse is charge pulse than the control parameter has the initial value of 1.

$X_{\text{ctrl}}(0)$=0; or if first pulse is found to be discharge pulse idle state (0 Current)
$X_{\text{ctrl}}(0) = 1$; If first pulse is found to be charge pulse
X_{ctrl} is a variable that determines the confidence level.

After initiation the control parameters perform as a probabilistic counter. This adds or subtracts the normalization of sampling time by either the transition time for discharge or the transition time for charge.

Then this control parameter, X_{ctrl} is governed by the following equations:

$$X_{\text{ctrl}}(i) = \max(X_{\text{ctrl}}(i-1) - \frac{\textbf{Time difference between two consecutive sample}}{\textbf{transition time for discharge}}, 0) \quad (1)$$

$$X_{\text{ctrl}} = \min\left(X_{\text{ctrl}}(i-1) + \frac{\textbf{Time difference between two consecutive sample}}{\textbf{transition time for charge}}, 1\right) \quad (2)$$

Obviously introduction of this control parameter ensures the smooth transition between the curves.

This X_{ctrl} parameter works like a weighting factor ($0 \leq X_{\text{ctrl}} \leq 1$) for determining the intermediate voltage level in between the transition of two curves. *Confidence Level* is a probabilistic parameter value that take the outline of an error function depending on the control parameter X_{ctrl}. $\text{erf}(z)$ is the "error function" encountered in integrating the normal distribution (which is a normalized form of the Gaussian function) [19]. It is an entire function defined by

$$\text{erf}(z) = \frac{2}{\sqrt{\pi}} \int_0^z e^{-t^2}\, dt \quad (3)$$

Confidence Level, CL=Error function (X_{ctrl});
Erf has the values $\text{erf}(0) = 0$ and $\text{erf}(\infty) = 1$.
So the intermediate voltage level will be determined by using the following formulae

$$\text{Intermediate Voltage} = CL * V_{charge} + (1 - CL) * V_{discharge} \tag{4}$$

This voltage can be used to find the SoC level of the battery. Because of the introduction of this confidence level parameter it circumvents computationally expensive Kalman filter as proposed by [16] to offset the error instigated by hysteresis effect. This parameter also functions as a smoother as it takes care of all the abrupt variations.

RESULTS AND DISCUSSION

The model centered SoC estimation of battery system significantly is subjected to different parameters such as the model selection, parameter estimation algorithms and so forth. Without appropriate mechanism this may lead to a fluctuant, even deviating result. For this goal, the stochastic principles behind Kalman Filters are very attractive as the filter finds the optimum between the model prediction and the measured data. A statistical predictor-corrector structure thereby offers obvious advantages over other alternatives. But only terminal quantities i.e., the input current and the measured terminal voltage of the battery are measured [20]. In this chapter, the IEC 62660-1 standard [21] for BEVs profile is used with a modification to accommodate the full charging and discharging of the battery cell. The effectiveness of the modified OCV curve and the hysteresis approach are shown in Figure 5.

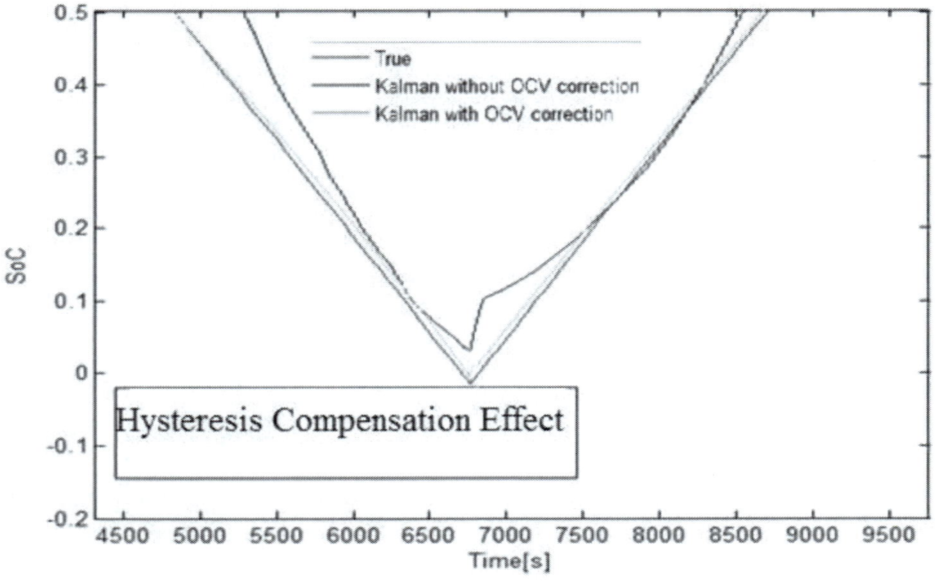

Figure 5. The figure proves the necessity of hysteresis compensation using SoC OCV curve. In case of no hysteresis presents (in blue graph) the graph may bend and jump into abrupt changes of SoC The usage of OCV correction will mitigate the problem (green curve).

The bump in SOC that would be visible if there is a change from discharge to charge, has disappeared. The output of Kalman filter is assumed to be modified OCV of the battery cell under test. The modified OCV level is depicted in Figure 4, plotted over SoC. Thereafter, the estimated bulk voltage is assigned to get the respective state of charge at different modified OCV level. Because of the introduction of this confidence level parameter it avoids computationally expensive EKF as proposed by [16] to offset the error instigated by hysteresis effect. This parameter also functions as a smoother as it takes care of all the abrupt variations. State of charge estimation is found only through the computation of the modified OCV. The voltage distribution over the time—this modified OCV voltage is used in co-ordination with hysteresis curve to determine the consequent SoC curve.

Conclusion

The hysteresis phenomena undoubtedly pose a great challenge in terms of battery modeling. The chapter elaborates different electrochemical and physical reasons behind hysteresis phenomena as well as a compensation structure. The inclusion of a hysteresis compensation scheme enhances the performances of the battery modeling. In the chapter a particular case on effect of hysteresis on battery modeling presented and explained. The consequence of introduction of hysteresis compensation scheme is that the model is now more robust, responsive and reliable for real life application.

References

[1] Khan M. R., Mulder G., Mierlo J.V., Kær, S. K., (2013),The Integration and Control of Multifunctional Stationary PV-Battery Systems in Smart Distribution Grid, *in EU PVSEC The 28th European Photovoltaic Solar Energy Conference and Exhibition Paris, France*, 3841 - 3851.

[2] Linden D., Reddy T. B., (2002), Handbook of Batteries, *Third Edition ed.: McGraw-Hill.*

[3] Sleigh A.K., Murray J.J., McKinnon W. R., (1991), Memory effects due to phase conversion and hysteresis in Li/LixMnO2 cells, *Electrochim. Acta.* 36, 1469-1474.

[4] Barker J., Pynenburg R., Koksbang R., Saidi M. Y.,(1996), An electrochemical investigation into the lithium insertion properties of LixCoO2, *Electrochim. Acta.* 41, 2481-2488.

[5] Carlier D., Saadoune I., Menetrier M., Delmas C.,(2002), Lithium electrochemical deintercalation from O2-LiCoO2 - Structure and physical properties, *J. of the Electrochem. Soc.* 149, A1310-A1320.

[6] Padhi A. K., Nanjundaswamy K. S., Goodenough J. B.,(1997), Phospho-olivines as positive- electrode materials for rechargeable lithium batteries, *J. of the Electrochem. Soc.* 144, 1188-1194.

[7] Dodd J. L., Yazami R., Fultz B.,(2006), Phase diagram of Li(x)FePO4, *Electrochem. and Solid State Letters.* 9, A151-A155.

[8] Srinivasan V., Newman J., (2004), Discharge model for the lithium iron-phosphate electrode, *J. of the Electrochem. Soc.* 151, 1517-1529.

[9] Srinivasan V., Newman J., (2006), Existence of path-dependence in the LiFePO4 electrode, *Electrochem. and Solid State Let.* 9, 110-114.

[10] Zhang Q., White R. E., (2007), Moving boundary model for the discharge of a LiCoO2 electrode, *J. of the Electrochem. Soc.* 154, 587-596.

[11] Zheng T., Liu Y. H., Fuller E. W., Tseng S.,Vonsacken U., Dahn J. R.,(1995),Lithium Insertion in High- Capacity Carbonaceous Materials, *J. of the Electrochem. Soc.* 142, 2581-2590.

[12] Zheng T., McKinnon W. R., Dahn J. R.,(1996),Hysteresis during lithium insertion in hydrogen- containing carbons, *J. of the Electrochem. Soc.* 143, 2137-2145.

[13] Yazami R., Reynier Y., (2006), Thermodynamics and crystal structure anomalies in lithium- intercalated graphite, *J. of Pow. Sou.* 153, 312-318.

[14] Gerschler J. B., Sauer D. U., (2009), Investigation of Open-Circuit-Voltage Behaviour of Lithium-Ion Batteries with Various Cathode Materials under Special Consideration of Voltage Equalisation Phenomena, *presented at the International Battery, Hybrid and Fuel Cell Electric Vehicle Symposium EVS24*, Stavanger, Norway.

[15] Roscher M. A., Sauer D. U.,(2011),Dynamic electric behavior and open-circuit-voltage modeling of LiFePO4-based lithium ion secondary batteries, *J. of Pow. Sou.* 196, 331-336.

[16] Vasebi A., Partovibakhsh M., Bathaee S. M. T.,(2007),A novel combined battery model for state-of- charge estimation in lead-acid batteries based on extended Kalman filter for hybrid electric vehicle applications, *J. of Pow. Sou.* 174, 30-40.

[17] Khan M. R., Mulder G., Mierlo J. V.,(2014),An online framework for state of charge determination of battery systems using combined system identification approach, *J. of Pow. Sou.* 246, 629-641.

[18] Ecker M., Gerschler J. B.,Vogel J.,Käbitz S.,Hust F.,Dechent P.,(2012),Development of a lifetime prediction model for lithium-ion batteries based on extended accelerated aging test data, *J. of Pow. Sou.* 215, 248-257.

[19] Acton F. S., (1990), Numerical Methods That Work., 2nd ed.: Washington, DC: *Math. AsSoC.* Amer.

[20] Plett G. L., (2004), Extended Kalman filtering for battery management systems of LiPB-based HEV battery packs - Part 1. Background, *J. of Pow. Sou.* 134, pp. 252-261.

[21] Commission I. E., (2010), IEC 62660-1 ed1.0 Secondary lithium-ion cells for the propulsion of electric road vehicles - Part 1: Performance testing, *Int. Electrotech. Comm.*

In: Hysteresis
Editor: José Carlos Dias

ISBN: 978-1-63321-336-4

Chapter 11

BIOLOGICAL HYSTERESIS ILLUSTRATED FROM DIFFERENT STUDY CASES: CONTROL FACTORS, RESPONSE PARAMETERS AND FUNCTIONAL MEANINGS

***Nabil Semmar*[1,2,*] *and Asma Hammami-Semmar*[3]**
[1]Institut Supérieur des Sciences Appliquées de Tunis (ISSBAT).
University of Tunis El Manar, Tunis, Tunisia
[2]Institut Méditerranéen de Biodiversité et d'Ecologie Marine et Continentale (IMBE),
Aix-Marseille Université, Marseilles, France
[3]Institut National des Sciences Appliquées et de Technologie (INSAT),
Tunis, Tunisia

ABSTRACT

Several natural systems are characterized by some resiliency due to the existence of three fundamental properties. The first one, called metastability, consists of the system's ability to oppose to internal variations despite those of external conditions. In this sense, metastability represents an energy-barrier favoring steady-states and protecting the system against perturbations. When the system reaches an extreme limit of metastability, it jumps to a new equilibrium state to avoid functional failure; this transition represents the second resiliency property. After jumping, the system enters into a new metastable field from which it begins a reverse (cyclic) variation (third resiliency property) until returning to its initial state which results in a hysteresis loop; and so on. To sum up, there is a dynamic stability which results in alternated trigger, imprinting, decay and annihilation of system states through a transition loop called hysteresis phenomenon. Beyond these basic functional traits, hysteresis is characterized by its rotation direction, size (loop area) and shape. These parameters provide useful information on structures, functions and evolutions of complex dynamic natural systems. Such hysteretic systems are observed at different biological scales including molecular, cellular, tissue or physiological level, organisms' behaviors, population, ecosystem level, etc. This chapter

* Corresponding Author: nabilsemmar@yahoo.fr.

presents several study cases to illustrate biological hysteresis where systems manifest varied cyclic equilibriums characterized by different trajectory shapes, sizes, locations and rotation directions.

Keywords: acid-base equilibrium, cell transport, CO_2 exchange, Diurnal cycle, drug effect, heart rate, lung volume, memory, metabolic regulation, molecular conformation, vision perception, pH, repelor

INTRODUCTION

Hysteresis is a Greek term meaning "to lag behind", and is used to designate systems where internal or dependent variable (*Y*) lags behind in its attempts to follow the changes in independent or control variable (*X*) [1]. Starting from initial system state A_0, "lag behind" process is at the origin of temporary metastable phases in which *Y* opposes to change while *X* pursues its variation (Figure 1a); this opposition to change of *Y* despite variation of *X* is called metastability. Metastability occurs when sufficiently high energy barriers prevent a spontaneous equilibrium transition between two states of system. At the limit of such resiliency, the continuous variation of control factor *X* makes that the system leaves the metastable state by jumping into a new *Y*-state (Figure 1b). From the new *Y*-state, the control variable *X* follows a reverse variation in which *Y* will be subject to a new metastability (Figure 1c). At the limit of such resiliency, the system returns to its initial state A_0 through a second abrupt non-equilibrium transition (jumping) (Figure 1d) [2-15].

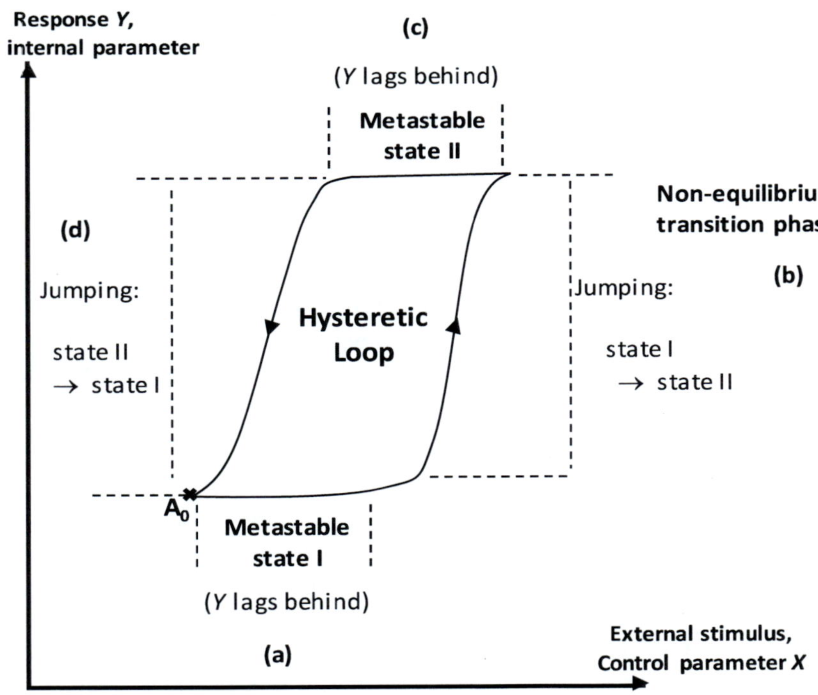

Figure 1. Illustration of hysteresis general shape with two metastable (a, c) and two non-equilibrium transition (b, d) phases.

Through the bistability-jumping loop, continuous external stimulus is converted into discrete internal response (of the system). Bistable states (I and II) and transition between them are responsible for hysteretic loop characterized by (i) a lag in evocation and (ii) perseverance in recovery. The majority of systems are rather in metastable states than in perfect ideal equilibrium.

In contrast to equilibrium curve, hysteresis implies two values for *Y* at a given value of *X*. The value of *Y* will depend on the history of tracing (states: *I*→*II* or *II*→*I*) between the two stability fields *I* and *II* (Figure 2a). This makes that the system has two different internal states *Y* in response to a given external variable *X*, and the occurrence of one state or the other in immediate future depends on the current positive or negative trend of *X*.

Thus, the outcome of a hysteresis process is related to (i) the current input value and (ii) its immediate previous state. This makes that hysteresis is a unique type of dynamics with rate-independent memory: regardless of how slow or fast the input values appear, only the previous extreme values determine the response *Y*, even if the input remains for a long time or merely flashes. Subsequently, hysteretic system remembers the history of tracing, i.e. the path by which the actual state change has been reached.

Under internal state or output response aspect, bistability in biological systems occurs at different scales (molecular, cellular, physiologic, behavioral, ecological, etc.), and concerns several response processes or internal parameters *Y*. These processes include molecular conformations, chemical transformations, conductance and opening levels of cell channels, contraction-relaxation of physiological organs, drug effects on living compartments, perceptions and behaviors in biological organisms, exchanges of matters in ecological environments, etc. [16-31].

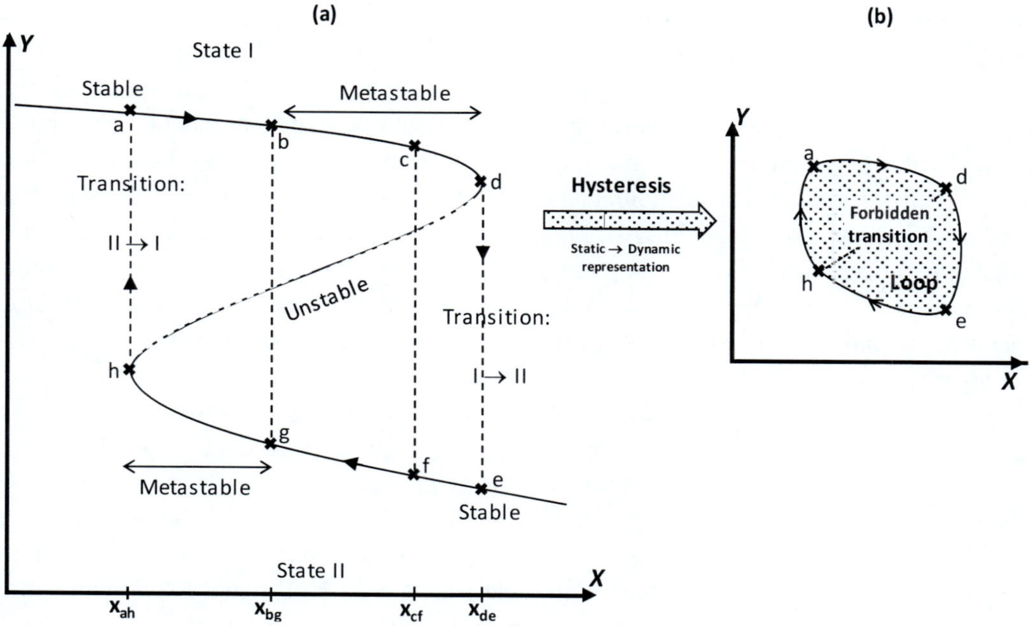

Figure 2. (a) Bifurcation diagram of a bistable system showing the existence of two *Y*-values (two system responses) associated to each level of control parameter *X*. The *Y*-values are irreversibly reached through a cyclic trajectory linking each state to its immediately previous one. (b) Hysteretic loop representing trajectory linking the set of authorized states (stable and metastable) of a system.

Under external conditions or input stimulation aspect, hysteretic loops are governed by several control factors *X* including temperature, pressure, radiation, pH, voltage, chemical concentration, angular effects, etc. The variation of control parameter *X* shows two reverse processes which maintain the dynamic stability (hysteretic loop) of the functional system. Coupled reverse processes include: heating-cooling (for temperature), inflation-deflation (for pressure), acid-base variation (for pH), activation-inactivation (for voltage), precursor-product and free-bound forms (for biochemical compounds), among many others such as magnetization-demagnetization, loading-unloading, adsorption-desorption, protraction-retraction, strain-stress, disappearance-saturation, drying-welting, etc. [7, 22, 32-37].

Geometrically, hysteretic loops are characterized by different parameters including (i) rotation direction, (ii) loop size or area and (iii) shape. Rotation direction can be of clockwise or anti-clockwise types which are fundamentally determined by the entropy increase accompanying unidirectional (irreversible) system state changes. In terms of system response, clockwise/anti-clockwise loops are associated to different regulatory processes such as acceleration/lag, attenuation/amplification, activation/inhibition, etc.. (ii) Loop size depends on input factors influencing the cycling rate. These factors can be intrinsic or extrinsic including biological population types, physiological or clinical states, experience time range, physical-chemical stimulus levels/types, etc.. (iii) Hysteretic loop shape can markedly change by the variations of external stimulus traits (e.g. amount, frequency, duration of governing factor) as well as those of system's intrinsic characteristics (e.g. healthy vs disease, non-stressed vs stressed systems, etc.).

Finally, functional properties of hysteresis revealed to be fundamental bases of several complex biological processes: for instance, uni-directionality (irreversibility) attributes to the system a memorization ability providing mechanistic bases for information recording and learning [38-39]. Bistability and switching are crucial concepts in cellular regulations. More generally, multistability is an important recurring theme in cell signaling [40].

This chapter provides illustrative presentation on different hysteretic biological systems in relation to different types of control parameters. Control parameters include temperature, pressure, voltage, radiation, pH, chemical concentration (of metabolites or drugs) and angular factors. Hysteretic responses of biological systems controlled by the above factors include: heart rate (contraction-relaxation), lung volume (inspiration-expiration), molecular protonation degree (acid-base titration), cell membrane channel state (opening-closure), environmental gas exchanges (uptake-release), effects and metabolic variations of drugs (tolerance, sensitization, cascade mechanisms), metabolite-enzyme interactions (precursor-derivative regulations), etc..

Introductive and intuitive illustration of hysteresis can be given by visual cases in which oscillatory perceptions link between two opposite image states (Figure 3a, c) [41-49]. A first well-known example of perception hysteresis was given by Haken [44] (Figure 3a): when we scan the sequence of pictures in figure 3a from left to right, the transition from a male face to a female form will appear after some a picture. Then, reversing the direction of scanning, the switching from the perception of female to male occurs through the successive points. This hysteresis phenomenon results from the co-existence of two stable "attractors" of some sort in the visual system (Figure 3b).

In a second visual example, figure 3c shows a young woman who is perceived for a "short" time to be followed by old woman perception; then pattern of young woman returns again, and so on. To explain these reversions, Köhler [49] assumed that once a pattern has been transmitted from the retina to the visual cortex, it triggers an electrochemical current which in turn enhances the resistance of the neural medium and the neurons become saturated. Subsequently to saturation, the current has to change its direction.

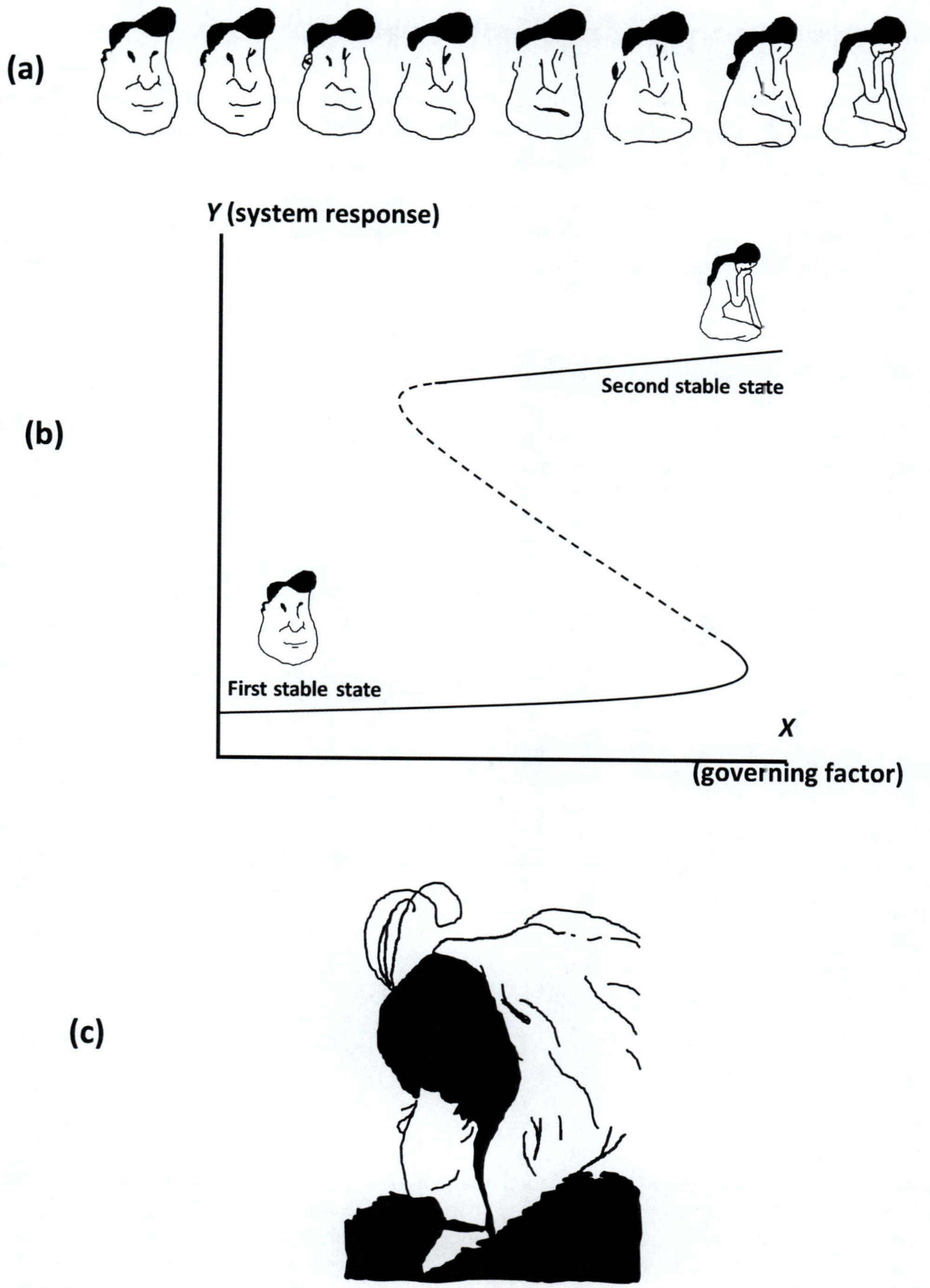

Figure 3. Two intuitive examples (a-b and c) of visual perceptions illustrating the concept of hysteresis.

Computational and Graphical Highlighting of Hysteresis

Let's consider a function *Y* describing a system in relation to a control variable *t*. System's equilibriums correspond to states where the rate of change of *Y* is null. Such states can be highlighted by means of graphical and analytical ways:

- Graphical way consists in plotting the rate of change (d*Y*/d*t*) as a function of the state variable *Y*. Then, equilibrium points will be identified from the intersection(s) of the derivative function d*Y*/d*t* with the *Y*-axis, i.e. where the rate of change is equal to 0 (Figure 4).
- Analytical way consists in finding the solution(s) for which the differential equation(s) of the system is (are) null.

Graphical Analysis of Equilibrium Points

The signs of the rate of change around a null derivative point determine whether corresponding equilibrium is stable or not (Figure 4).

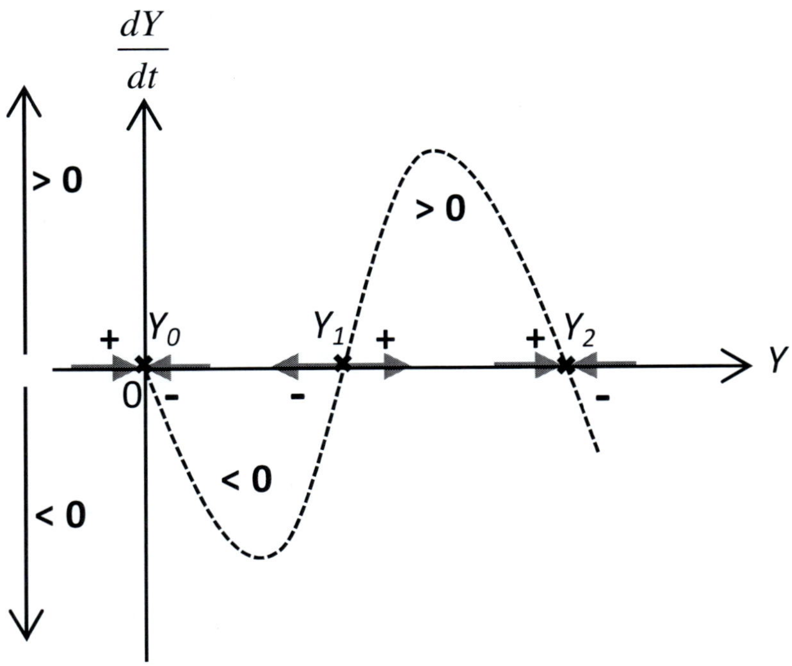

Figure 4. Illustration of graphical determination of equilibrium states as intersection points between state variable *Y* and its derivative d*Y*/d*t*. From the curve d*Y*/d*t*, positive (>0; +) and negative (<0; -) derivative areas can be defined. Therefore, points in positive area will be associated with right arrows, and those in negative area will be associated with left arrows. It results that the points at the intersection with *Y*-axis (equilibrium points) will be identified as attractor or repeller according to the convergence or divergence of their two associated arrows, respectively. Attractors (Y_0 and Y_2) and repeller (Y_1) correspond to stable and unstable equilibrium points, respectively.

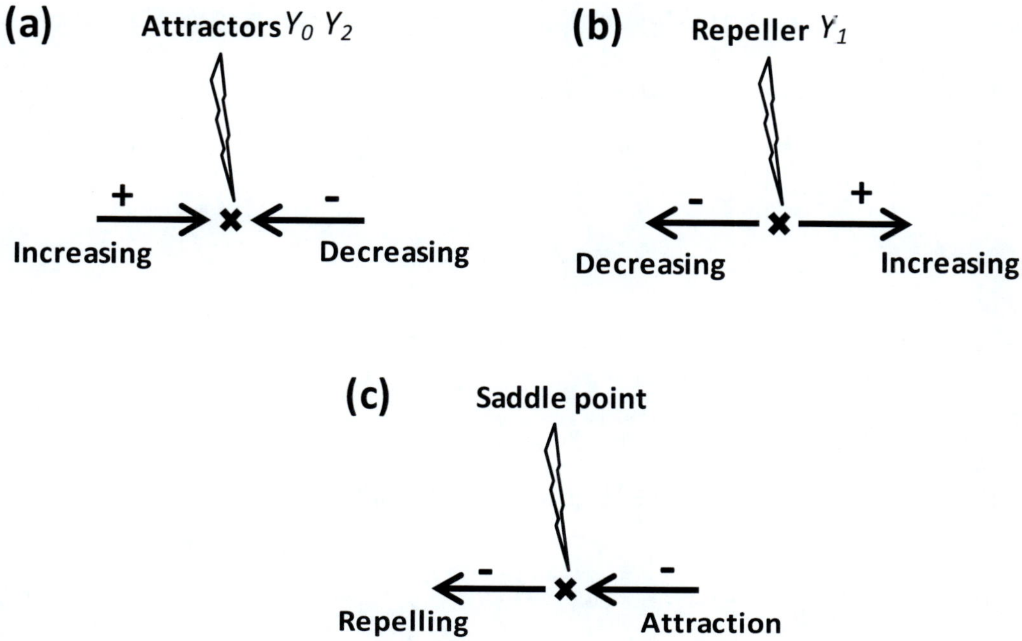

Figure 5. Different types of equilibrium points depending on the signs of derivative function (rate of change sign) at left and right sides.

In the illustrative plot of figure 4, there are three equilibrium points where the rate of change (dY/dt) is equal to 0: Y_0, Y_1, Y_2. Positive and negative rates of changes are considered around each equilibrium point Y_i to identify whether this point (Y_i) behaves as attractor or repeller:

Attractor points have negative rates of change (decreasing Y) at the right side and positive rates of change (increasing Y) at the left side (Figure 5a). Consequently, these points correspond to stable equilibrium (Y_0, Y_2 in figure 4). Repeller points have opposite properties since the solution will move away from the point leading to an unstable equilibrium (Y_1) (Figure 5b).

Apart from attractor and repeller points, there is a third type of equilibrium point called saddle point showing the properties of attractor on one side and repeller on the other side (Figure 5c). A summary of the three types of equilibrium points is presented in the following table:

Signs of the rates of change		Left		
		+	-	
Right	+	Saddle	Repeller	Equilibrium points
	-	Attractor	Saddle	

Analysis of System's Equilibrium Points by Solving Differential Equation

Let's consider the following derivative function:

$$\frac{dY}{dt} = Y\left[r\left(1-\frac{Y}{K}\right) - q\cdot\frac{1}{Y+k}\right] \tag{1}$$

where:

Y is the state variable defined in the variation field [0, +∞[.

r, q, K and k are control parameters.

$\frac{dY}{dt}$ is the rate of change of the system.

By setting the rate of change dY/dt equal to 0, we can write:

$$Y\left[r\left(1-\frac{Y}{K}\right) - q\cdot\frac{1}{Y+k}\right] = 0 \tag{2}$$

The system reaches steady state (equilibrium) at Y-points which satisfy (2). This equation is satisfied for:

$$Y = 0 \tag{a}$$

or

$$r\left(1-\frac{Y}{K}\right) - q\cdot\frac{1}{Y+k} = 0 \tag{b}$$

The later case (b) can be rearranged to give (3):

$$\begin{aligned}
r\left(1-\frac{Y}{K}\right) - q\cdot\frac{1}{Y+k} = 0 \quad &\Leftrightarrow \quad r\left(1-\frac{Y}{K}\right) = q\cdot\frac{1}{Y+k} \\
&\Leftrightarrow \quad r(K-Y) = \frac{qK}{Y+k} \\
&\Leftrightarrow \quad r(K-Y)(Y+k) = qK \\
&\Leftrightarrow \quad r(K-Y)(Y+k) - qK = 0 \\
&\Leftrightarrow \quad (K-Y)(Y+k) - \frac{qK}{r} = 0 \\
&\Leftrightarrow \quad KY + Kk - Y^2 - kY - \frac{Kq}{r} = 0
\end{aligned}$$

$$\Leftrightarrow \quad Y^2 + Y(k-K) + \frac{Kq}{r} - Kk = 0 \tag{3}$$

Solutions of quadratic equation 3 are given by calculating the discriminant Δ:

$$\Delta = (k-K)^2 - 4\left(\frac{Kq}{r} - Kk\right)$$

$$\Delta = \frac{r(k-K)^2}{r} - 4\left(\frac{Kq}{r} - \frac{rKk}{r}\right)$$

$$\Delta = \frac{r(k-K)^2 - 4(Kq - rKk)}{r}$$

$$\Delta = \frac{r(k^2 - 2kK + K^2) - 4Kq + 4rKk}{r}$$

$$\Delta = \frac{rk^2 - 2rkK + rK^2 - 4Kq + 4rKk}{r}$$

$$\Delta = \frac{rk^2 + 2rkK + rK^2 - 4Kq}{r}$$

$$\Delta = \frac{r(k+K)^2 - 4Kq}{r}$$

$$\Delta = (k+K)^2 - \frac{4Kq}{r}$$

The existence and the type of solution of the derivative function depend both on the Δ-sign and value. To explicitly illustrate this, let's set the parameters q, k and K to be equal to 0.1, 1 and 10, respectively, by varying the r-value:

- If $\Delta < 0$, (3) has not real solution, and the studied system has only one global equilibrium state at the point $Y^* = 0$ (solution (a) of (2)) (Figure 6a).
- If $\Delta = 0$, there are two equilibrium points at $Y^* = 0$ and at the point:

$$Y^* = \frac{-(k-K) + \sqrt{\Delta}}{2} = \frac{-(k-K) - \sqrt{\Delta}}{2} = \frac{-(k-K)}{2} = 0.5(K-k) = 4.5$$

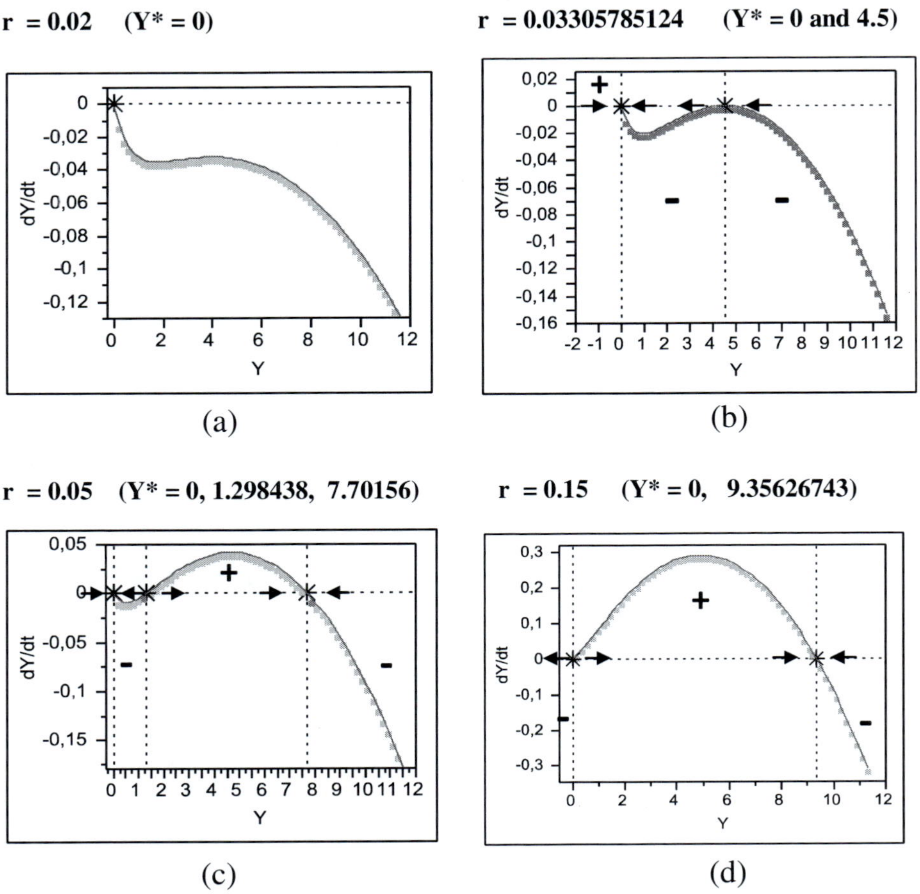

Figure 6. Variation of derivative function d*Y*/d*t* in relation to control parameter *r* by fixing the three other ones to $q = 0.1$, $k = 1$ and $K = 10$. (a), (b) unique solution cases Y^* for discriminant $\Delta < 0$ and $\Delta = 0$, respectively; (c), (d) solution cases for $\Delta > 0$ varying with *r*.

Note that for fixed values of *q*, *k*, *K* (0.1, 1, 10), $\Delta = 0$ implies that $r = 0.03305785124$ (Figure 6b).

The first equilibrium point $Y^*=0$ corresponds to a globally stable state whereas the second one $Y^*=0.5(K-k)$ corresponds to a saddle point where the equilibrium is not stable (Figure 6b). As shown in figure 6b, the saddle point $Y^* = 0.5(K-k)$ is due to the fact that:

(i) *Y* converges to Y^* when the initial conditions are higher than Y^*.
(ii) *Y* diverges from Y^* when starting from values lower than Y^*.

- If $\Delta > 0$, the quadratic equation 3 has two solutions given by:

$$Y^* = \frac{-(k-K)+\sqrt{\Delta}}{2} \tag{4}$$

and

$$Y^* = \frac{-(k-K)-\sqrt{\Delta}}{2} \tag{5}$$

These two cases can be distinguished in relation to the parameter r which directly influences Δ (Figure 6c, d):

(i) If $\sqrt{\Delta} < (K-k)$ (4, 5) (i.e. for $r < 0.1$), we obtain three equilibrium points including one unstable and two stable states (Figure 6c).

(ii) If $\sqrt{\Delta} > (K-k)$ (4, 5) (i.e. for $r > 0.1$), two relevant equilibriums are obtained including one globally stable and another unstable (Figure 6d).

Highlighting of Hysteresis through Bifurcation Diagram

The two previous sections showed that changes in the values of a parameter (e.g. r) can result in drastic changes in the number of equilibrium points and their stability properties. It has been shown that different equilibrium values of Y (i.e. Y^*) are associated to different values of control parameter r. More gradual variations of equilibrium point Y^* are illustrated in figure 7 by considering more values of r.

By considering the continuous variations of Y^* vs r, we obtain a bifurcation diagram showing the qualitative change of system behavior (Figure 8): by fixing the parameters q (=0.1), k (=1) and K (=10), and by continuously varying r from 0.01 to 0.15, the set of all the solutions (Y^*) are given by $Y^*=0$ and equations (4, 5). After calculation of Y^* for different values of r, the bifurcation diagram Y^* vs r can be visualized to analyze system transitions through its different stability states (Figure 8). The plot shows four successive phases through which the equilibrium states Y^* vary in relation to the control parameter r:

Starting from low values of r (e.g. 0.01), the system shows a stable equilibrium at the value $Y^*=0$ (point a in Figure 8a, b). Despite the increase of r, the system opposes or resists to variations by maintaining stability at the level $Y^* = 0$ (Figure 8b, step I). The resistance phase where Y shows no variation or small variation in relation to r leads to metastable state of the system. Beyond a certain value of r, the system is subjected to a variation which constraints it to leave the point (r, $Y^*= 0$) (Figure 8: point b). Therefore, the system acquires a new stable equilibrium by jumping along the Y-axis from point b to point c (Figure 8b: step II).

From point c, the system can enter into a new stable equilibrium despite the decreasing values of control parameter r (Figure 8a): therefore, the system shows some resilience around a high value of Y^* leading to a second metastability state (Figure 8b: step III). The metastable state will be left at a critical low r-value (saddle point d) from which the system abruptly returns to the initial equilibrium point $Y^*=0$ (for a new stability cycle) (Figure 8b: step IV); and so on.

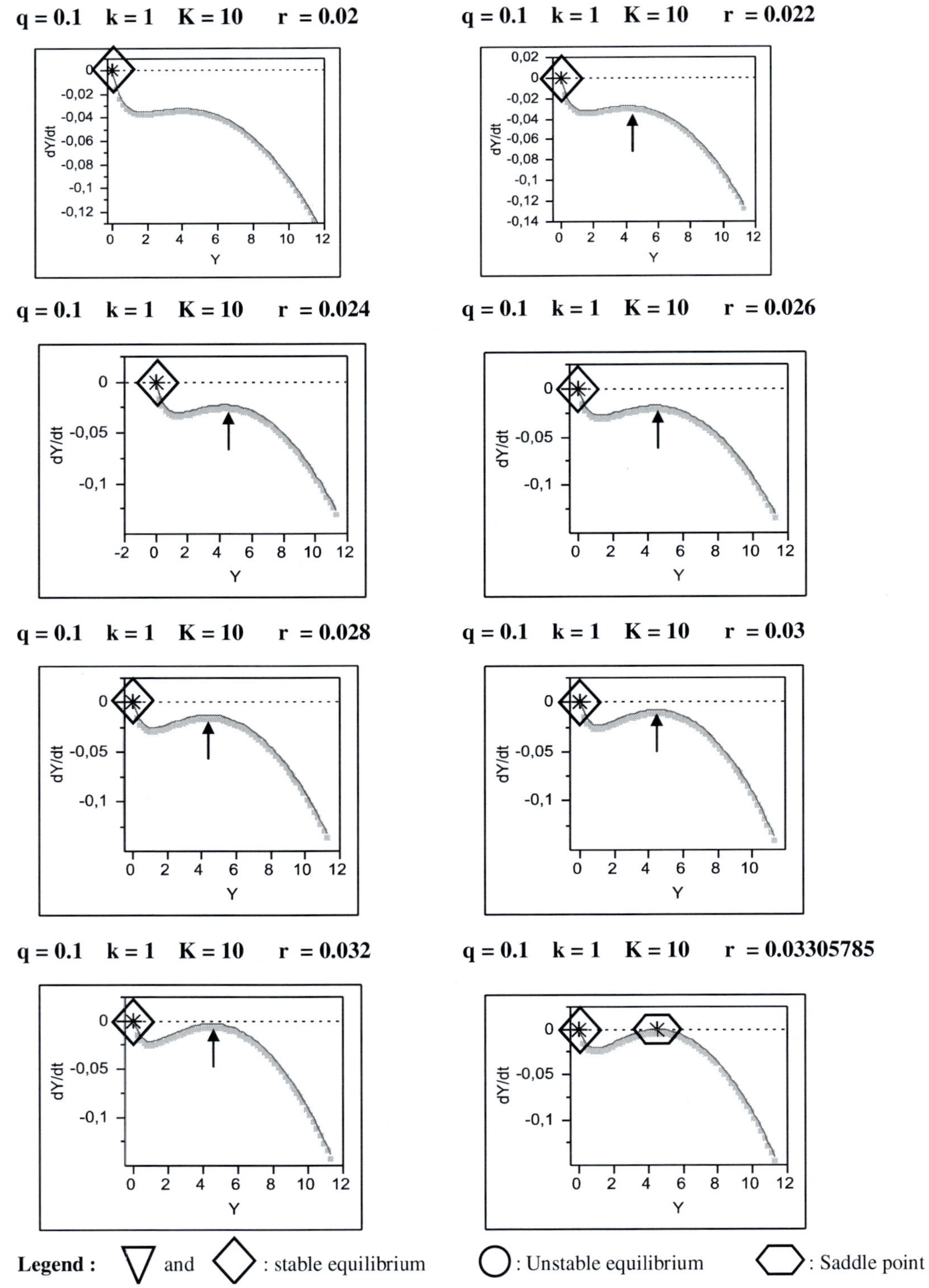

Figure 7. (Continued).

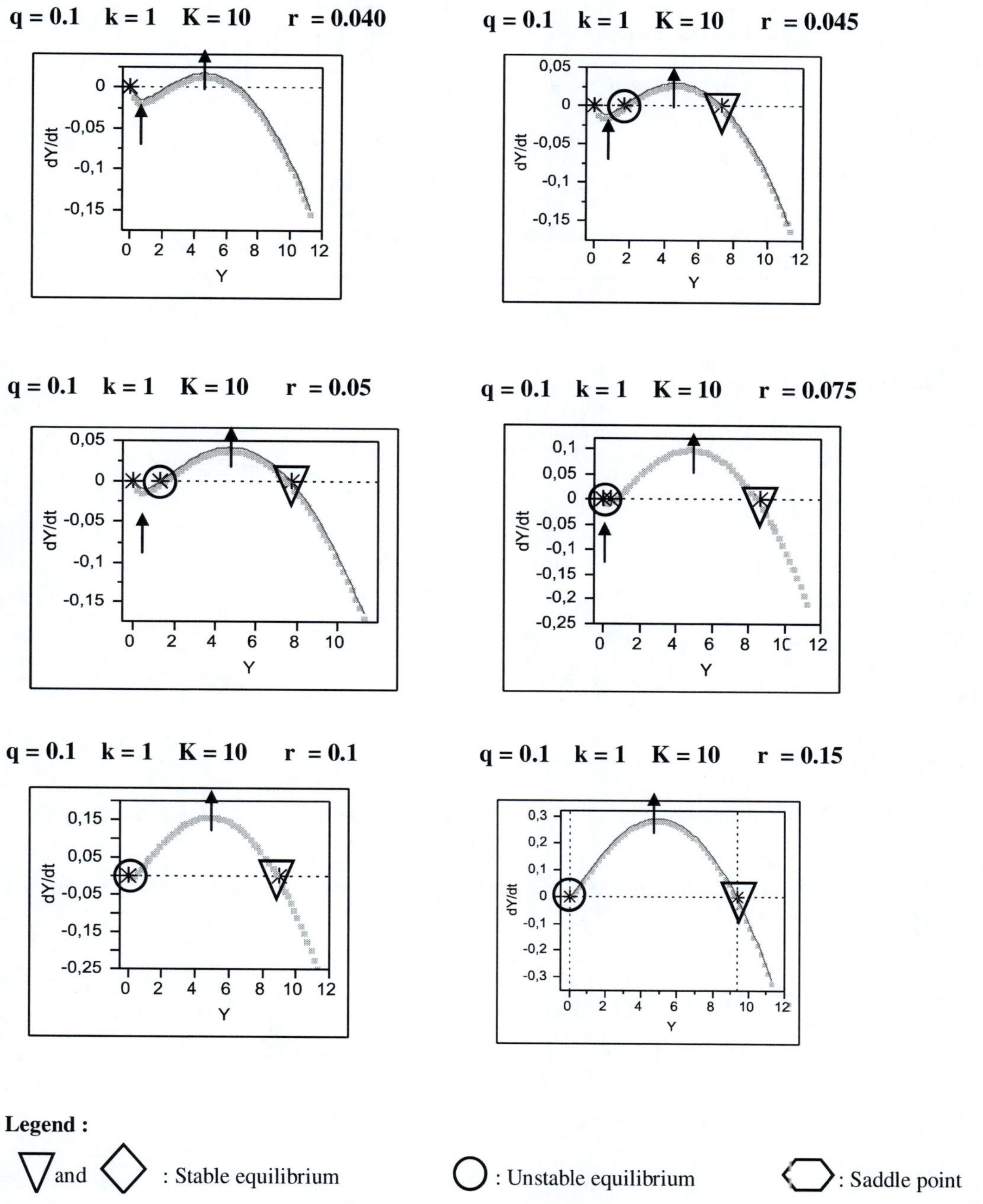

Figure 7. Illustration of gradual variation of equilibrium states (Y^*) of a differential system in relation to control parameter r.

This multi-step cyclic pathway governing the stability states of a dynamical system is called hysteresis. In summary, a hysteretic system shows two stable states, and it transits from the first state to the second one according to a path that is different from the next one which leads to the initial state from the second.

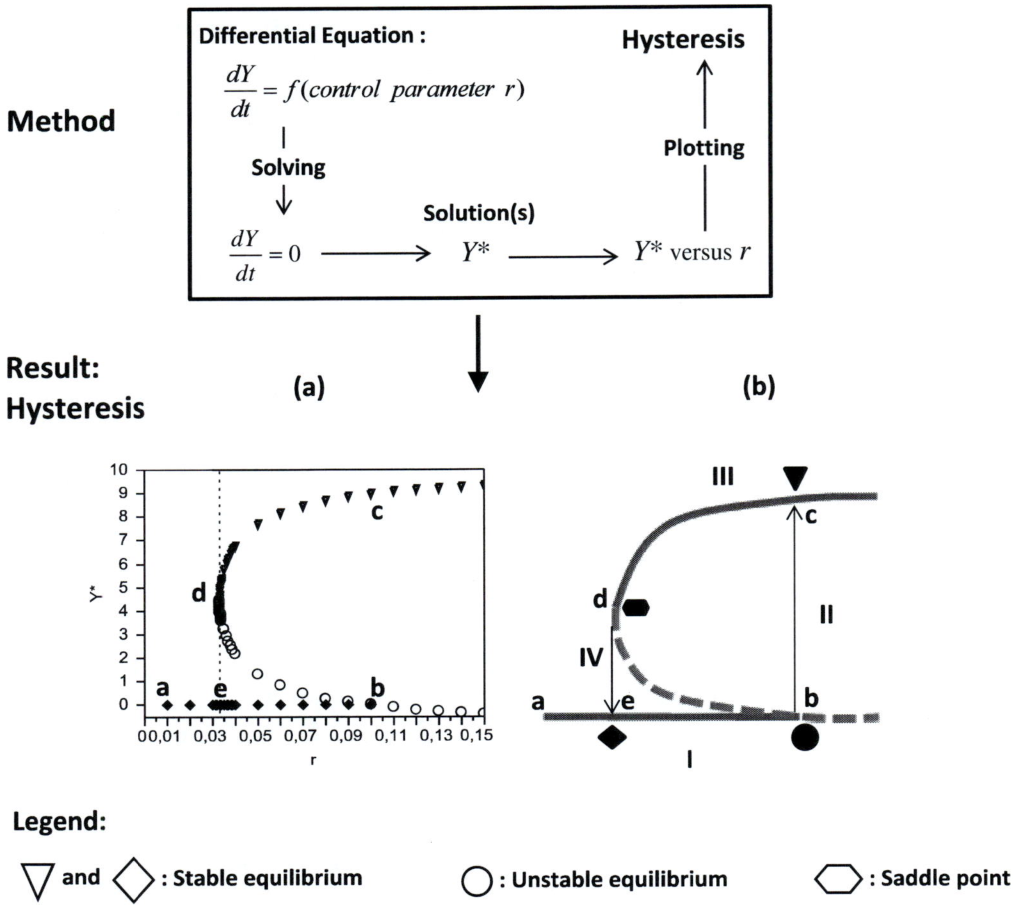

Figure 8. Bifurcation diagram showing loop trajectory governing the stability of a system analyzed by differential equation (1).

BIOLOGICAL HYSTERESIS DEPENDING ON DIFFERENT CONTROL PARAMETERS

Temperature Dependent Hysteresis: The Heart Rate Case

Biological hysteresis depending on temperature will be illustrated by study cases on heart rate (HR) in reptilians and crustaceans [50, 51]. Heating-cooling experiences in these animals were revealed to be followed by differential heart rates which can be graphically highlighted by a hysteresis between HR and body temperature (BT) (Figures 9, 10).

Heart rate Hysteresis in Reptiles

Variations of HR result in differences at the level of peripheral blood flow enabling reptiles to control heat transfer between their body core and their environment [52, 53]. Several studies on reptilian species showed that HR during heating is significantly faster than the one occurring during cooling [50, 52, 54, 55]. This regulatory process was shown to be at the origin of hysteresis in lizards and crocodiles.

HR hysteresis is illustrated here by a study on HR variation in the estuarine crocodile *Crocodylus porosus* in response to irradiation heating over 60 minutes [50] (Figure 9):

Heating was performed by irradiation from a heat lamp under dry conditions. The heat lamp was switched on until BT reached 32-33°C, then switched off (Figure 9a). At the beginning of heating, HR increased sharply (Figure 9b), then it continued to increase until the removal of the heat source (figure 9b: rhombus points). When heat source was switched off, HR decreased instantaneously and sharply (figure 9b: asterisk points). The plot of HR vs BT highlighted a hysteresis consisting of a clockwise loop where HR was significantly faster during heating phase (phase B) than during cooling one (phase D) (Figure 9c).

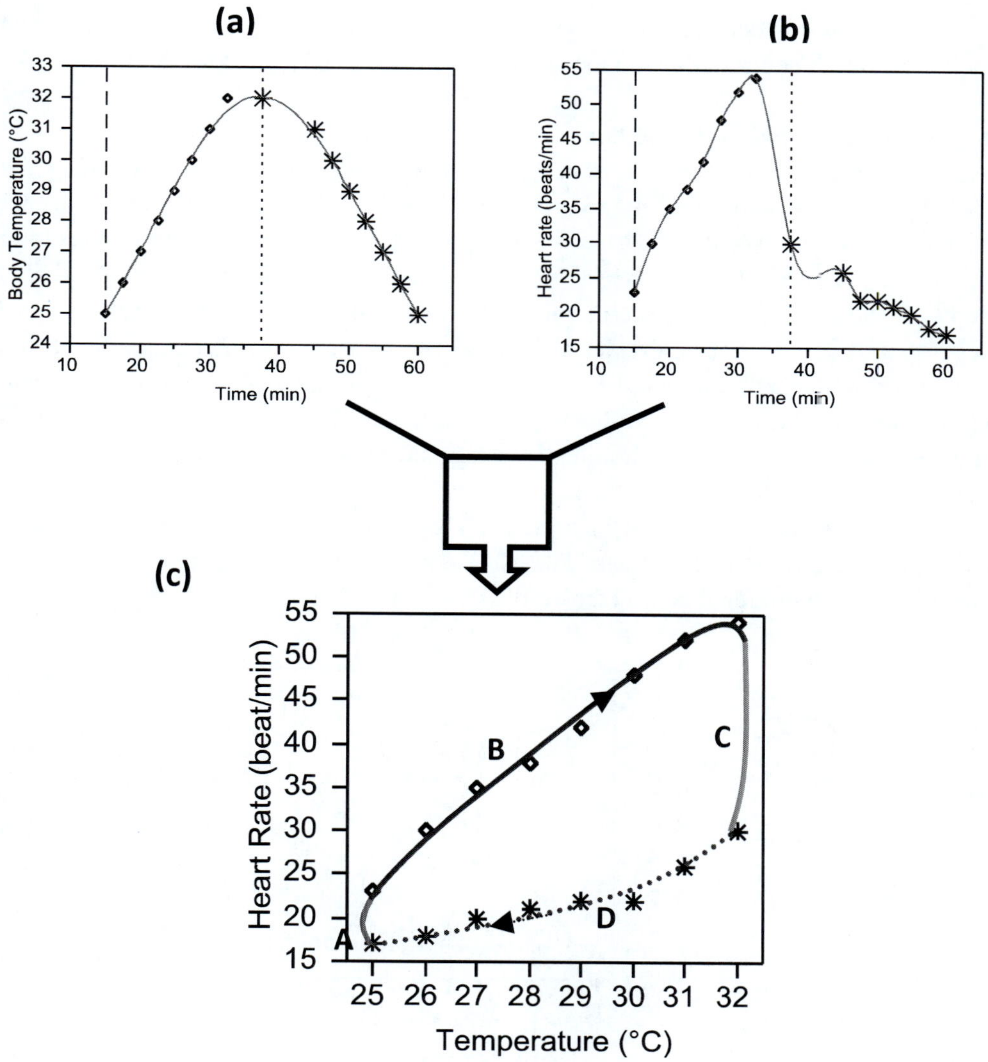

Figure 9. Heart rate hysteresis highlighted in the crocodile *Crocodylus porosus* exposed to body heating up to 32°C followed by cooling phase. Variation of body temperature (a) and heart rate (b) with heating-cooling time. (c) Clockwise hysteresis resulting from the variation of heart rate relatively to that of body temperature.

During the initial phases, after heat was applied or removed, HR changed dramatically despite the fact that BT remained stable (Figure 9c: phases A, C). During these two rapid response phases, HR is controlled independently of BT [50]. A similar but less pronounced hysteretic pattern has been reported for the lizard *Pogona barbata* and was partly explained by the action of the cholinergic and β-adrenergic nervous systems [56]. Moreover, HR hysteresis may indicate the presence of a heat-sensitive control mechanism that triggers a HR response before BT changes. Prostaglandins are possible mechanism that may control cardiac response during heating and cooling via the baroflex, by contraction or dilatation of capillary beds, and/or by directly stimulating the heart [57]. Baroflex is likely to play a role in modulating heart rate, particularly in crocodiles [58].

The thermoregulatory benefits of HR hysteresis are that it enables a reptile to maximize the time during which BT remains within its preferred range by optimizing heat exchanges with external environment [59]. This hysteretic behavior shows high thermal sensitivity of physiological rate functions allowing the crocodile to be active for a longer time.

Heart Rate Hysteresis in Crustacean

Second illustration of heart rate hysteresis is given by a study on the crustacean *Cherax destructor* which was subjected to successive heating-cooling phases [51]: eight crustacean individuals were initially subjected to increasing water (environment) temperature resulting in variations in their body temperatures (BT) from 20 to 30°C. After BT reached 30°C, the heating was stopped leading to cooling phase (BT: 30→20°C). During these two successive phases (observed in the eight individuals), several HR measures were obtained for different BT values. From the eight measured HR values for each BT (in heating and cooling phases), mean standardized heart rates (MSHR) were calculated. Then, a plot of MSHR vs BT highlighted a clockwise loop which indicated a regulatory shifting of heart rate in relation to BT variation (Figure 10):

The hysteresis showed that HR increased to a peak in response to the heat provided by flowing hot water (Figure 10: phase B). The removal of heat source was followed by instantaneous and sharp regulatory decrease of HR (Figure 10: phase C).

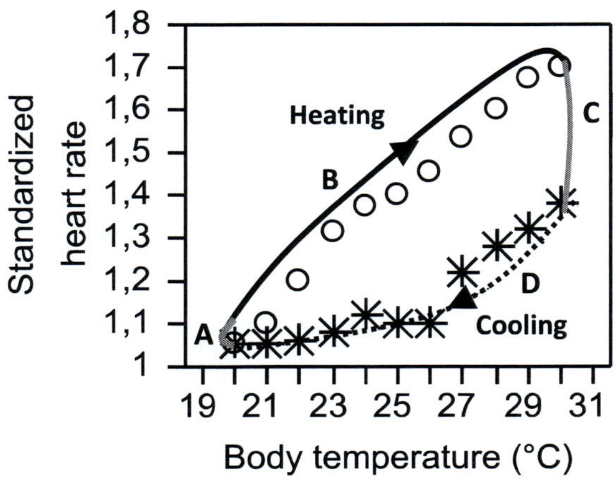

Figure 10. Heart rate hysteresis highlighted in the crustacean *Cherax destructor* exposed to body heating until 30°C followed by cooling phase.

Comparison between the two phases showed that HR was markedly faster during heating than cooling phase at any BT value (Figure 10). During the cooling phase following the sharp decline, HR steadied at a rate slightly elevated from the pre-determined resting HR (measured prior to treatment) (Figure 10D). This HR hysteretic pattern is compatible with the ability of crustacean to rapidly acclimate to changes in temperature [60, 61]; it has been explained by the presence of thermoreceptors or thermosensitive neurons which may mediate the cardiac response [51]. However in reptiles, HR hysteresis was reported to be associated with cholinergic and β-adrenergic receptors of the autonomic nervous system [50].

Pressure Dependent Hysteresis: The Lung Volume Case

General Presentation of the Hysteresis Pattern

Biological hysteresis depending on pressure will be illustrated by the lung which is an elastic body dissipating energy through a hysteresis process. Lung hysteresis is an anti-clockwise loop highlighted by plotting the lung volume vs lung pressure. During inflation (inspiration), the lung volume increases by the opening of distal air units whereas in deflation (expiration), it decreases due to decruitment (downsizing). Energy dissipation mainly occurs in the alveolar recruitment process in which forces of molecular adhesion (such as surface tension) are at work [62]. In hysteresis, the energy applied to the lung in inflation (inspiration) is not recovered in deflation (expiration).

Different lung hysteresis phenomena (Figure 11) have been linked to several mechanisms including [63-65]:

- True tissue hysteresis associated to plastic behavior of tissue. Elastic and collagen fibers are grouped in the interstices forming a network that endows the lung with significant properties relating to stiffness, camping distortion and elasticity [66, 67].
- Surface hysteresis associated to air-liquid surface forces in alveoli [68]. Surface tension can act either directly, by providing an additional contractile force, or indirectly by distorting the alveolar geometry and by increasing the force in the tissue elements (smaller bubbles) [69].
- Differences in the sequence of recruitment and decruitment of lung units between inflation and deflation.
- Considerable trans-mural pressure needed in inflation phase for opening peripheral airways that were closed during deflation.

The different parts of pressure-volume hysteresis have been associated with different lung processes consisting of the opening and closing of the peripheral airspaces that take place during the respiratory cycle [64, 70-72]. Moreover, the loop shows several characteristic points and phases which have different physiological meanings (Figure 11): increasing and decreasing curves of lung hysteresis correspond to inflation and deflation phases, respectively. These two distinct limbs make that for a given lung pressure, the lung volume is higher during deflation than during inflation.

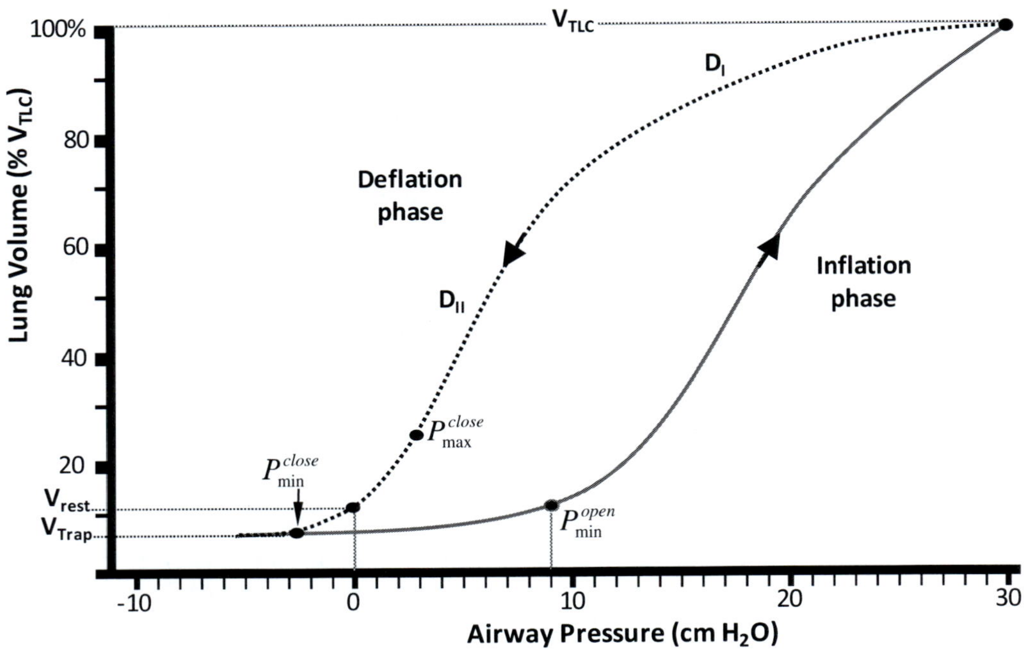

Figure 11. General illustration of lung hysteresis (volume vs pressure) with different characteristic points and phases.

In the initial phase of inflation, the lung size increases because the number of peripheral airspace (PAS) increases linearly. Inflation begins from a minimal volume of air which was previously trapped by the closure of non-cartilaginous airways at minimal pressure corresponding to the closure of all pulmonary pathways [17, 73, 74]. This minimal gas volume is referred to as trapped-gas volume V_{Trap} (Figure 11).

From V_{Trap}, the initial ascending part of hysteresis results in very little change in volume until a critical inflection point pressure $P_{\min}^{open}$ is reached (Figure 11). This pressure was interpreted as the minimum airway opening pressure [74]. The point $P_{\min}^{open}$ shows that lung pressure necessary to reopen a closed airway would be expected to be greater than required to close it ($P_{\min}^{close}$); this can be explained by differences in surface forces under the two circumstances [75].

Beyond the "knee" point ($P_{\min}^{open}$), the slope of inflation curve increases exponentially indicating the opening of previously closed PAS [64, 76]. In this transition phase of hysteresis, tissue forces predominate [77-79]. The exponential increasing of lung pressure-volume curve during inflation leads to a high end point corresponding to the total lung capacity (TLC) (V_{TLC}, Figure 11).

From the TLC point, deflation curve occurs according to two descending phases which are less and more rapid, respectively [62] (Figure 11): in the first descending phase (D_I), the lung volume decreases without PAS closing. In the second phase (D_{II}), the lung retracts when the PAS closes; this second descending phase manifests as a decreasing exponential function of the lung volume in relation to lung pressure (Figure 11). In the rapid decreasing phase (D_{II}), the limb of hysteresis deviates to the left leading to inflection point $P_{\max}^{close}$ which corresponds to the pressure at which the airway closure begins [80]. Peripheral airways are closed with a liquid plug as a result of the surface tension forces operating on the bronchioles liquid lining.

From $P_{\max}^{close}$ point, airway closure continues until minimal volume (V_{Trap}) which is associated to pressure $P_{\min}^{close}$. A difference between $P_{\max}^{close}$ and $P_{\min}^{close}$ was reported to be an index of inhomogeneity within the lung [64, 74]. The minimal lung pressure $P_{\min}^{close}$ has been found to reach negative values indicating some resistance of the peripheral airways to collapse at low volumes [75]. Different closing pressures $P_{\min}^{close}$ revealed to be intrinsic characteristics of different lung compartments:

They are less negative for airways than for alveoli. Experiments on lung resistance to atelectasis (entirely degassed lung state) in rabbits, cats and dogs showed that $P_{\min}^{close}$ are in the order of −1 to −2 cm H_2O for the airways, and −3 to −6 cm H_2O for alveoli [81]. Subsequently, during lung deflation, V_{Trap} is usually reached before onset of atelectasis. This attribute to V_{Trap} a high functional value in respiratory system because V_{Trap} makes that smaller pressures are required to reopen closed airways compared to the pressures needed for the reopening of closed alveoli. Apart from V_{Trap}, resting lung volume (V_{rest}) is defined as the volume at zero pressure (Figure 11). The V_{rest} can be influenced by several factors including animal species, variability of pulmonary blood content, previous volume history of lungs, etc. [74].

In addition to V_{Trap} and V_{rest}, recycling lung volume is defined as the volume characterizing the end or the beginning of a given respiratory cycle (ordinary, short or long cycle). It has significant effect on next hysteretic pattern. This point will be illustrated in the two next sections through high variability of loop's area, shape and position governed by intrinsic and extrinsic lung conditions.

Effects of Recycling Pressure Conditions on Hysteresis Pattern

Inflation volume represents a key memorial parameter in lung hysteresis. In experiences on isolated dog's lung [64], hysteresis was obtained by initiating inflation phase from different deflation-end points corresponding to different recycling pressures and volumes (Figure 12). These inflation starting points varied between 2.5 and 20cm H_2O and corresponded to pressure values lower or higher than maximum airway closing pressure $P_{\max}^{close}$.

Loops occurring between inflation end pressure (30cm H_2O) and end-deflationary pressure > $P_{\max}^{close}$ were inlaid because of a common deflation limb and fairly separate straight inflation limbs (Figure 12). However, as soon as the end-deflationary pressure falls below $P_{\max}^{close}$, the volume-pressure characteristics are altered leading to a right shifted hysteresis. Such hysteresis exhibits a distinct "knee" from which there is a rapid reopening of the breathing units so far closed during deflation (Figures 11, 12). Below the "knee", the initial inflation curve should result from (i) inflation of the breathing units still open at the end of the deflation (Figure 12: point V_{04}) which are (ii) possibly coupled with the reopening of a relatively small number of closed units.

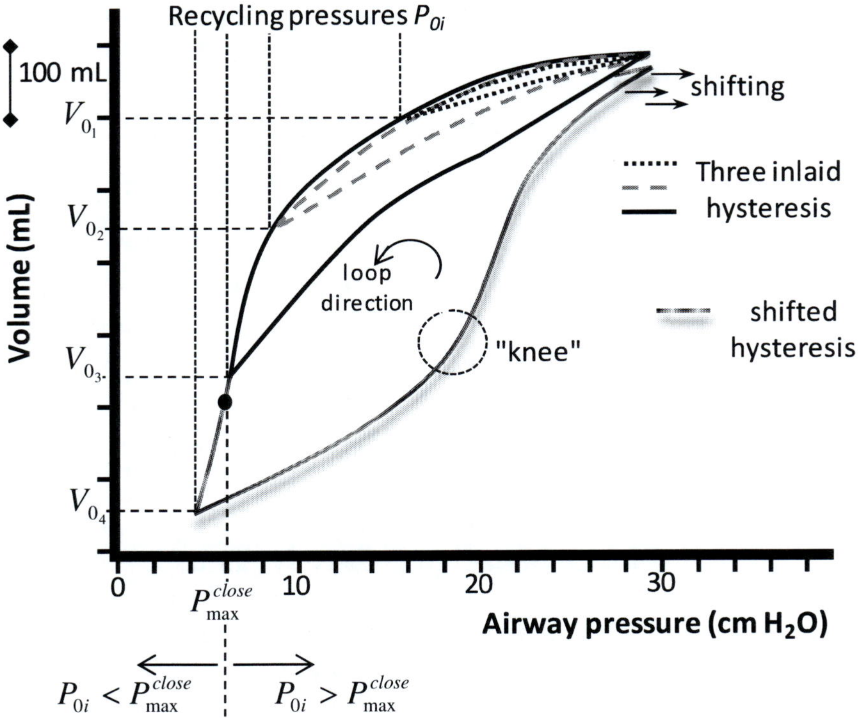

Figure 12. Different hysteresis loops observed in isolated dog's lung subjected to different recycling pressure P_{0i} and volumes V_{0i} (V_{01}, V_{02}, V_{03}, V_{04}). The recycling pressures were either lower or higher than $P_{\max}^{close}$ (pressure at which the airway closure begins).

Effects of Respiratory Environment on Lung Hysteresis Pattern

Different respiratory matrices revealed to be responsible for important variability of shape, area and location of lung hysteresis (Figure 13). This can be illustrated from four experience types:

- Normally air inflated lung (natural physiologic situation)
- Completely degassed lung (absence of air in alveoli) (simulating asphyxia conditions)
- Lung subjected to small volume cycles (simulating respiration difficulties)
- Saline filled lung (e.g. simulating drowning)

Completely degassed lung is clinically known by atelectasis. Analysis of hysteresis corresponding to atelectasis (Figure 13b: curve 1) showed that atelectatic lungs need a high transpulmonary pressure for inflation beginning. Pressure and volume measurements in excised rabbit's lung under atelectatic state showed that opening of breathing units begins at critical pressures in the order of 30cm H_2O (Figure 13a, b: curve 1) [63].

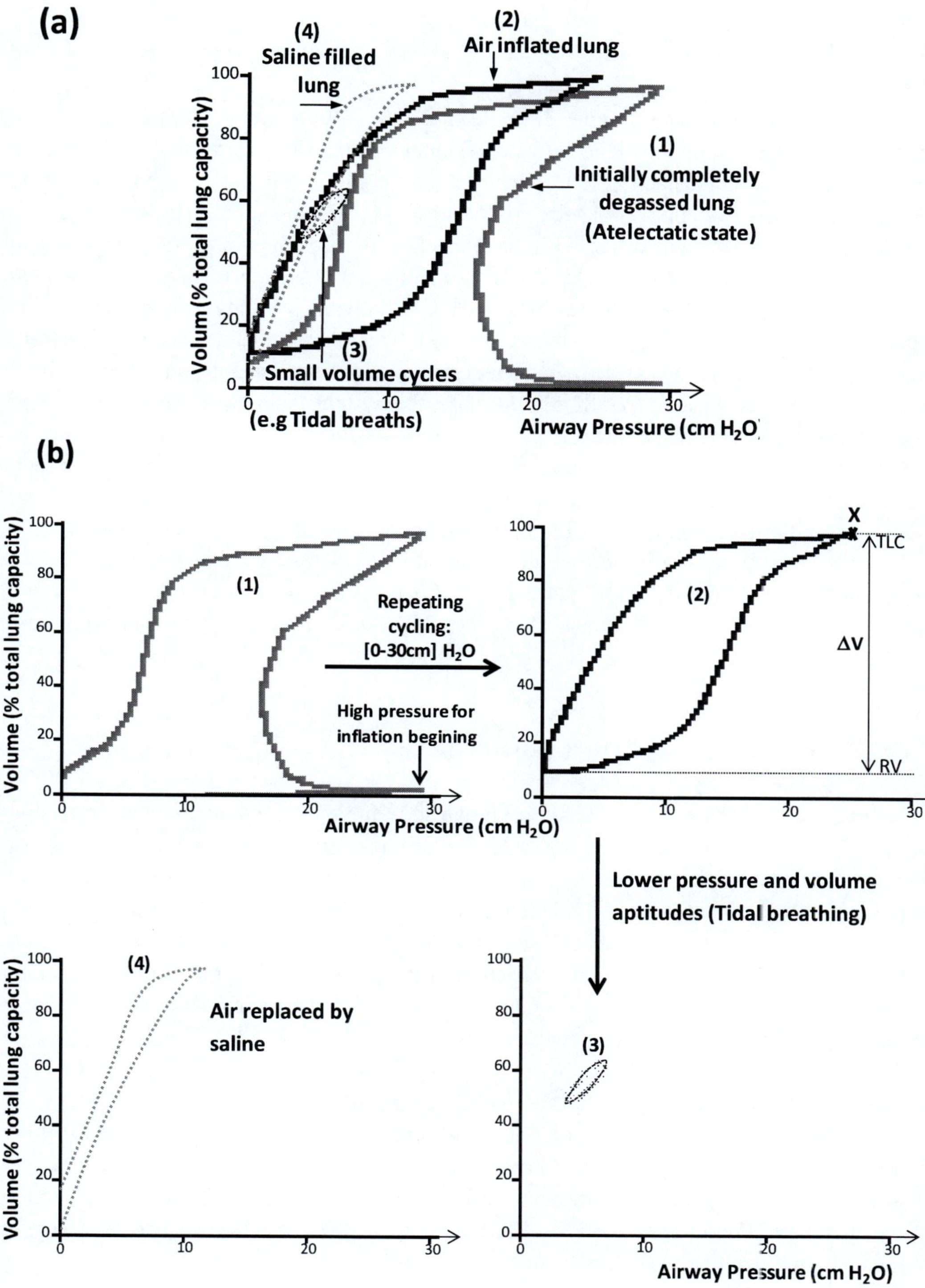

Figure 13. Variation of lung hysteresis shape, area and position due to different intrinsic and extrinsic respiratory factors.

Starting from completely degassed state, repeated cycling of airway pressure between 0 and 30cm H_2O makes pronounced hysteresis to become consistent (Figure 13: curves (1)→(2)) [63].

In the resulting hysteresis (Figure 13b: curve 2), the volume range (ΔV) represents the volume that a subject would achieve between residual volume (RV) and a maximal inflation effort to reach total lung capacity (TLC). The high end point (X) of hysteresis corresponds to stiffen lung resulting from both connective tissue stiffness and increasing surface tension.

Respiratory disorders are observed in several clinical cases including tidal breathing in vivo or Cheyne-Stokes syndrome. This consists of transient cessation of respiration (apnea) for several seconds (until 20s or more) followed by initially slow inflation returning then by gradual increase in rhythm causing noisy respiration. This *in vivo* disturbance of respiratory system implies small volume cycles which require small pressure changes leading to small hysteresis (Figure 13: curve 3) [63, 82]. In this case, the hysteresis does not seem to be related to surface forces and may be an intrinsic property of the tissue [83].

Surface forces are lost when the lung is filled with saline (liquid) rather than with air (e.g. in drowning case) (Figure 13b: curve 4). Saline filling results in much less recoil of lung (deflation vs inflation) compared to its amplitude under air filling (over the same volume range). This leads to narrow hysteresis along pressure axis (small hysteresis area) [63, 84]. This suggests that, as the air-liquid surface is removed from the lung, there is no surface tension to oppose distension, so that lung reaches TLC with lower transpulmonary pressure [78, 82, 83, 85].

pH Dependent Hysteresis: The Case of Nucleic Acid Conformation

Several polynucleotide complexes are known to exhibit thermodynamically metastable states and hysteresis [86-89]. Hysteresis is particularly pronounced in acid-base titrations of the three-stranded complex poly(A)-2 poly(U) [90]. Association between adenine (A) and uracil (U) is characteristic trait of RNA. This triple helix (Figure 17) is obtained if poly(A) and poly(U) are mixed at neutral pH and sufficiently neutral salt concentration [90, 91]. From neutral pH, acid-base titration results in variation of the protonation degree in polynucleotide by a pH-dependent hysteretic way. Characteristics and biological interests of such potentiometric hysteresis will be presented in the two next sections.

Characteristic Components of Potentiometric Hysteresis

Acid-base titration of poly(A)-2poly(U) shows hysteresis indicating that the protonation and deprotonation of macromolecular complex are irreversible processes which succeed within a loop containing two different and opposite trajectories (Figure 14). Under isothermal-isobaric conditions, every equilibrium state of the (A-2U) system is completely determined by the two parameters: pH and ionic strength.

Under variable pH (control parameter) ($3 \leq pH \leq 7$), the internal potentiometric state of RNA can be spectrophotometrically monitored by its absorbance at the wavelength of 280nm (A_{280}) (Figure 14).

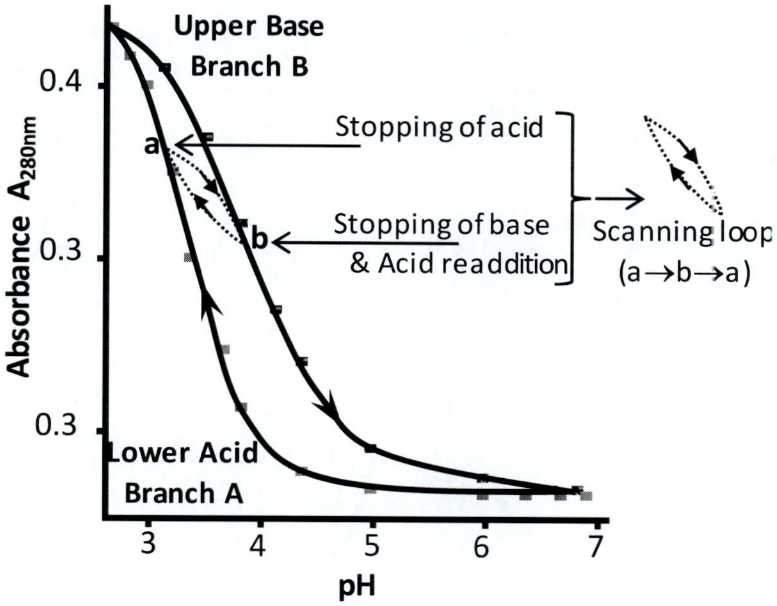

Figure 14. Potentiometric hysteresis resulting from acid-base titration of a triple helix (2A-U) and highlighted from spectrophotometric measure (absorbances at 280nm) versus pH [39, 92].

Other internal parameters can be used to highlight the hysteretic behavior of RNA in relation to pH, viz. the degree of protonation α (Figure 15) and the extent of the conformational change ξ (Figure 16) [39, 92]. Degree of protonation α corresponds to the ratio between the numbers of bounds' and protonation partners' moles.

Starting from pH = 7, the lower branch of hysteresis results from the titration of the neutral solution with acid (Figures 14, 15, 16). However, by stopping acid titration followed by subsequent titration of the acid solution with base, the macromolecular system returns to higher pH via another trajectory represented by the upper branch of the loop:

Potentiometric hysteresis showed that the degree of proton uptake occurs as irreversible and bi-stable function of pH (Figure 15). The internal parameters (A_{280} and α) increased monotonically as pH decreased from 7 to 3 (Figures 14, 15). Reversing the direction of pH variation by adding alkali showed that the lower branch is not retraced, but a new curve is obtained as the base (upper) branch of hysteresis (Figure 15).

Within the potentiometric hysteresis, elementary scanning loop can be obtained by stopping acid titration at an intermediate point *a* followed by alkali addition up to intermediate point *b*. The repetition of this short acid-base cycle makes an intermediate local loop ($a \rightarrow b \rightarrow a$) to be observed (Figure 14). This scanning loop shows a good resiliency property of the macromolecular titration system. Moreover, addition of acid after stopping brings back to the boundary curves of the global loop. Such global return provides further observations on the resiliency property of the macromolecular hysteresis ($AB \rightarrow ab \rightarrow AB$).

By considering the degree of protonation as internal parameter of hysteresis, the overall reaction occurring during the titration can be written in terms of reactive chain residues or segments (Figure 15):

$$2\,(A\text{-}2U)\text{-}H^{+} \leftrightarrows (A\text{-}A) + 4(U)$$

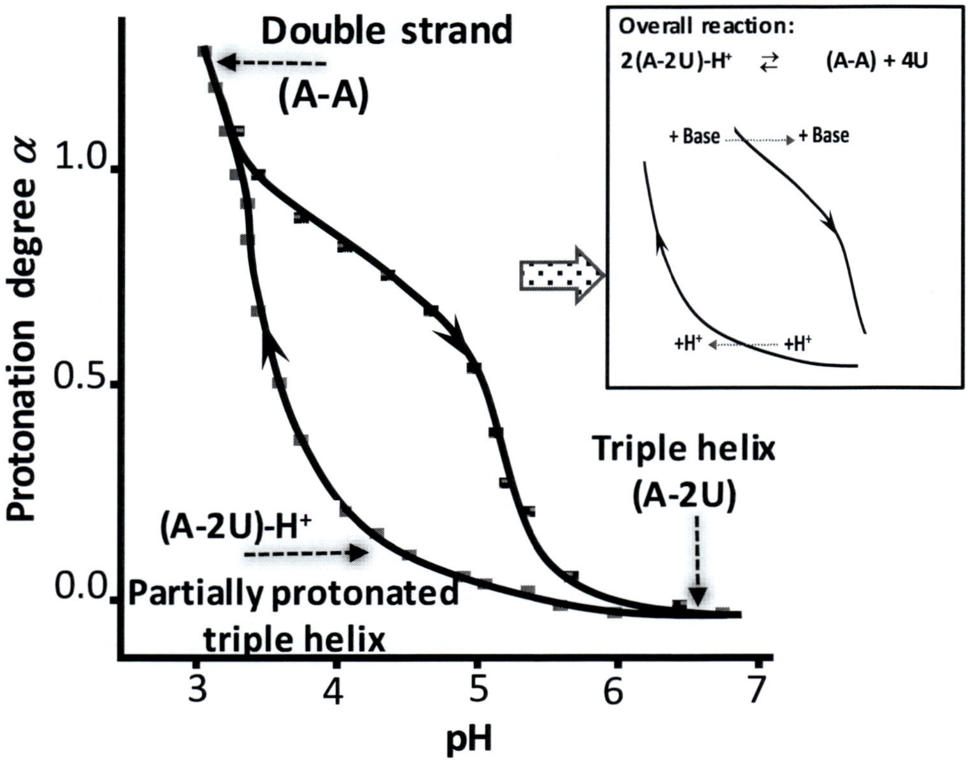

Figure 15. Potentiometric hysteresis resulting from acid-base titration of a triple helix (2A-U) and highlighted by plotting the degree of protonation α vs pH [39, 92].

The measured α-value can be considered as a sum of two contributions (Figure 17):

- Protons bound to the adenine-adenine double helix (A-A): → degree of protonation α_1.
- Protons bound to (A) in triple helix (adenosine - 2 uraciles) (A-2U): → degree of protonation α_2.

In polynucleotide complex, long lived metastability is observed neither in the (A-U) double helix nor in the (A-A) complex, but rather in the (A-2U) triple strand (Figure 17) [90]. The partial protonation of the triple helix (A-2U) represents a metastable pre-equilibrium (Figure 15) leading to two states:

- Opening of (A-2U) segment: (A-2U) → A + 2U
- Protonation of open segments A according to: A + H^+ → (AH^+)

The metastability of (A-2U) triple helix provides some energy barrier opposing to the opening of its segments within a partially protonated state (Figure 15). Moreover, the increased protonation in the acid titration compensates a part of the electrostatic repulsion energy arising from the phosphate residues of opposite strands. It results that opening of (A-2U) segments becomes more difficult before a critical low pH is reached. Therefore, the opening of (A-2U) triple helix is forced leading to double helix (A-A). This abrupt phenomenon is more clearly observed in the hysteresis of the conformational change ξ versus pH (Figure 16) [92]. Relationship between conformational change ξ and protonation degree α is given by:

$$\alpha = \alpha_1\xi + \alpha_2(1-\xi) \quad (6)$$

The pH-dependent values of the conformational change ξ were calculated from UV-spectrophotometric data [90]. By plotting ξ vs pH, one obtains a hysteresis which revealed to be sensitive to some key processes in acid-base titration system, viz. (Figure 16):

- Along the acid branch, (A-A) formation from (A-2U) occurs abruptly at pH values near to 3.5.
- Along the base branch, conversion of the double helix (A-A) to triple helix (A-2U) is a relatively slower process.

Apart from these mechanistic processes, the areas of the loops (α vs pH) and (ξ vs pH) provide indications on the amount of Gibbs free energy dissipated in one isobaric-isothermal titration cycle [39].

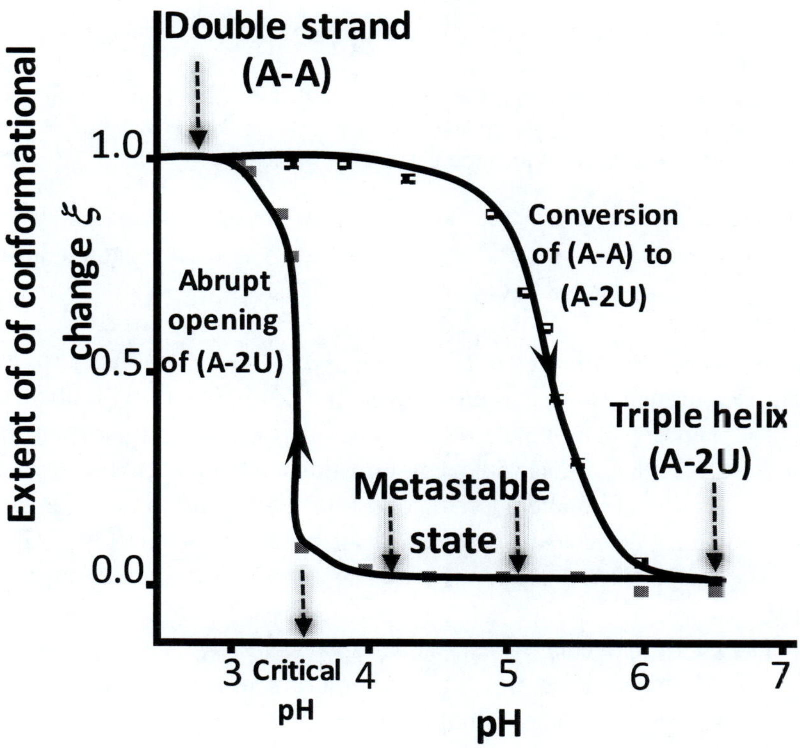

Figure 16. Potentiometric hysteresis resulting from acid-base titration of a triple helix (2A-U) and highlighted by plotting the extent of conformational change (or structural transition) ξ versus pH [39, 92].

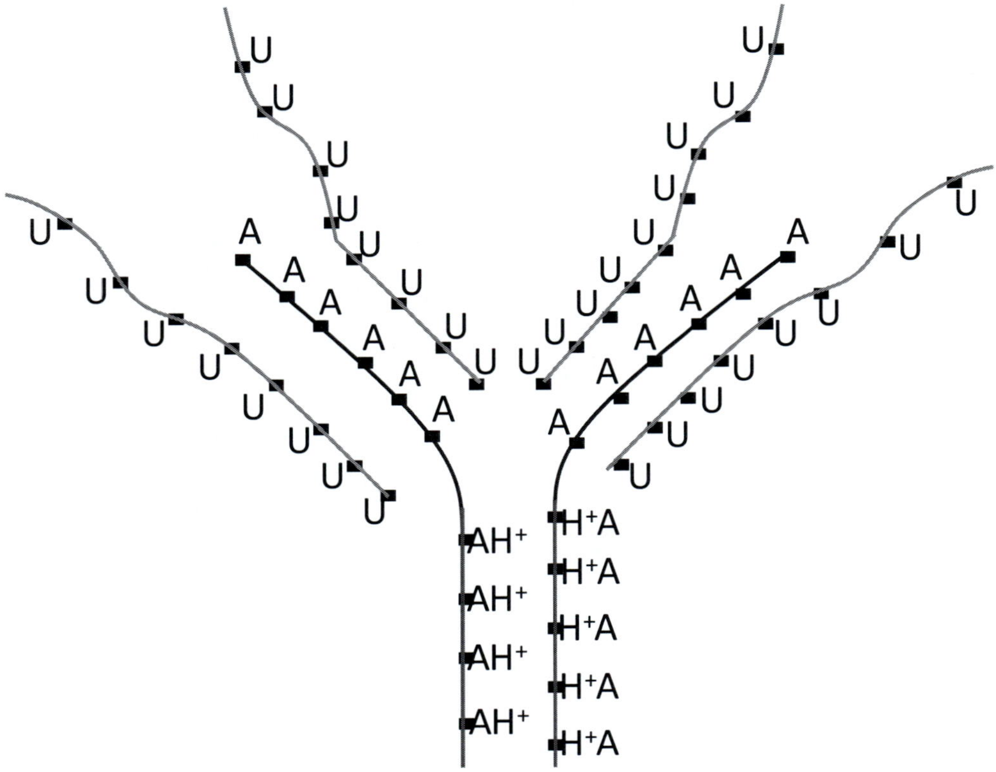

Figure 17. Schematic representation of junction between (A-2U) and (A-A) stretches in RNA system at an intermediary pH. Almost all the single stranded (A) segments are deprotonated ($\alpha_2 \approx 0$) and almost all the (A) residues in (A-A) sequences are protonated ($\alpha_1 \approx 1$) [93].

These different results show the usefulness of hysteresis to follow the "trans-crystallization" reaction between (A-2U) triple helix and (A-A) double strand during the acid-base titration of the A-U system (RNA system).

Biological Interests of Potentiometric Hysteresis

Potentiometric hysteresis has been shown to be implied in biological memory mechanisms [39]. There was accumulative evidence that the first step of the memory is a physical process consisting of plausible conformational changes in biopolymers (polyA, polyU) of RNA. Such conformational changes could lead to the production of long-lived metastable states readily detected in molecular hysteresis phenomena [39]. Apart from the pH, ionic strength and external electrical fields have been found to influence molecular conformational responses.

The first example of molecular hysteresis was discovered by Cox and co-workers [94] in the potentiometric acid-base titration of RNA. Conformation, size and position of RNA domains in the macromolecule are topological variables implied in molecular memory systems [90]. Domains are structured nucleotide stretches with different lengths and stabilities forming "microcrystalline" regions distributed along a single chain molecule.

Changes in the environment of metastable macromolecules or subcellular macromolecular organizations (such as membranes) could induce conformational variations playing a role of reproducible imprints of a memory nature [39]. The imprinting process itself would be based on the release of a metastable state followed by a sharp non-equilibrium structural change. Changes in conformation or location of macromolecules have been reported to be responsible for permeability alteration of nerve membranes and release of suitable transmitter substances [95].

A previous study on memory mechanism reported that if a pattern has to occur for a very long time, the pathways have to be given an additional property akin to magnetic hysteresis; therefore, the informon becomes a permanent discriminator of the pattern to which it is adapted [96].

Voltage Dependent Hysteresis: The States of Cell Membrane Channels

Voltage dependent hysteresis refers to conductance delays in response to voltage change. Delayed conductance was found to be a structural property of cell ion channels; it is a voltage-independent switch of kinetics causing delayed voltage-dependent transitions [97]. This hysteretic effect can be observed at both "macroscopic" and "microscopic" levels. In biological field, conductance hysteresis has been revealed to be a process controlling the opening and closing of ion channels. Channels consist of nanoscopic transmembrane pore-forming proteins maintaining and regulating ionic currents through cell membrane. They are key dynamic elements implied in membrane excitability and neural signals (action potentials) [98, 99]. They can (i) sustain a membrane potential around [-80, -60] mV, and (ii) generate action potentials about [+80, +100] mV in amplitude.

"Microscopic" and "macroscopic" systems concern studies of single channel (molecular scale) and population of channels (whole cell scale), respectively. Macroscopic system is due to aggregation of microscopic currents through thousands of ion channels, smoothing out the details of single channel activity [100]. However, single channel records reveal the gating process itself, i.e. discrete changes in conductance over time. At single channel level, the gating process involves conformational changes of the protein, resulting in alternations between open and closed states of the protein channel.

Protein channels (pores) inserted into cell bilayer lipid membrane were found to manifest gating bi-stability in response to variable external voltages (control factor) [101-103]. Among voltage-gated protein channels, those based on lysenin have been revealed to manifest hysteretic behaviors [104]. Hysteresis consisted of clockwise loop where current versus voltage showed two phases (Figure 18c):

- Increasing voltage phase associated to system inactivation due to channels' closing.
- Decreasing voltage phase corresponding to reactivation due to channels' reopening.

Lysenin is a 297 amino acids-pore-forming protein in which hysteresis represents an intrinsic characteristic of gating mechanism [105]. In its native environment, lysenin is neither an ion channel nor a transmembrane protein [106, 107]; however, laboratory experiments showed that this protein has cytolytic activity suggesting a pore-forming toxin having important feature of ion channels like voltage and high transport rate [105, 108, 109]: when inserted into artificial bilayer lipid membranes, lysenin replicates fundamental properties of ion channels. Some of these properties have been highlighted by hysteretic pattern of current response (*I*) in relation to applied voltage (*V*) [104]. Such conductance hysteresis at molecular level has been shown to play a fundamental role in data storage and emerging memory application [110-112].

Conductance Hysteresis at "Macroscopic" Level

At macroscopic level, conductance hysteresis arises in response to periodic voltage that matches ion channel kinetics [102, 113-116]. Conductance hysteresis is highlighted by plotting the macroscopic current (*I*) measured in a population of uniform lysenin channels in response to linearly variable voltage (*V*) (external stimulus). Variability in hysteretic patterns was observed by varying the voltage at different sweep rates (from $1Vs^{-1}$ to $0.025\ mVs^{-1}$) (Figure 18a) [104]. These different sweep rates were categorized into rapid, slow and very slow voltages and resulted in very different shapes of conductance hysteresis (Figure 18):

Hysteresis Resulting from Rapid Voltage Rate (1 V.s^{-1})

Rapid voltage stimulation of lysenin channels' population resulted in a linear and symmetrical response (Figure 18b). Identical amperage–voltage curves over the entire voltage range were obtained for both ascending and descending voltage applications. This superimposition between ascending (inactivation, i.e. channel's closing) and descending voltage curve (reactivation, i.e. channel's reopening) means that a rapid voltage variation did not allow the channels to adjust their conformation in response to instantaneous electric tension.

Hysteresis Resulting from Slow Voltage Rate (0.01 V.s^{-1})

Decrease of voltage rate from 1 to 0.01 Vs^{-1} (one hundred-fold decrease) broke the previous symmetry of amperage-voltage curve resulting in a clear clockwise hysteresis (Figure 18c): a slow equilibration of conformational changes that switch the ion channels from the conducting to the non-conducting state in response to external voltage stimuli (*V*) is an essential condition for observing dynamic hysteresis in macroscopic currents (*I*) [115]. Such conductance hysteresis seemed to be introduced by lags:

(i) Between the voltage variation and the dynamic conformational equilibrium.
(ii) Between close and open states of channels.

Like many other hysteretic processes, conductance hysteresis originates from a dynamic break in symmetry leading to competing time scales between the period of an oscillating external perturbation (e.g. oscillating applied voltage) and the system's typical relaxation time (e.g. conformational transitions between open and close states) [101-103].

Although the effective voltage threshold to promote channel switching is ~20mV (Figure 18a), the *I-V* curve remained in its pre-switching state until *V*~40mV (channels' closure delay) (Figure 18c). This switching bias (delay) is due to the slow voltage (0.01 V.s^{-1}) which makes that the application time is shorter than required to reach equilibration. Consequently, the opened state of channels persists at population scale. This switching bias is overcome by a voltage exceeding 40 mV; in hysteresis, this is shown by a decreasing current (*I*) while increasing the applied voltage (V) (Figure 18c: segment ab). This transition phase (a→b) of hysteresis indicated promoted channel inactivation by positive voltage effects.

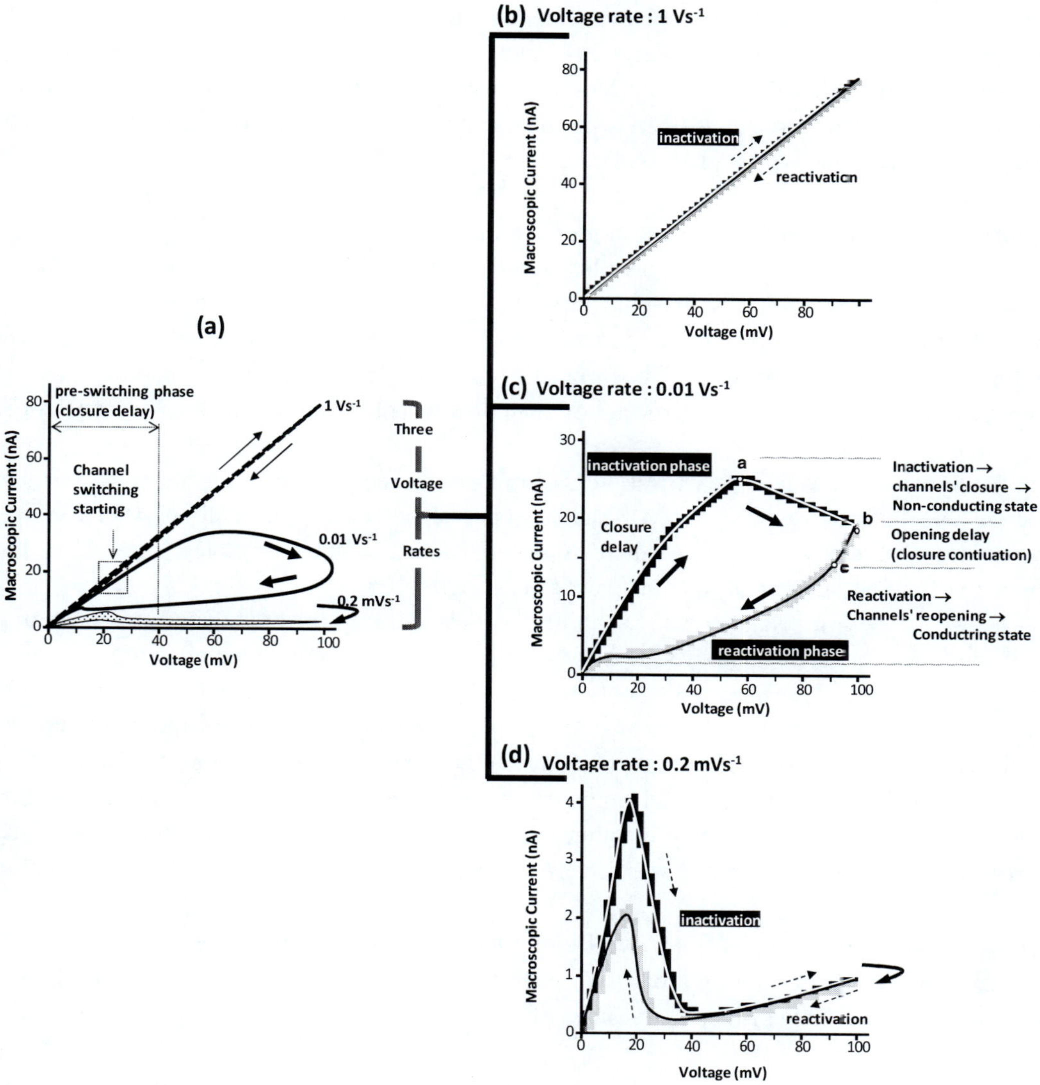

Figure 18. Different conductance hysteretic patterns of closure-reopening lysine channel highlighted by plotting measured current vs applied voltage at different sweep rates.

From maximal voltage application (Figure 18c: point b), descending I-V curve (transition b→0) showed a different trajectory from that of ascending phase (transition 0→a→b). At the beginning of this phase (b→c), the macroscopic current I decreases more rapidly than voltage; this process indicates a continuation of channel closure (opening delay). From point c, the I-V curve showed some linear decreasing phase until 0 corresponding to protein channels re-opening (transition c→0).

Hysteresis Resulting from Very Slow Voltage Rate (0.2 and 0.05 mV.s^{-1})

A further reduction of voltage rate (from 0.01 V.s^{-1} to 0.2 and 0.05 mV.s^{-1}) accentuated non linearity in the hysteresis pattern (Figure 18d). Moreover, the transition between conducting and non-conducting phases showed steeper aspect. Although the time-scale (period = 1/sweep rate) of the voltage was much larger (because of very slow voltage rate) than the characteristic relaxation time of the lysenin channels, the non-symmetrical response remained, which leads to a small persisting hysteresis (i.e. despite the unfair competing time scales of the input and output processes).

Conductance Hysteresis at "Microscopic" Level

At lower scale than channels' population, conductance hysteresis has been shown to be an intrinsic property of single lysenin channel. At this "microscopic" level, ion channel hysteresis refers to an enduring conduction delay that arises in response to voltage change despite steady-state conditions following the change [102, 115, 117]. Such hysteresis represents elementary or basic process of that observed at population level (macroscopic currents).

At single channel level, each individual unit can be described as a nano-scale bistable system able to achieve one of two possible states (inactivation-reactivation) (i.e. closing-reopening) characterized by different resistances (*U*/*I*) or conductances (*I*/*U*) linked to the applied voltage and electrical state history of the system [104].

Hence, a lysenin channel shows ability to "remember" its previous state indicating intrinsic memory which manifests by a hysteretic behavior at biomolecular (nano-structure) levels. Such memory-resistance elements simulate biological memory mechanisms [118]. On this basis, they may constitute the explanatory components of intelligent abilities and memory displayed by unicellular organisms [119] or other cells without brain functions. This hysteretic concept of biomolecular memory provides an understanding of how channel proteins found in living cells may form an immediate record of the molecular history which then determines their future responses to various stimuli.

Synthesis on Conductance Hysteretic Behaviors under Variable Voltage

Analysis of hysteresis loop area (nW) (electrical power) in relation to sweep rate (V.s^{-1}) helped to conclude three principal attributes of dynamic voltage hysteresis (Figure 19):

- Disappearance of hysteresis for fast voltage variations, i.e. when the characteristic relaxation time of lysenin channels greatly exceeds the period of oscillatory voltage stimulus.
- Slightly asymmetrical bell shape of loop area distribution versus sweep rates.
- Asymptotic decrease of the loop area to a small but not null value for very slow voltages (sweep rates < 0.2 mV.s^{-1}).

Alternatively to current variable (*I*) (Figure 18), the probability (*P*) of channel being in open state was plotted versus the voltage (*V*) to highlight a hysteresis with more classical shape (Figure 20). This internal state parameter (*P*) was extracted from *I-V* curves of macroscopic current recorded during inactivation and reactivation at 0.2 mV.s^{-1} [120-122].

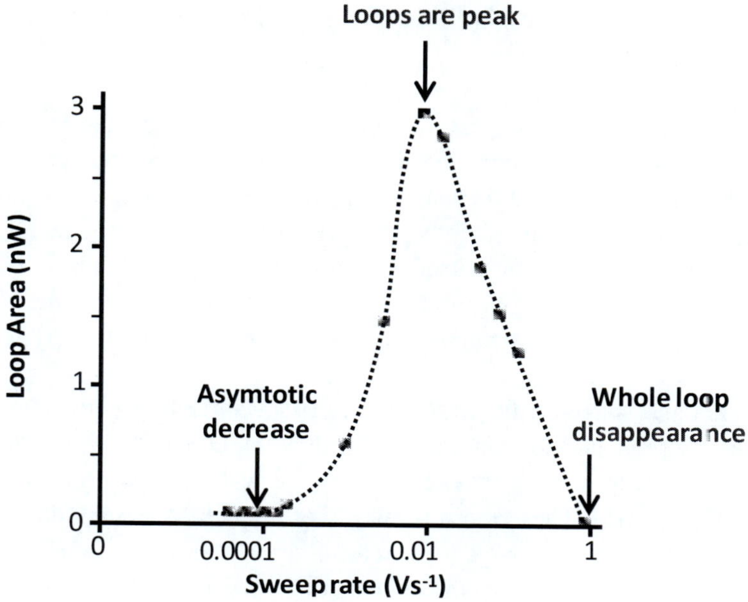

Figure 19. Asymmetrical bell shape giving the distribution of current-voltage hysteresis area in relation to sweep rate.

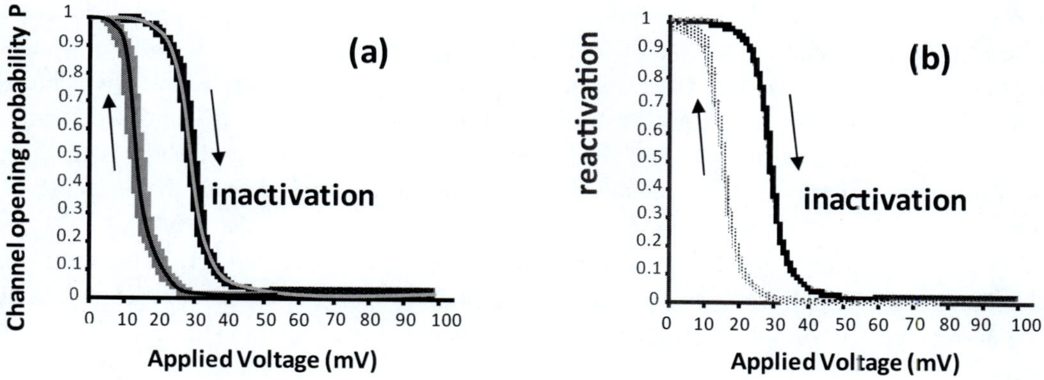

Figure 20. Hysteretic loops of opening probability of single channel (a) and channels' population (b) of lysenin in response to voltage-induced gating at the sweep rates in the range [0.1, 0.2] mVs^{-1} [104].

At the scale of a single channel, the probability P (1 for open, 0 for close) was assigned to each voltage value by observing the current I and determining the conducing state. Therefore, multiple runs at 0.1 $mV.s^{-1}$ were applied to cumulate a serial of 1 or 0 values from which average probability was calculated under each current (Figure 20a). Similar hysteretic pattern was obtained for average opening probability calculated from several runs applied at population level, i.e. on a system of lysenin channels (Figure 20b). In both loops, the average opening probability showed a bistability of lysenin channel(s) with a clear leftward shift during reactivation (Figure 20a, b).

Radiation-Dependent Hysteresis: The Case of Plant-environment Gas Exchanges

Current climatic changes led to research about the question of how drought influences carbon exchanges between ecosystems and the atmosphere. The net ecosystem exchange of CO_2 (NEE) relies on the balance between CO_2 uptake and emission through (i) plant photosynthesis and (ii) ecosystem (plant and soil) respiration, respectively [123]. Recent agricultural studies on southeastern US ecosystems showed that photosynthetically active radiation (PAR) was the primary control factor of photosynthesis at day time scale [124]. A daily study on the peanut *Arachis hypogaea* highlighted an anti-clockwise hysteresis of NEE in relation to PAR:

Moreover, area and shape of this hysteresis varied significantly with water stress. Very low soil water content resulted in significant NEE depression despite high PAR values. The magnitude of the hysteresis was much larger during the water-stress days (Figure 21b) than non-stress days (Figure 21a). These loops showed that a same PAR value does not cause the same NEE response whether it belongs to increasing or decreasing phase (Figure 21): in the morning, increasing PAR resulted in the increase of carbon uptake (more negative NEE values); however in the afternoon, carbon uptake was lower than during the morning at a given PAR value. Beyond these general characteristics of hysteresis, the two loops' patterns showed marked differences in shape as direct effect of water-stress conditions [124]:

On non-stress days, NEE increased (gets more negative) in response to increasing PAR in the morning (Figure 21a). This variation reaches NEE peak at the highest PAR of 2080 μmol photons $m^{-2}.s^{-1}$. In the afternoon (as PAR decreased), NEE declined. However, during water-stress conditions, NEE showed two variation phases in response to increasing PAR (Figure 21b): NEE reached a peak at PAR of 1100 μmol.photons.$m^{-2}.s^{-1}$, then it decreased rapidly until reaching almost zero at the end of the morning (PAR > 2000 μmol photons $m^{-2}.s^{-1}$). Throughout the afternoon, NEE remained constant at nearly zero despite the decrease of PAR.

The diurnal pattern of NEE provides an output picture on the interaction between the atmosphere and plant physiological response. Asymmetric shape of NEE hysteresis at diurnal scale during dry periods with high vapor pressure deficit was attributed to [125]:

- Control of photosynthesis via stomatal closure at high vapor pressure deficit.
- Enhanced respiration induced by high temperature. Ecosystem respiration has been reported to exhibit hysteresis which may be attributed to [126]:
    - Out of phase temperature between the different biological and ecological compartments (from leaf to deep soil) [124, 127, 128].

  and/or
    - Lagged response of root respiration to light [129].

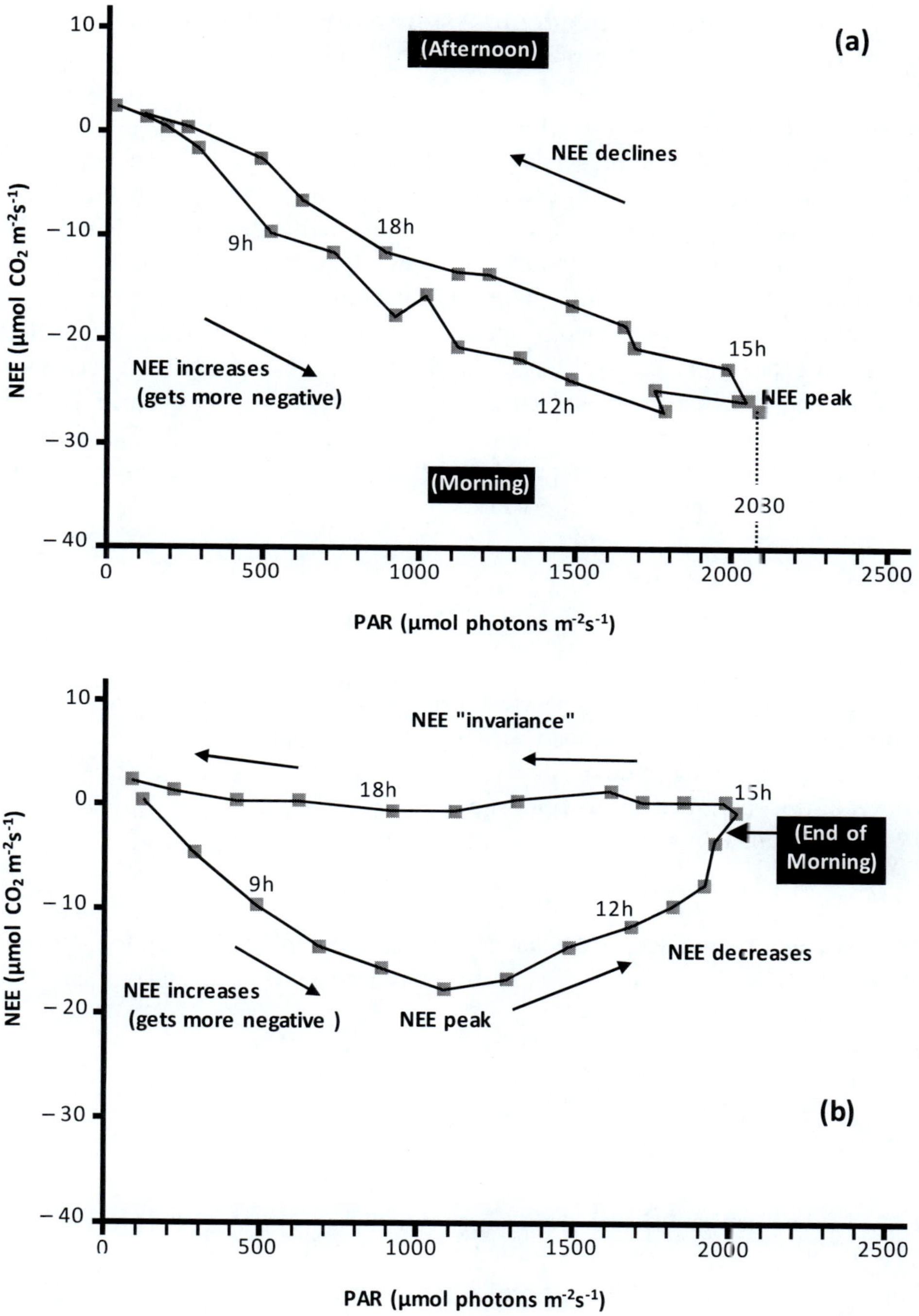

Figure 21. Hysteretic relationships between net ecosystem CO_2 exchange (NEE) and photosynthetically active radiation (PAR) during a diurnal cycle in the peanut *Arachis hypogaea* L.. (a) Hysteresis under non-stress conditions; (b) hysteresis under water-stress conditions.

Chemical Concentration Dependent Hysteresis: Pharmacological Effects of Drugs

Introduction to Pharmacodynamic Models

Administrated drugs (and food compounds) in living body are quantitatively shared between blood compartment and physiological site of action. Between these two compartments, drugs have intermediate states including their binding to protein receptors and their distribution into peripheral compartment (tissues) (Figure 22). If drug levels in blood are in equilibrium with those in the site of action, relationship between drug effect and free circulating quantity can be formulated by means of a classic pharmacodynamic model known as Michaelis-Menten equation (7). If steady state equilibrium is established between blood compartment and drug receptor, any change in circulating drug levels will be instantly mirrored by a change in receptor binding and therefore drug effect. Thus for a drug interacting reversibly with a receptor and giving a proportional effect to the number of occupied receptors, we can write (Figure 22):

$$\mathrm{CP} \rightleftarrows \mathrm{C} \rightleftarrows \mathrm{D} + \mathrm{R} \rightleftarrows \mathrm{DR} \rightleftarrows \mathrm{E}$$

with:

CP: drug-protein form in central compartment (blood or plasma)
C: free drug level in central compartment (blood or. plasma)
D: free drug level at the receptor site *R*
R: level of free (not filled up) receptor site
DR: level of bound drug to receptor site (target tissue)
E: drug effect

This rearranges to produce the Michaelis-Menten equation relating the drug effect (*E*) to its free level (*D*) at the receptor site (*R*) by considering the binding constant (K_D) and maximal effect (E_{max}):

$$E = \frac{E_{\max} \cdot D}{K_D + D} \tag{7}$$

where K_D (binding constant) is the drug concentration producing 50% maximal effect (50% E_{max}).

Graphically, this equation is represented by a hyperbola showing that as drug level increases, receptor occupancy increases rapidly at first but slows down thereafter as the receptors fill up (Figure 22: model M_1). Drug concentration responsible for maximal effect (E_{max}) may be many folds higher than that producing the half of response (K_D). This attributes to K_D a clinical usefulness helping to evaluate effects, reactivity and biological activities of administrated chemical compounds.

Assuming that the drug is in equilibrium, i.e. *D* is proportional to *DR*, it is possible to relate drug levels in central compartment (*C*) (blood or plasma) with the measured effect (*E*).

However, this relationship can be changed if the binding of the drug to plasma protein (CP) changes between individuals due to different physiological and/or disease states.

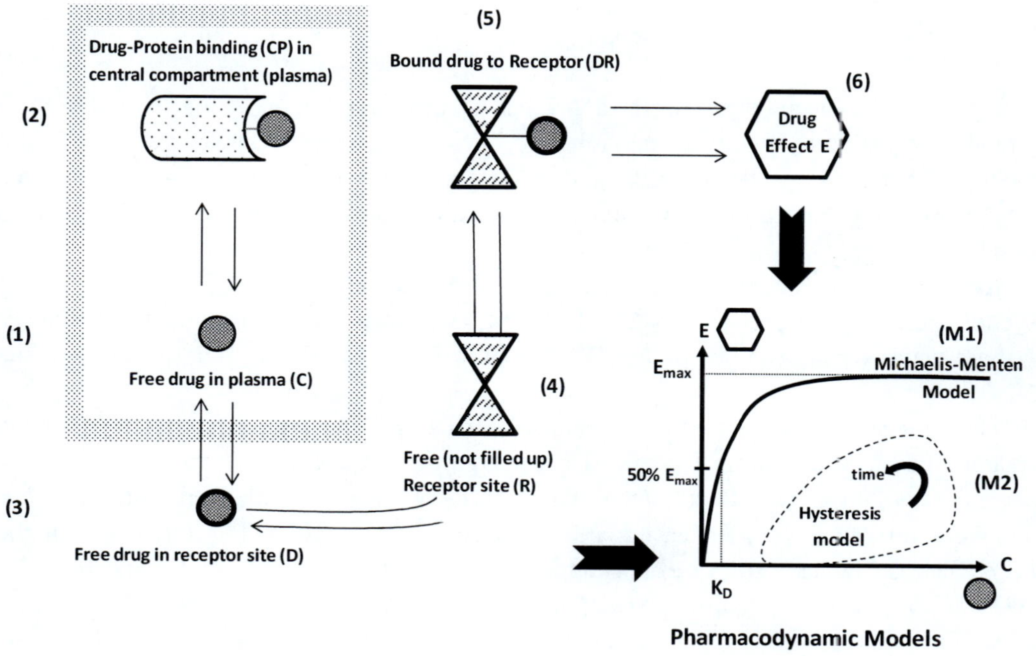

Figure 22. Illustration of different compartments (1-5) controlling drug concentration from systemic circulation (1-2) (blood or plasma compartment) to drug effect (6). Graphical representation of different pharmacodynamic models (M1, M2) showing different types of relationships between drug concentration *C* and effect *E*.

Moreover, not all drug receptor interactions can be described simply by (7) (Figure 22: model M_1) for many possible cases: irreversible binding, multiple binding, down or up regulation, sparse receptors, partial agonist, etc. These factors can be at the origin of more complex relationships between drug level and drug effect, including hysteretic relationships (Figure 22: model M_2). Finally, unless the drug works within blood compartment (e.g. antibiotics), application of (7) needs some assumptions and simplifications for many drugs acting on not directly accessible sites of action.

Types and Origins of Pharmacological Hysteresis

When a drug is not in equilibrium with its site of action, hysteresis occurs and drug levels are out of phase with activity. Hysteresis can be conceived as a process implying delays or discrepancy between drug activity and its free circulating level *C* (Figures 23-26). By reference to rotation direction, one distinguishes two types of pharmacological hysteresis: anti-clockwise and clockwise (called proteresis).

Anti-Clockwise Pharmacological Hysteresis

Anti-clockwise hysteresis represents a pharmacological process where a given drug level has a higher effect in more advanced time (Figure 22: model M2) [130, 131]. This shifting between drug level and effect can be caused by several regulatory factors:

- Uptake into an active site
- Active metabolites deriving from administrated drug form (called pro-drug)
- Cascade activity
- Sensitisation or reverse tolerance

Hysteresis has been coined for the delay that it takes for a drug to be transported from the systemic circulation to its site of action then to illicit a measurable response (effect E). Anti-clockwise hysteresis has been highlighted in different clinical cases: for instance, different psychotropic drugs take a finite time before being equilibrated with their sites of action in the brain [132, 133].

An illustration is given for delta-9-tetrahydrocannabinol (THC) known for its psychological "high effect" on subjects [132, 134]. The drug was administrated to volunteers by three ways: (i) intravenous bolus (IV), (ii) oral or *per os* administration (PO), smoking (SK).

Then, the subjective high effect was plotted versus THC plasma concentration taking into account the measure time order. By considering sequential observations of drug concentration-subjective effect, anti-clockwise loops were highlighted for the three administration routes, with decreasing area in the sense: IV → SK → PO (Figure 23): higher plasma THC levels and greater effects appeared at earlier time points for IV injection. The declining portions of the loops (after approximately 30 min) occur together for IV injection and smoking. This hysteretic phenomenon was explained by the fact that THC takes a finite time before it is equilibrated with its site of action in the brain. This phenomenon supports the fact that an intermediate compartment (called effect compartment) needs to reach preliminary equilibrium with the central one (plasma) before the intensity of effect reaches a peak. Note that PO administration results in very small hysteresis, since more time is allowed for drug penetration into systemic circulation (transit from intestine to plasma).

Apart from the administration route, administrated dose revealed to be a significant factor governing the hysteresis area (Figure 24): three increasing levels of THC were administrated following smoking route (1.3, 2.0, 2.5% THC by cigarette). Anti-clockwise loops were obtained indicating that drug effect is related to its plasma concentration through a delay mechanism. The loops' areas were larger for higher potency cigarettes. It can be seen that a given plasma concentration results in a higher subjective effect at later time compared to initial time. The declining phase of the loops, occurring around 45 min after beginning of smoking, were not distinguishable for the three potencies. This suggests that the decrease of the subjective high effect occurs independently from its potency.

Apart from the delay process linked to drug uptake into active site, pharmacological anti-clockwise hysteresis were observed for active metabolic derivatives released after drug transformation. In fact, metabolic transformations can result in a lag time between drug effect and its blood concentration. For example, fenfluramine (a psychotropic pro-drug) is converted into a metabolite which is active for up to 15h because of its long half-life (five-fold longer than that of its precursor) [135].

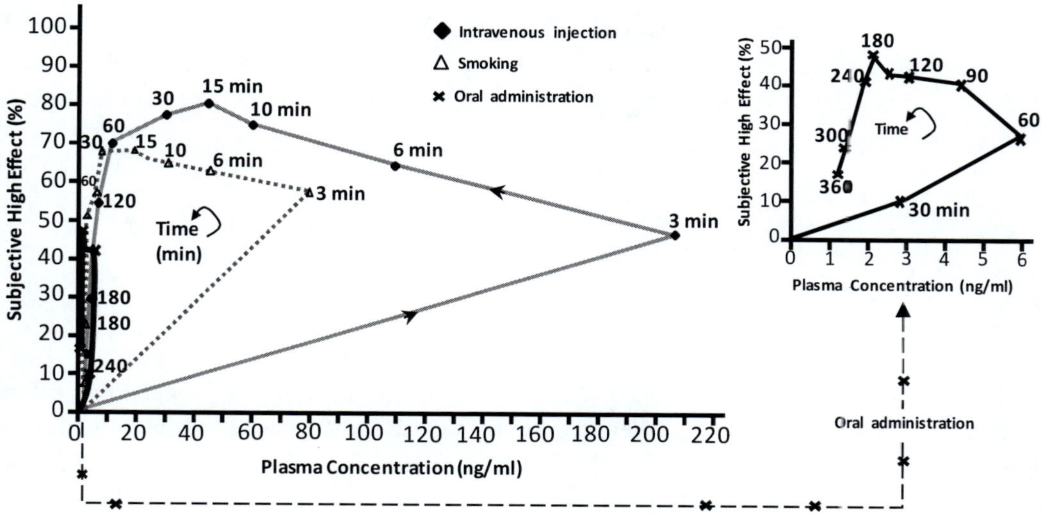

Figure 23. Three hysteresis loops showing the delayed subjective effect of psychotropic drug (THC) administrated by three different ways (intravenous, oral and smoking routes) [132].

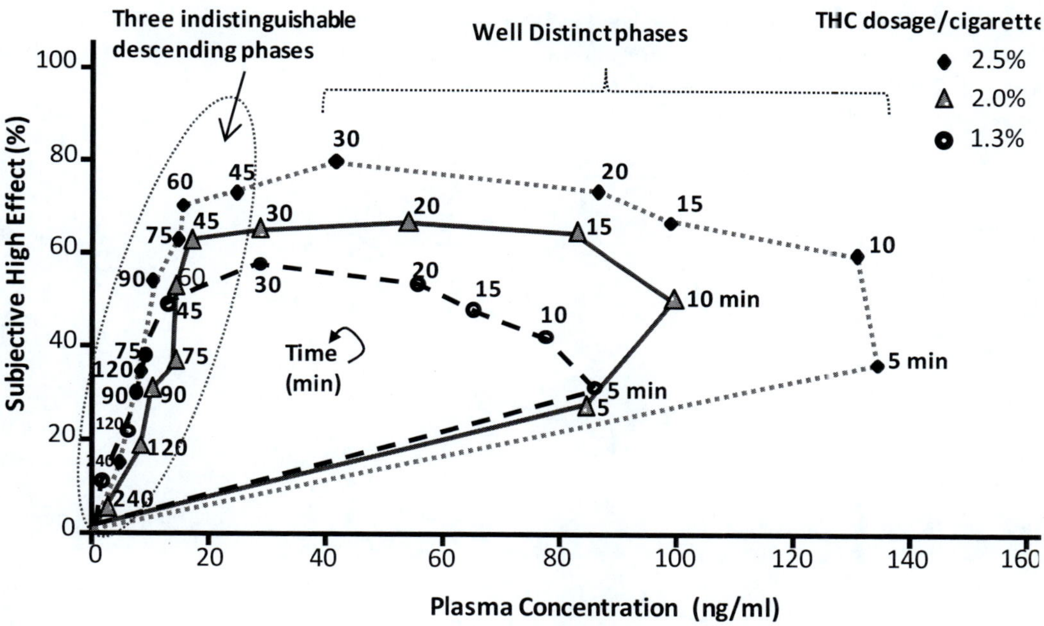

Figure 24. Three hysteresis showing increasing subjective effect (increasing loop's area) of psychotropic drug (THC) with administrated dose in smoked cigarette. After 45 min, the effect decreases by a similar pattern independently from the received dose [132].

Also for some drugs, anti-clockwise hysteresis of pharmacological effects result from progressive series of events which lead chronologically to a delayed or indirect activity. Examples can be found in antidepressant activity of some centrally acting drugs where several weeks may be needed to produce a maximal effect [136]. Also, cascade reactions leading to delayed drug effects concern anticoagulant coumarin drugs [137, 138].

Clockwise Pharmacological Hysteresis (Proteresis)

Pharmacological clockwise hysteresis (called proteresis) is graphically identified by a decreasing drug effect with time for any given drug concentration. This can be caused by different mechanisms:

- Tolerance
- Active antagonistic metabolites
- Learning effects
- Feedback regulations

Tolerance is defined as a reduced effect of xenobiotic at the same concentration that earlier produced a greater effect. Tolerance proteresis can be illustrated by the reduction of euphoria with consumed cocaine over time (due to tolerance) (Figure 25) [139, 140]. Also tolerance proteresis was reported for several centrally acting products including the barbiturates [141], alcohols [142], morphine [143], etomidate [144] and some benzodiazepines [145, 146].

Ellinwood and co-workers [145] showed marked tolerance for both alprazolam and diazepam (benzodiazepines) on the cognitive task of tracking (Figure 26). In the clockwise loop, the contrast between the onset and offset areas provides a robust indicator of acute tolerance: for a blood concentration around of 600ng/ml, the diazepam effect (tracking delta scores) at 25min was shown to be four times the level at 155min (Figure 26).

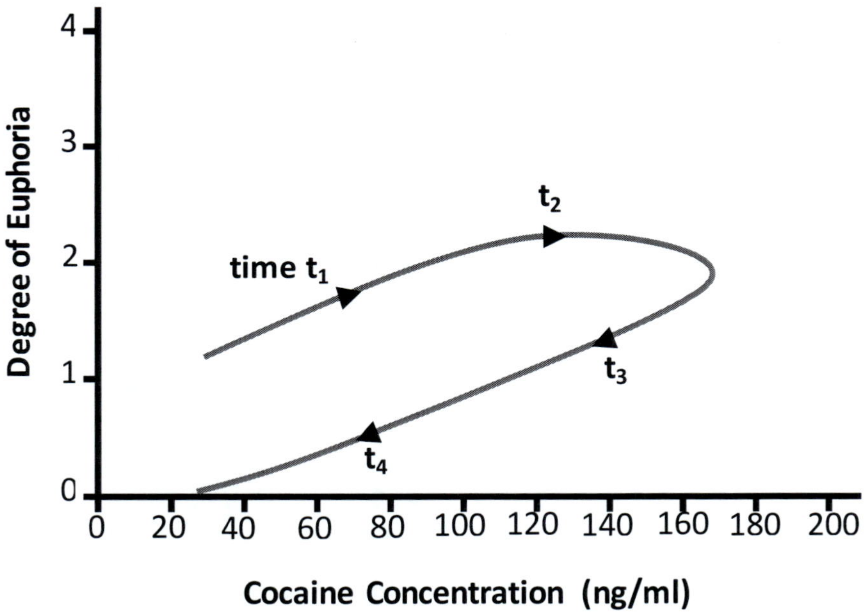

Figure 25. Clockwise hysteresis showing the reduction of degree of euphorbia for a given concentration of cocaine due to tolerance process [139].

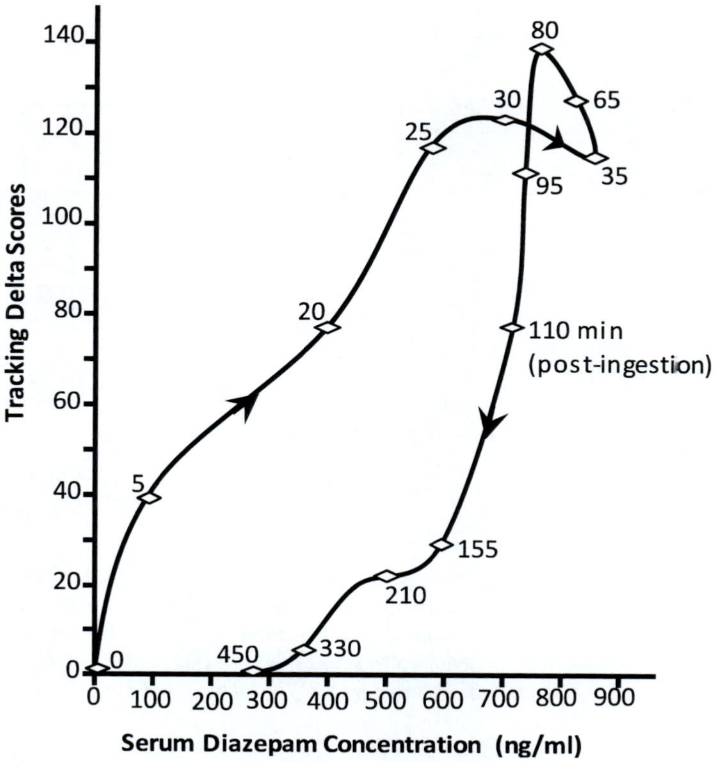

Figure 26. Hysteresis showing tolerance of clinical population toward diazepam. Tracking performance is reduced with time leading to a reduction of drug efficiency for a same blood concentration [145].

Chemical Concentration Dependent Hysteresis: The Case of Enzyme Level Control

In biochemical field, hysteresis represents metabolic shifts that may be due to different processes such as competitions between metabolites for substrates or enzymes, activation of some enzyme reactions (pathways), etc. [21, 147]. These processes can be experimentally measured by metabolic parameters including enzymatic reaction rates, enzyme concentrations, products' or ligands' regulation levels, etc. . In more general terms, two types of phenomena can summarize the origins of metabolic hysteresis:

- The presence of different metabolic responses of a system depending on a ligand concentration being increased or decreased.
- The presence of conformation changes of enzymes.

Taking into account these two hypotheses, Ricard and Buc [21] simulated enzyme hysteresis from a simple mnemonical metabolic network showing the property to generate meta-steady-state (MSS) kinetics (Figure 27). The concept of MSS refers to an enzyme reaction process which implies the presence of metastable enzyme conformations. As a consequence of MSS, enzyme reaction rate and enzyme conformation level display hysteresis.

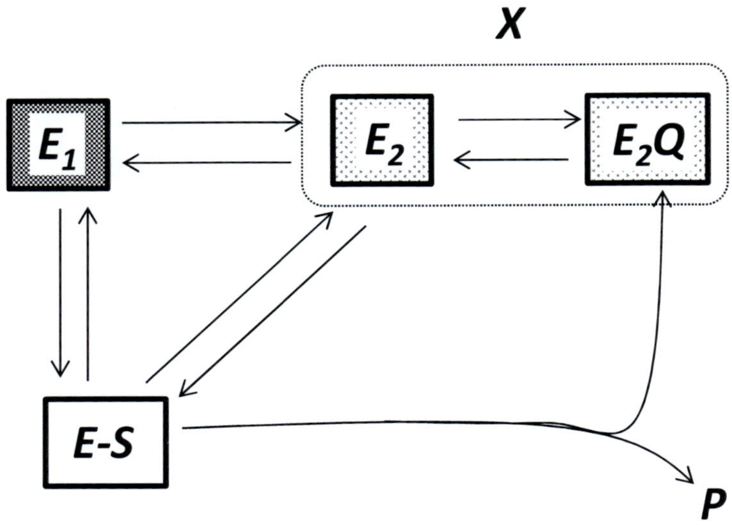

Figure 27. Simple mnemonical metabolic network used as a basic model for simulation of hysteretic behaviors of enzyme conformation levels (E_1 and X) in relation to two variation directions (increase and decrease) of the ligand Q. E_1 and E_2 are two possible conformations of the enzyme E; E-S is the bound form of enzyme E to substrate S; X represents both free and bound forms of E_2; E_2Q is the bound form of enzyme E_2 to ligand Q; P is the second product (with Q) of enzyme reaction.

Mnemonical model used by Ricard and Buc [21] consisted of an enzyme (E) having two possible conformations E_1 and E_2 interacting with substrate S at different rates k_1 and k_2, respectively (Figure 27). The metabolic reaction is assumed to release two products, P and Q. The mnemonical metabolic model shows three states of the enzyme E: E_1, X ($E_2 \leftrightarrow E_2$-Q) and E-S. The conformational state X ($E_2 \leftrightarrow E_2$-Q) is stabilized by the product Q. Moreover, the production of Q results in shifting of the equilibrium between the initial concentrations of the conformational forms E_1 and E_2. Initial concentrations of E_1 and X are interdependent according to the unit sum relationship:

$$\frac{E_1(t=0)}{E_{Tot}} + \frac{X(t=0)}{E_{Tot}} = 1 \quad \Leftrightarrow \quad \frac{E_1(0)}{E_{Tot}} + \frac{E_2(0) + E_2Q(0)}{E_{Tot}} = 1$$

where: E_{Tot} is the total concentration of enzyme (E) by considering all its conformational forms with their free and bound states.

At a given time t, the dependent concentrations $E_1(t)$ and $X(t)$ will be different according to whether the actual concentration of Q is reached after a decrease or an increase of a previous concentration of that ligand. The MSS is therefore dependent on both the concentration of Q and the direction of the concentration change.

Simulations showed that levels of enzyme conformations (E_1, X and E-S) can manifest hysteresis in relation to ligand concentration ([Q]) [21]. Such hysteretic behaviors were linked to:

- Differential ability of enzymes to bind the ligand.
- Slowness of the conformational transition.

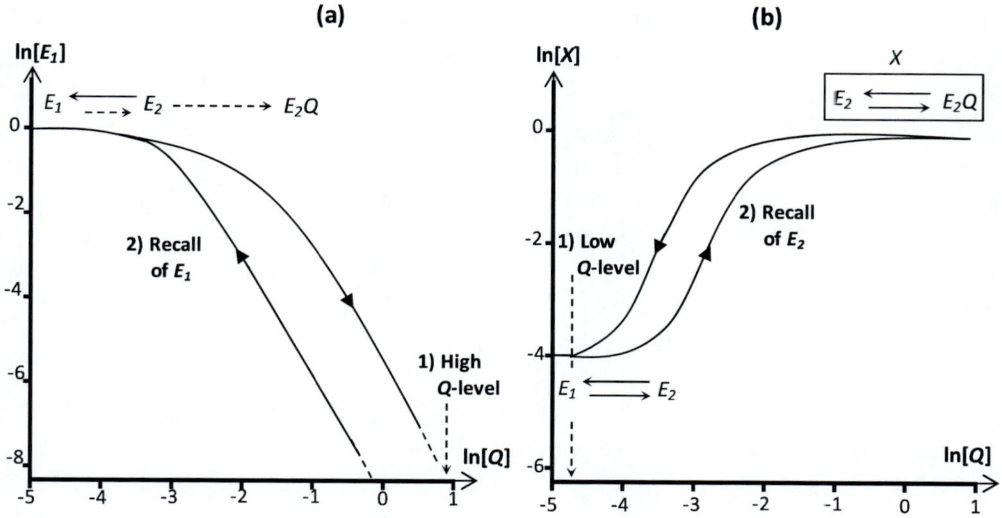

Figure 28. Hysteresis of the levels of enzyme conformations E_1 (a) and X (b) in relation to ligand level (Q) acting as regulator of biochemical reactions [21].

Enzyme conformation hysteresis (Figure 28a, b) makes that enzyme states are implied in detection of both Q-concentration and its variation direction. In this sense, enzyme states may be considered as molecular sensors. The opposite directions of hysteresis loops for alternative conformations (E_1 and X) indicate that a change of chemical potential of the two enzymes (E_1 and E_2) is associated with the specific sensing of the variation direction for a given ligand concentration ([Q]).

Owing to the slowness of the transition between enzyme conformation forms ($E_1 \rightleftarrows E_2 \rightleftarrows E_2Q$), the system tends to recall the conformations E_1 and E_2 under high and low levels of Q, respectively (Figure 28). In this sense, the change of ligand concentration ([Q]) acts as a trigger of the system by generating a thermodynamic force which builds up a difference of chemical potentials of the various enzyme forms. Non-null value of such difference of potential represents a memory basis for enzyme system [21]. Enzyme memory means that the response of the metabolic system depends on its own history. This concept helps the system to detect the direction of the force which is expected subsequently to the variation of ligand concentration ([Q]).

Finally, the enzyme conformations E_1 and X may be considered as molecular sensors, whereas the enzyme-substrate complex (E-S) may be viewed as a convertor of information (e.g. into chemical reactions or biological signals).

Statistically Highlighted Hysteresis

Statistically Highlighted Hysteresis in Metabolomics

Anti-clockwise hysteresis has been statistically highlighted between homovanilic acid (HVA) and its precursor, 3,4-dihydroxyphenyl-acetic acid (DOPAC), in a population of 34 Parkinson disease suffering patients [147]. This work provided a new computational approach to highlight hysteresis from kinetic metabolic datasets. The approach is known by the terms "Weighted Metabolic Profiles Analysis" (WMPA) [147, 148]:

The 34 patients received L-dihydroxy-phenylalanine (L-dopa) as a pro-drug which transforms into dopamine (dopa) (the active metabolite) and other derivatives in the human body (Figure 29) [149-152].

After oral administration of L-dopa, plasma samples were collected at seven times varying from 0.1 to 5, 6 or 7h. In each sample, L-dopa and different derivatives were quantitatively analyzed. With 34 patients and seven samples per patient, the whole dataset consisted of (34 x 7) = 248 metabolic profiles (HPLC profiles) representing L-dopa and its derivatives. From this dataset, four (q = 4) metabolic trends (MbTr I, II, III, IV) were highlighted by statistical multivariate analysis. These MbTrs showed that some metabolites are highly regulated at the expense of the others (Figures 29, 30):

- MbTr I corresponded to high metabolic regulation of 3-O-methyldopa (3-OMD) at the expense of all the others (Figure 30: MbTr I).
- MbTr II was mainly shared between 3-OMD and L-dopa (Figure 30: MbTr II)
- MbTr III was clearly characterized by high regulations of L-dopa (Figure 30: MbTr III)
- MbTr IV showed relatively high regulation of L-dopa and HVA in presence of cumulated levels of 3-OMD (Figure 30: MbTr IV).

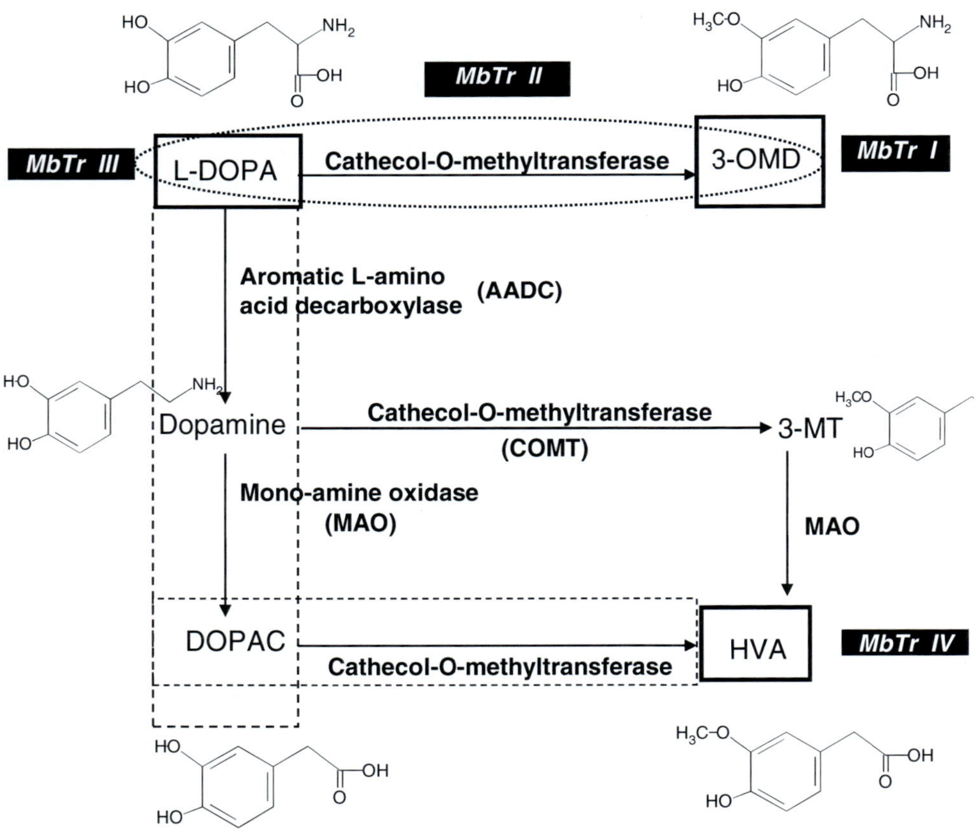

Figure 29. Metabolic chart showing biosynthesis pathways of different metabolites deriving from L-DOPA used as pro-drug against Parkinson disease.

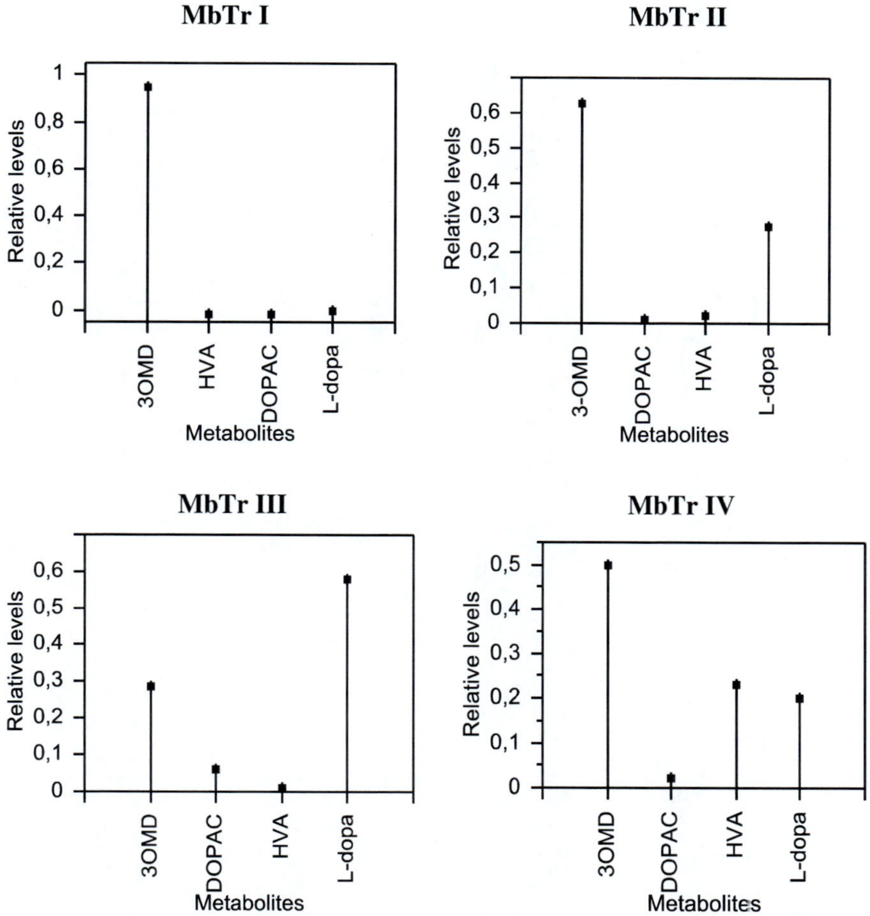

Figure 30. Four typical metabolic profiles representing the four metabolic trends (MbTrs I-IV) highlighted in Parkinson disease suffering patients after administration of L-dopa (pro-drug) [147].

This static picture of metabolism was completed by dynamic analysis taking into account the variation of metabolic concentrations at different sampling times:

The whole dataset was initially stratified into 7 time ranges going from]0-1h] to]6-7h]. For each time range, the four MbTrs were separated (Figure 31a), and their HPLC profiles were iteratively combined and averaged to obtain a complete set of smoothed metabolic patterns (Figure 31b, c, d). These smoothed results represented a population backbone from which different metabolic regulations were graphically highlighted (Figure 31e):

For that, the different HPLC profiles of the four MbTrs were mixed and averaged according to a complete set of mixture design called Scheffé's simplex design (Figure 31b) [147, 148, 153-155]. From each MbTr j, m_j HPLC profiles were randomly sampled such that $\sum_{j=1}^{q=4} m_j = m$. The variable m_j represents the contribution (weight) of MbTr j in the considered mixture (a row of Scheffé's matrix). The constant m is *a priori* fixed and represents the total number of HPLC profiles which are combined to carry out a given entire mixture. In this study, the N different mixtures are quadruplets (m_1, m_2, m_3, $m_{q=4}$) which distinguish the ones from the other by the four variable contributions (m_1-m_4) of the four combined MbTrs.

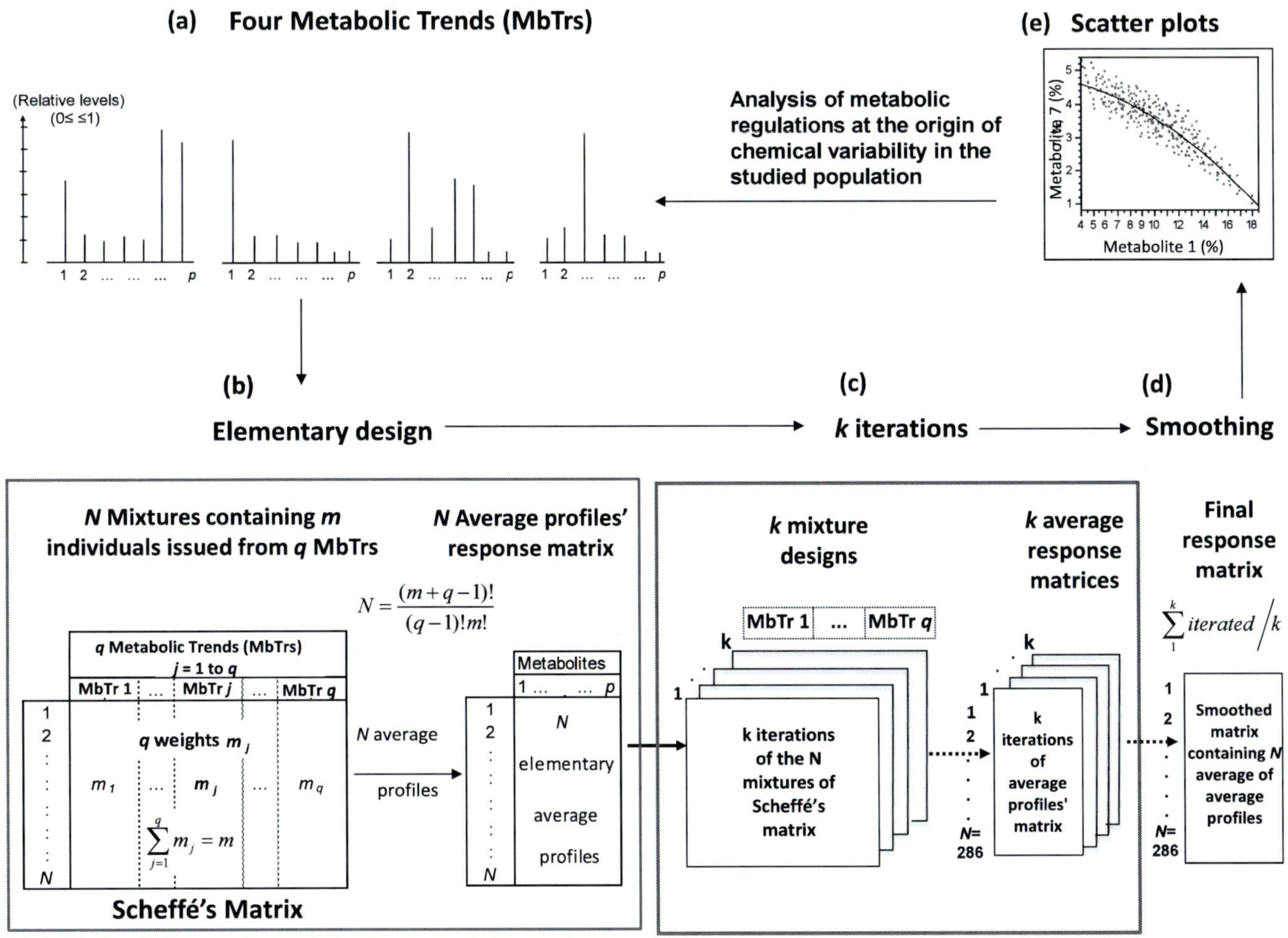

Figure 31. Algorithm applied to analyze regulatory relationships between several metabolites from a stratified population. Application of this algorithm on different time dependent data subsets revealed hysteretic relationship between some metabolites [147].

The total number (N) of mixtures can be calculated by a combinatorial formula depending on m and q (Figure 31b):

$$N = \frac{(m+q-1)!}{(q-1)!m!} \tag{8}$$

From the m randomly sampled profiles of a given mixture (m_1, m_2, m_3, m_4; $\Sigma m_j = m$), an average profile is calculated as a weighted response of the q (=4) MbTrs (Figure 31b). The set of the N average profiles (resulting from an elementary complete mixture design) was smoothed by iterating it several times to explore the variability within each MbTr (Figure 31c).

After k iterations of the mixture design, k response matrices containing N average metabolic profiles were obtained (Figure 31c). Finally, the k elementary response matrices were averaged to obtain a single smoothed response matrix (Figure 31d). In the final response matrix, the N smoothed metabolic profiles provide a set of N gradually weighted states from which central regulatory relationships between metabolites can be highlighted (Figure 31e).

Application of WMPA on the seven separated data subsets (corresponding to the seven time ranges) revealed to be useful for dynamic analysis of regulations between different metabolites. Especially, the pair of metabolites (DOPAC, HVA) showed a hysteretic relationship in which HVA regulation had delayed regulation by reference to its precursor DOPAC [147]: after separated applications of WMPA on the seven time ranges, the seven smoothed data subsets (HVA vs DOPAC) were plotted along discontinuous time axis (Figure 32). The increasing time (from 1 to 7) showed that the seven scatter plots were spatially organized according to an anti-clockwise loop. This lagged regulation of HVA by reference to that of DOPAC could be explained by considering both (i) enzymatic competitions between precursors (DOPAC and 3-MT) and (ii) the downstream position of HVA in the metabolic network (Figure 29):

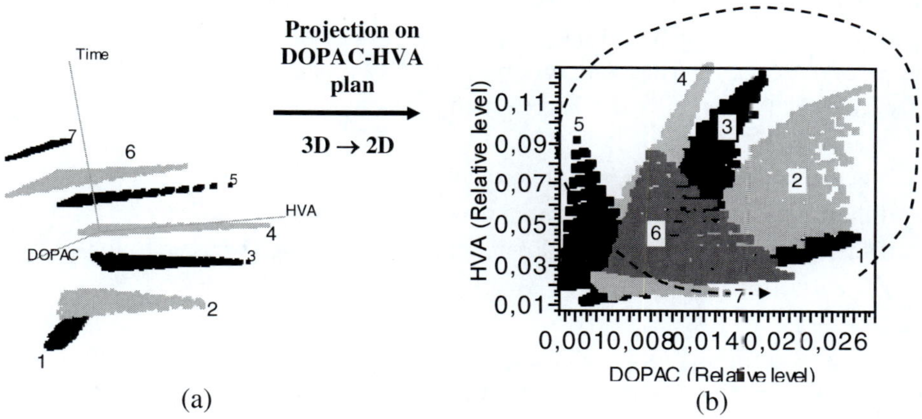

Figure 32. Anti-clockwise hysteretic relationship between HVA and its precursor (DOPAC) highlighted in Parkinson disease suffering population after application of WMPA approach on seven time range-data subsets and projection of the seven smoothed results along discrete time axis. (a) Tridimensional plot HVA vs DOPAC vs Time; (b) bidimensional plot HVA vs DOPAC.

HVA derives either from DOPAC or 3-MT according to two metabolic pathways competing for the same enzymes. The first pathway (Dopamine → DOPAC → HVA) consists of two metabolic steps depending respectively on MAO and COMT. The order of this dependence is reversed in the second metabolic pathway (Dopamine → 3-MT → HVA). This metabolic organization can be at the origin of competitions between the two metabolic pathways leading to a delayed regulation of the final and common product (HVA).

Statistically Highlighted Hysteresis in Visual Perception Field

In vision field, hysteresis provides a context to study perceptual comparisons within an experimental setting. Perception of an object at a given time directly influences its perception ways in the immediate future. In other words, the continuation of an initial perception decision is governed by the immediately preceding perceptual states. Such phenomena imply dynamical systems in which the hysteresis results from a competitive interaction between currently stimulated detectors and those that will be stimulated in the immediate future [156, 157]. Intuitive illustration was presented in figure 3.

Several studies [158-161] used motion quartets as perceptual stimulus: the system consists in simultaneously presenting two small elements in the diagonally opposite corners of an imaginary rectangle. This diagonal corresponds to the hypotenuse of a right triangle. Then the two elements in the other diagonally opposite corners are simultaneously presented. Then the first pair again, and so on.

In experimental response, elements are perceived to move either vertically or horizontally, though never in both directions at the same time. The observer had to mention which motion is perceived: the horizontal or the vertical one. The ratio of vertical distance on the horizontal one (called aspect ratio) was calculated, and several values of this parameter were obtained by gradually increasing or decreasing the vertical distance under fixed horizontal one [24]. This ratio can be associated to a tangent value specific to the visualized right triangle.

For ascending trials, each trial is composed of a sequence of display cycles with increasing aspect ratio's values, the last of which was used to characterize the trial (called End-Point Aspect Ratio : EPAR) (Figure 33a).

The set of n ascending trials with different EPAR is repeated q times.

After each trial, observers (A, B and C) mentioned:

- Whether the initial percept was vertical or horizontal.
- If there was change from the initial percept to the alternative one.
- If there was a loss of the initial percept anytime during the trial (followed by a well-defined percept or not).

As illustrated in figure 33b, q elementary response matrices were obtained. Each matrix contains n rows representing the n trials and three columns representing binary responses to the three perception questions.

Then $3nq$ responses were classified by EPAR (Figure 33c) to allow the calculation of proportions of (Figure 33d):

- Change to new percept

- Loss of the initial percept

These calculations were carried out for each observer.

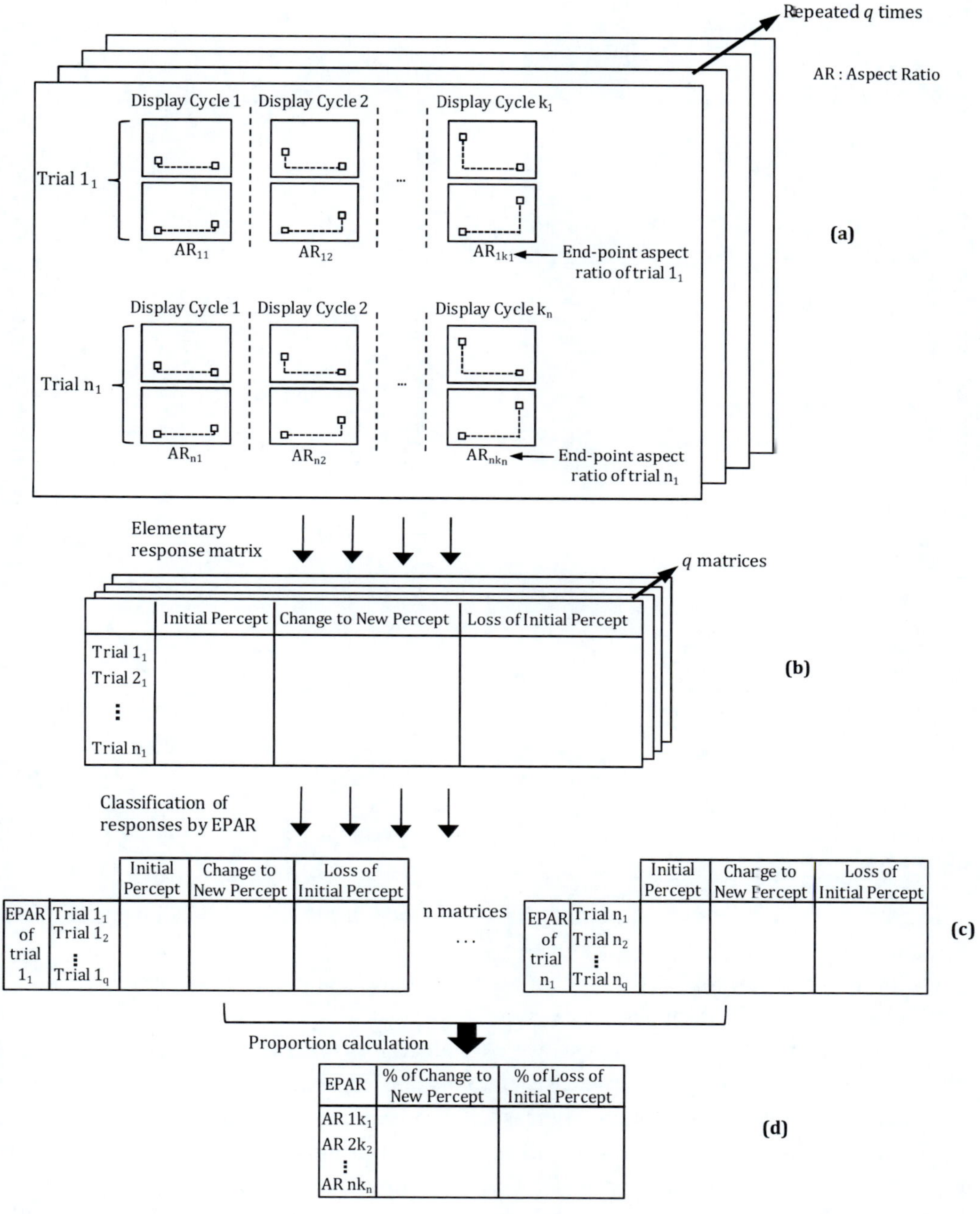

Figure 33. Multi-step methodology based on iterated visual experiences leading to statistically calculate two hysteretic variables: (i) proportions of changes to new percept and (ii) proportions of loss of initial percept. Trials were indexed by two numbers *ij* corresponding to i^{th} trial (i=1-*n*) and its jth iteration (*j*=1-*q*).

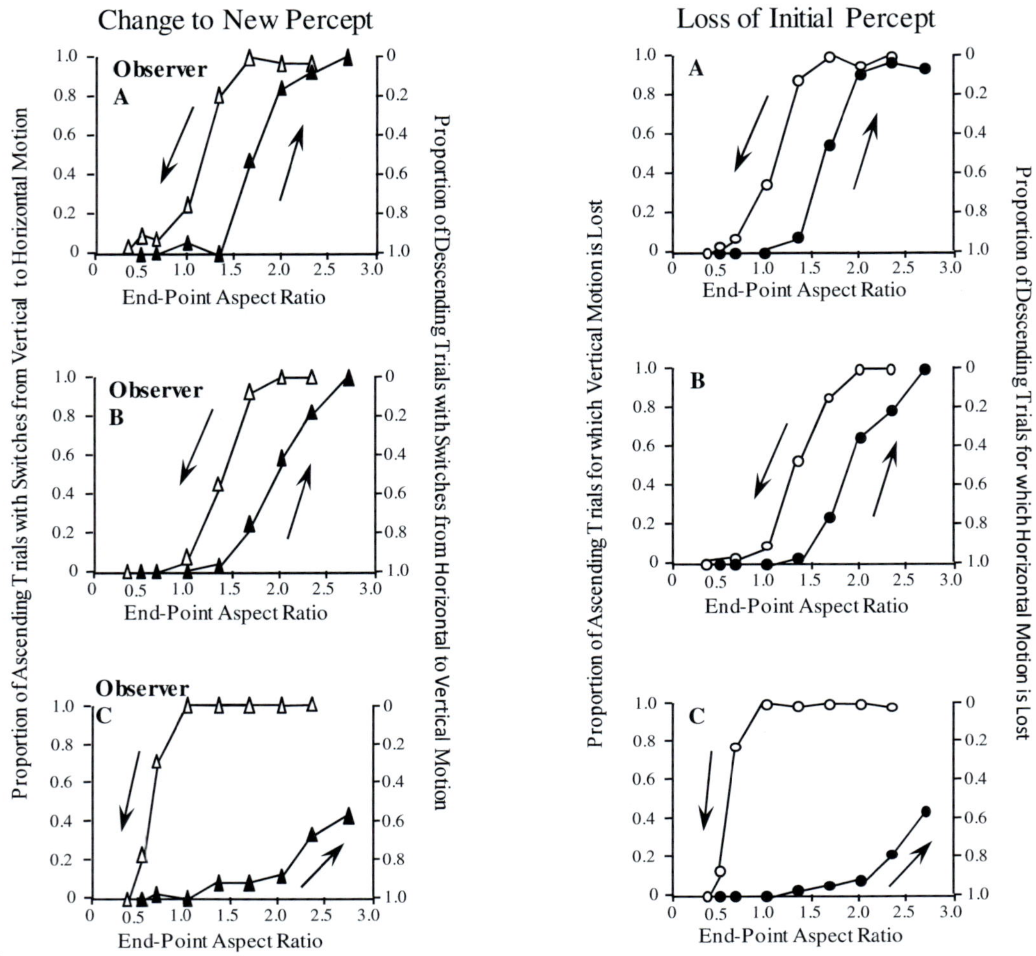

Figure 34. Perception hysteresis concerning three observers A, B, C and highlighted by plotting the proportions of change to new percept (at left) and the loss of the initial percept (at right) versus the EPAR values.

The same procedure was done for the descending trials where display cycles were ordered by decreasing aspect ratio's values.

The different proportions calculated through ascending and descending trials were plotted versus EPAR variable for each observer separately.

As illustrated in figure 34, hysteresis were obtained for (i) the change to new percept and (ii) the loss of initial percept. Large aspect ratios are compatible with higher sensitivity toward horizontal motion perception whereas small ones favor perception of vertical motion [24]. For intermediate values, there were difficulties to discriminate a motion pattern. For this zone, the likelihood of perceiving vertical versus horizontal motion pattern significantly differed between ascending and descending trials. This zone is characterized by a bistability phenomenon: when a percept was lost, it was immediately replaced by an alternative percept [24]. Within-observer analysis showed that the two hysteresis (change to new percept; loss of the initial percept) had globally similar aspects indicating a common phenomenon governing the two progressive perceptions. This phenomenon is linked to a pre-disposition to continue with a previous perceptual decision linked to the inhibitory influence of currently stimulated detectors on those that will be stimulated in the immediate future [156]. However, between-observers analysis highlighted significant difference between hysteresis: among the three observers, the most naïve (C) showed the biggest loop area. Thus, the loop area provides some evaluation of contrast level of individual perception (Figure 34).

ACKNOWLEDGMENT

The authors are grateful toward Ms. Ines Semmar-Houas for the chapter reading.

REFERENCES

[1] Ewing, A., 1885. Experimental Researches in Magnetism. *Phil. Trans. Roy. SOC. London* 176, 523-640.

[2] Enderby, A. J., 1955. The domain model of hysteresis. *Transactions of the Faraday Society* 51, 835-848.

[3] Preisach, F., 1935. Uber die magnetische Nachwirkung. *Z. Physik.* 94, 277-302.

[4] Néel, L., 1942. Théorie des lois d'aimantation de Lord Rayleigh, 1. *Cahiers de Physique* 12, 1-20.

[5] Néel, L., 1943. Théorie des lois d'aimantation de Lord Rayleigh, 2. *Cahiers de Physique* 13, 18-30.

[6] Everett, D. H., 1954. A general approach to hysteresis – part 3: a formal treatment of the independent domain model of hysteresis. *Transactions of the Faraday Society* 50, 1077-1096.

[7] Everett, D. H., 1955. A general approach to hysteresis – part 4: an alternative formulation of the domain model. *Transactions of the Faraday Society* 51, 1551-1557.

[8] Poulovassilis, A., 1962. Hysteresis of pore water – an application of the concept of independent domains. *Journal of Soil Science* 92, 405-412.

[9] Philip, J. R., 1964. Similarity hypothesis for capillary hysteresis in porous materials. *Journal of Geophysical Research* 69, 1553-1562.

[10] Topp, G. C., 1971. Soil–water hysteresis: The domain theory extended to pore interaction conditions. *Soil Science Society of America Proceedings*, 35, 219-225.

[11] Topp, G. C., 1971. Soil water hysteresis in silt loam and clay loam soils. *Water Resources Research* 7, 914-920.

[12] Mualem, Y., 1973. Modified approach to capillary hysteresis based on a similarity hypothesis. *Water Resources Research* 9, 1324-1331.

[13] Parlange, J.-Y., 1976. Capillary hysteresis and the relationship between drying and wetting curves. *Water Resources Research* 12, 224-228.

[14] Parlange, J.-Y. 1980. Water transport in soils. *Annual Revision of Fluid Mechanics* 12, 77-102.

[15] Pang, W. C., Rosowsky, D. V., Pei, S. and van de Lindt, J. W., 2007. Evolutionary Parameter Hysteretic Model for Wood Shear Walls. *Journal of Structural Engineering* 133 (8), 1118-1129.

[16] Chang, H. H., Oh, P. Y., Ingber, D. E. and Huang, S., 2006. Multistable and multistep dynamics in neutrophil differentiation. *BMC Cell Biology*, 7 (11), 1-12.
[17] Hughes, J. M. B; Rosenzweig, D. Y.; Kivitz, P. B., 1970. Site of airway closure in excised dog lungs: histologic demonstration. *J. Appl. Physiol.* 29, 340-344.
[18] Hughes, J. M. B., Jones, H. A. and Wilson, A. G., 1975. Bronchial Hysteresis in excised lungs. *J. Physiol.*, 249, pp. 435-443.
[19] Poisson, P. and Bhalerao, K. D., 2013. Hidden hysteresis – population dynamics can obscure gene network dynamics. *Journal of Biological Engineering* 7 (16), 1-12.
[20] Noori, H. R., 2014. *Hysteresis Phenomena in Biology*. Springer, New York, 45p.
[21] Ricard, J. and Buc, J., 1988. 1. Enzyme memory and sensing chemical signals. *Eur. J. Biochem.* 176, 103-109.
[22] Sutton, G. P, Macknin, J. B., Gartman, S. S., Sunny, G. P., Beer, R. D., Crago, P. E., Neustadter, D. M., Chiel, H. J., 2004. Passive hinge forces in the feeding apparatus of Aplysia aid retraction during biting but not during swallowing. *J. Comp. Physiol. A*, 190, 501-514.
[23] Marck, C., Thiele, D., Schneider, C. and Guschlbauer, W., 1978. Protonated polynucleotides structures - 22.CD study of the acid-base titration of poly(dG).poly(dC). *Nucleic Acids Research* 5 (6), 1979-1996.
[24] Hock, H. S., Bukowski, L., Nichols, D. F., Huisman, A. and Rivera, M., 2005. Dynamical vs. judgmental comparison: hysteresis effects in motion perception. *Spatial Vision* 18 (3), 317-335.
[25] Ricklefs, R. E. and Williams, J. B., 2003. Metabolic responses of shorebird chicks to cold stress: hysteresis of cooling and warming phases. *The Journal of Experimental Biology* 206, 2883-2893.
[26] Brewer, K. K., Sakai, H., Alencar, A. M., Majumdar, A., Arold, S. P., Lutchen, K. R., Ingenito, E. P. and Suki, B., 2003. Lung and alveolar wall elastic and hysteretic behavior in rats: effects of in vivo elastase treatment. *J. Appl. Physiol.* 95, 1926-1936.
[27] Ben-Tal, A., 2006. Simplified models for gas exchange in the human lungs. *Journal of Theoretical Biology* 238, 474-495.
[28] Ganzinger, U. and Nenmann, K., 1992. Quantification of tissue distribution of antibiotics by kinetic hysteresis analysis. *Eur. J. Clin. Pharmacol.* 43, 517-522.
[29] Pleuvry, B. J., 2005. Hysteresis in drug response. *Anaesthesia and Intensive Care Medicine* 6 (8), 286-287.
[30] Pleuvry, B. J., 2008. Hysteresis in drug response. *Anaesthesia and Intensive Care Medicine* 9 (8), 372-373.
[31] Neff, R., Blasius, B., Beck, F. and Lüttge, U., 1998. Thermodynamics and energetics of the tonoplast membrane operating as a hysteresis switch in an oscillatory model of crassulacean acid metabolism. *J. Memb. Biol.* 165, 37-43.
[32] Krasnosel'skii, M. A. and Pokrovskii, A. V., 1989. *Systems with Hysteresis*, Springer, New York, 410p.
[33] Okatan, M. B., Mantese, J. V., Alpay, S. P., 2010. Effect of space charge on the polarization hysteresis characteristics of monolithic and compositionally graded ferroelectrics. *Acta Materialia* 58, 39-48.
[34] Pham, H. Q., Fredlund, D. G. and Barbour, S. L., 2005. A study of hysteresis models for soil-water characteristic curves. *Can. Geotech. J.* 42, 1548-1568.

[35] Nguyen, P. T. M, Do, D. D. and Nicholson, D., 2013. Simulation study of hysteresis of argon adsorption in a conical pore and a constricted cylindrical pore. *Journal of Colloid and Interface Science* 396, 242-250.

[36] Shi, Y. X., Seto-Poon, M., and Wheatley, J. R., 1998. Hysteresis of the nasal pressure-flow relationship during hyperpnea in normal subjects. *J. Appl. Physiol.* 85, 286-293.

[37] Stoeckel, D., 1995. The shape of memory effect. Phenomenon, Alloys and Applications. *Proceedings: Shape Memory Alloys for Power Systems EPRI, Nitinol Devices & Components*, pp. 1-13. www.nitinol.com.

[38] Neumann, E. and Katchalsky, A., 1972. Long-Lived Conformation Changes Induced by Electric Impulses in Biopolymers. *Proc. Nat. Acad. Sci. USA* 69 (4), 993-997.

[39] Katchalsky, A. and Neumann, E., 1972. Hysteresis and molecular memory record. *Intern. J. Neuroscience* 3, 175-182.

[40] Angeli, D., Ferrell, J. E., Jr. and Sontag, E. D., 2004. Detection of multistability, bifurcations, and hysteresis in a large class of biological positive-feedback systems. *Proc. Natl. Acad. Sci.* 101 (7), 1822-1827.

[41] Fisher, G. H., 1967. Measuring ambiguity. *Am. J. Psychol.* 80, 541-547.

[42] Marbe, K., 1893. Die Schwankungen der Gesichtsempfindungen. *Phil. Stud.* 8, 615-637.

[43] Eichler, W., 1930. Der rhythmische Wechsel in der Auffassung räumlich-zweideutiger geometrischer Figuren. *Z. Sinnesphysiol.* 61,154-193.

[44] Haken, H., 1983. *Synergetics. An Introduction*, Vol. 1, Springer-Verlag, Berlin.

[45] Orbach, J., Ehrlich, D., Heath, H. A., 1963. Reversibility of the Necker cube. I. An examination of the concept of "satiation of orientation". *Percept Mot. Skills* 17, 439-458.

[46] Stadler, M. and Erke, H., 1968. Über einige periodische Vorgänge in der Figuralwahrnehmung. *Vision Res.* 8, 1081-1092.

[47] Pöppel, E., 1982. *Lust und Schmerz. Neuronale Grundlagen menschlichen Verhaltens*. Severin & Siedler, Berlin.

[48] Gräser, H., 1977. *Spontane Reversion sprozesse in der Figuralwahrnehmung*. Dissertation, Trier.

[49] Köhler, W., 1940. *Dynamics in psychology*. Liveright, New York.

[50] Frankin, C.E. and Seebacher, F., 2003. The effect of heat transfer mode on heart rate responses and hysteresis during heating and cooling in the estuarine crocodile *Crocodylus porosus*. *J. Exp. Biol.* 206, 1143-1151.

[51] Goudkamp, J. E., Seebacher, F., Ahern, M., Franklin, C. E., 2004. Physiological thermoregulation in a crustacean? Heart rate hysteresis in the freshwater crayfish *Cherax destructor*. *Comparative Biochemistry and Physiology*, Part A 138, 399-403.

[52] Grigg, G. C., Alchin, J., 1976. The role of the cardiovascular system in thermoregulation of *Crocodylus johnstoni*. *Physiol. Zool.* 49, 24-36.

[53] Bartholomew, G. A., 1982. Physiological control of body temperature. In: Gans, C., Pough, F. H. (Eds.), *Biology of the Reptilia*, vol. 12. Academic Press, New York, 167-212.

[54] Grigg, G. C., Seebacher, F., 1999. Field test of a paradigm: hysteresis of heart rate in thermoregulation by a free-ranging lizard (*Pogona barbata*). *Proc. R. Soc. Lond.* B266, 1291-1297.

[55] Seebacher, F., Grigg, G. C., 2001. Changes in heart rate are important for thermoregulation in the varanid lizard *Varanus varius*. *J. Comp. Physiol.* B171, 395-400.

[56] Seebacher, F., Franklin, C. E., 2001. Control of heart rate during thermoregulation in the heliothermic lizard, *Pogona barbata*: importance of cholinergic and adrenergic mechanisms. *J. Exp. Biol.* 204, 4361-4366.

[57] Robleto, D. O. and Herman, C. A., 1988. Cardiovascular effects of prostaglandin I2 and prostaglandin F2□ in the unanesthetised bullfrog, *Rana catesbeiana*. *J. Exp. Zool.* 246, 10-16.

[58] Altimiras, J., Franklin, C. E. and Axelsson, M., 1998. Relationship between blood pressure and heart rate in the salt water crocodile *Crocodylus porosus*. *J. Exp. Biol.* 201, 2235-2242.

[59] Seebacher, F., 2000. Heat transfer in a microvascular network: the effect of heart rate on heating and cooling in reptiles (*Pogona barbata* and *Varanus varius*). *J. Theor. Biol.* 203, 97-109.

[60] Forward, R. B., 1990. Behavioural responses of crustacean larvae to rates of temperature change. *Biol. Bull.* 178, 195-204

[61] Jury, S. H., Watson III, W. H., 2000. Thermosensitivity of the lobster, *Homarus americanus*, as determined by cardiac assay. *Biol. Bull.* 199, 257-264.

[62] Escolar and Escolar, 2004. Lung hysteresis: a morphological view. *Histology and Histopathology* 19, 159-166.

[63] Hoppin F. G. Jr, Stothert J. C. Jr, Greaves I. A., Lai, Y. L. and Hildebrandt, J., 1986. Lung recoil: elastic and rheological properties. In: *Handbook of physiology: the respiratory system*. Vol. III, Part I. Macklem P, Mead J, (Eds). Bethesda, MD: American Physiological Society, 195-215.

[64] Glaister, D. H., Schroter, R. C., Sudlow, M. F. and Milic-Emili, J., 1973. Bulk elastic properties of excised lungs and the effect of transpulmonary pressure gradient. *Respir. Physiol.* 17, 347-364.

[65] Hoppin F. G., 1999. Sources of lung recoil. In: Respiratory mechanics. Milic-Emili J. (Ed.). *Respir. Monogr.* 12, 33-53.

[66] Yuan, H., Kononov, S., Cavalcante, F. S. A., Luctchen, K. R., Ingenito, E. P. and Suki, B., 2000. Effects of collagenase and elastase on the mechanical properties of lung tissue strips. *J. Appl. Physiol.* 89, 3-14.

[67] Stamenovic, D. and Smith, J. C., 1986. Surface forces in lung.III. Alveolar suface tension and elastic properties of lung parenchima. *J. Appl. Physiol.* 60, 1358-1362.

[68] Von Neergaard, K., 1929. Neue Auffassungen über einen Grundbegriff der Atemmechanik; Die Retraktionskraft der Lunge, abhängig von der Oberflächenspannung in den Alveolen. *Z. Ges. Exp. Med.* 66, 373-394.

[69] Wilson, T. W., 1981. Relations among recoil pressure, surface area and surface tension in the lung. *J. Appl. Physiol.: Respir. Environ. Exerc. Physiol.* 50, 921-926.

[70] Frazer, D. G., Weber, K. C. and Franz, G. N., 1985. Evidence of sequential opening and closing of lung units during inflation of excised rat lungs. *Respir .Physiol.* 61, 277-288.

[71] Cheng, W., DeLong, D. S., Franz, G. N., Petsonk, E. L. and Frazer, D. G., 1995. Contribution of opening and closing of units to lung hysteresis. *Respir. Physiol.* 102, 205-215.

[72] Ingimarsson, J., Björklund, L. J., Larsson, A. and Werner, O., 2001. The pressure at the lower inflexion point has no relation to airway collapse in surfartant-treated premature lambs. *Acta Anaesthesiol. Scand.* 45, 690-695.
[73] Kleinman, L. E., Poulos, D. A., Siebens A. A., 1964. Minimal air in dogs. *J. Appl. Physiol.* 19, 204-206.
[74] Milic-Emili, J. and D'Angelo, E., 2005. Statics of the lung. In: *Physiologic Basis of Respiratory Disease.* Hamid, Q., Shannon, J., Martin, J. (Eds). BC Decker Inc, Hamilton, Ontario, 27-34.
[75] Macklem, P. T., Proctor, D. F., Hogg, J. C., 1969. The stability of peripheral airways. *Respir. Physiol.* 8, 191-203.
[76] Milic-Emili, J., Henderson, J. A. M, Dolovich, M. B., Trop D., Kaneko K., 1966. Regional distribution of inspired gas in the lung. *J. Appl. Physiol.* 21, 749-759.
[77] Fredberg, J. J. and Stamenovic, D., 1989. On the imperfect elasticity of lung tissue. *J. Appl. Physiol.* 67, 2408-2419.
[78] Bachofen, H., Hildebrandt, J. and Bachofen, M., 1970. Pressure-vclume curves of air- and liquid-filled excised lung-surface tension in situ. *J. Appl. Physicl.* 29, 422-431.
[79] Bachofen, H., Schürch, S., Urbinelli, M. and Weibel, E. R., 1987. Relations among alveolar surface tension, surface area, volume, and recoil pressure. *J. Appl. Physiol.* 62, 1878-1887.
[80] Glaister, D. H., Schroter, R. C., Sudlow, M. F. and Milic-Emili, J., 1973. Transpulmonary pressure gradient and ventilation distribution in excised lungs. *Respir. Physiol.* 17, 365-385.
[81] Cavagna G. A., Stemmler E. J., DuBois A. B., 1967. Alveolar resistance to atelectasis. *J. Appl. Physiol.* 22, 441-52.
[82] Mead J., Whittenberger J. L. and Radford E. P., 1957. Surface tension as a factor in pulmonary volume-pressure hysteresis. *J. Appl. Physiol.* 10, 191-196.
[83] Bachofen, H. and Hildebrandt, J., 1971. Area analysis of pressure-volume hysteresis in mammalian lung. *J. Appl. Physiol.* 30, 493-497.
[84] Radford E. P., 1965. Static mechanical properties of mammalian lungs. In: *Handbook of physiology*. Fenn W. O. (Ed.). Bethesda: American Physiological Society 1964-1965, 429-449.
[85] Jones, T. A., Petsonk, E. L. and Frazer, D. G., 1996. Effect of temperature on pressure-volume hysteresis of excised lung. *Respir. Physiol.* 106, 47-55.
[86] Warner, R. C., 1956. Ultraviolet spectra of enzymatically synthesized polynucleotides. *Fed. Proc* 15, 379.
[87] Beers, R. F. and Steiner, R. F., 1958. Spectrophotometric, light scattering, and titration studies upon the interaction of polyadenylic and polyuridylic acids. *Nature* 181, 30-32.
[88] Cox, R. K., 1963. Dissociation properties of ribonucleic acid I. Titration of rat-liver RNA and model polynucleotides. *Biochim. Biophys. Acta* 68, 401-410.
[89] Vetterl, V. and Cuschlbauer, W., 1972. Protonated polynucleotide structures: XI. Polyadenylic acid at high ionic strength. *Archives of Biochemistry and Biophysics* 148, 130-140.
[90] Neumann, E. and Katchalsky, A., 1970. Thermodynamische Untersuchung der Hysterese in System Polyriboadenyl-Polyribouridylsäure-Modell einer makromolekularen Gedächtnis-Aufzeichnung, Ber. Bunsenges. *Physik. Chem.* 74, 868-879.

[91] Felsenfeld, G., Davies, D. R. and Rich, A., 1957. Formation of a three-stranded polynucleotide molecule. *J. Am. Chem. Soc.* 79, 2023-2024.

[92] Neumann, E., 1973. Molecular Hysteresis and its Cybernetic Significance. *Angew. Chem. internat.* 12 (5), 356-369.

[93] Spodheim, M. and Neumann, E., 1975. Ionic strength of the hysteresis in the polyriboadenylate- polyribouridylate system. *Biophysical Chemistry* 3, 109-124.

[94] Cox, R. A., Jones, A. S., Marsh, G. E. and Peacocke, A. R., 1956. On hydrogen bonding and branching in a bacterial RNA. *Biochim. Biophys. Acta* 21, 576-577.

[95] Roberts, R. B. and Flexner, L. B., 1969. The Biochemical basis of long-term memory, *Quart Rev. Biophysics* 2, 135-173.

[96] Uttley, A. M., 1970. The informon: the network of adaptive pattern recognition. *J. Theor. Biol.* 27, 31-67.

[97] Andersson, T., 2010. Exploring voltage-dependent ion channels in silico by hysteretic conductance. *Mathematical Biosciences* 226, 16-27.

[98] Börjesson, S. I., Elinder, F., 2008. Structure, function, and modification of the voltage sensor in voltage-gated ion channels. *Cell Biochem. Biophys.* 52 (3), 149-174.

[99] Hille, B., 2001. *Ion Channels of Excitable Membranes* (3rd ed.). Sinauer Associates, Inc., Sunderland, USA.

[100] Hodgkin, A. L., Huxley, A. F., 1952. A quantitative description of membrane current and its application to conduction and excitation in nerve. *J. Physiol.* 117, 500.

[101] Kaestner, L., Christophersen, P., Bernhardt, I., Bennekou, P., 2000. The non-selective voltage-activated cation channel in the human red blood cell membrane: reconciliation between two conflicting reports and further characterization. *Bioelectrochemistry* 52, 117-125.

[102] Männikkö, R. Pandey, S. Larsson, H. P., Elinder, F., 2005. Hysteresis in the voltage dependence of HCN channels: conversion between two modes affects pacemaker properties. *J. Gen. Physiol.* 125, 305-326.

[103] Nowak, L. M. and Wright, J. M., 1992. Slow voltage-dependent changes in channel open-state probability underlie hysteresis of NMDA response in Mg^{2+}-free solutions. *Neuron* 8, 181-187.

[104] Fologea, D., Krueger, E., Mazur, Y. I., Stith, Ch., Okuyama, Y., Henry, R., Salamo, G. J., 2011. Bi-stability, hysteresis, and memory of voltage-gated lysenin channels. *Biochimica et Biophysica Acta* 1808, 2933-2939.

[105] Fologea, D., Krueger, E., Lee, R., Naglak, M., Mazur, Y., Henry, R., Salamo, G., 2010. Controlled gating of lysenin pores. *Biophys. Chem.* 146, 25-29.

[106] Shakor, A. B. A., Czurylo, E. A. and Sobota, A., 2003. Lysenin, a unique sphingomyelin-binding protein. *FEBS Lett.* 542, 1-6.

[107] Yamaji-Hasegawa, A., Makino, A., Baba, T., Senoh, Y., Kimura-Suda, H., Sato, S.B., Terada, N., Ohno, S., Kiyokawa, E., Umeda, M., Kobayashi, T., 2003. Oligomerization and pore formation of a sphingomyelin-specific toxin, lysenin. *J. Biol. Chem.* 278, 22762-22770.

[108] Ide, T., Aoki, T., Takeuchi, Y., Yanagida, T., 2006. Lysenin forms a voltage-dependent channel in artificial lipid bilayer membranes. *Biochem. Biophys. Res. Commun.* 346, 288-292.

[109] Kwiatkowska, K., Hordejuk, R., Szymczyk, P., Kulma, M., Abdel-Shakor, A.-B., Plucienniczak, A., Dolowy, K., Szewczyk, A., Sobota, A., 2007. Lysenin-His, a

sphingomyelin recognizing toxin, requires tryptophan 20 for cation-selective channel assembly but not for membrane binding. *Mol. Membr. Biol.* 24, 121-134.

[110] Lu, W., Lieber, C. M., 2007. Nanoelectronics from the bottomup. *Nat. Mater.* 6, 841-850.

[111] Waser, R. and Aono, M., 2007. Nanoionics-based resistive switching memories. *Nat. Mater.* 6, 833-840.

[112] Wurstbauer, U., Sliwa, C., Weiss, D., Dietl, T., Wegscheider, W.. 2010. Hysteretic magnetoresistance and thermal bistability in a magnetic two-dimensional hole system. *Nat. Phys.* 6, 955-959.

[113] Azene, E. M., Xue, T., Marbán, E., Tomaselli, G. F. and Li, R. A., 2005. Non-equilibrium behavior of HCN channels: insights into the role of HCN channels in native and engineered pacemakers. *Cardiovasc. Res.* 67, 263.

[114] Nayak, T. K., Sikdar and S. K., 2007. Time-dependent molecular memory in single voltage-gated sodium channel. *J. Membr. Biol.* 219 (1-3), 19.

[115] Pustovoit, M. A., Berezhkovskii, A. M. and Bezrukov, S. M., 2006. Analytical theory of hysteresis in ion channels: two-state model. *J. Chem. Phys.* 125 (19), 194907.

[116] Villalba-Galea, C. A., Sandtner, W., Starace, D. M., Bezanilla, F., 2008. S4-based voltage sensors have three major conformations. *PNAS* 105 (46), 17600.

[117] Xu, H., Ramsey, I. S., Kotecha, S. A., Moran, M. M., Chong, J. A.. Lawson, D., Ge, P., Lilly, J., Silos- Santiago, I., Xie, Y., DiStefano, P. S., Curtis, R., Clapham, D. E., 2002. TRPV3 is a calcium-permeable temperature-sensitive cation channel. *Nature* 418, 181-186.

[118] Pershin, Y. V., La Fontaine, S., Di Ventra, M., 2009. Memristive model of amoeba learning. *Phys. Rev.* E 80, 021926.

[119] Saigusa, T., Tero, A., Nakagahi, T., Kuramoto, Y., 2008. Amoebae anticipate periodic events. *Phys. Rev. Lett.* 100, 018101.

[120] Bezanilla, F., 2000. The voltage sensor in voltage-dependent ion channels. *Physiol. Rev.* 80, 555-592.

[121] Bezanilla, F., 2005. Voltage-gated ion channels. *IEEE Trans. Nanobiosci.* 4, 34-48.

[122] Nilius, B., Talavera, K., Owsianik, G., Prenen, J., Droogmans, G., Voets, T., 2005. Gating of TRP channels: a voltage connection? *J. Physiol.* 567, 35-44.

[123] Chapin, F., Woodwell, G., Randerson, J., Rastetter, E., Lovett, G., Baldocchi, D., Clark, D., Harmon, M., Schimel, D., Valentini, R., Wirth, C., Aber, J., Cole, J., Goulden, M., Harden, J., Heimann, M., Howarth, R., Matson, P., McGuire, A., Melillo, J., Mooney, H., Neff, J., Houghton, R., Pace, M., Ryan, M., Running, S., Sala, O., Schlesinger, W., and Schulze, E. D., 2006. Reconciling carbon-cycle concepts, terminology, and methods. *Ecosystems* 9, 1041-1050.

[124] Pingintha, N., Leclerc, M. Y., Beasley, J. P., Durden1, Jr., D., Zhang, G., Senthong, C. and Rowland, D., 2010. Hysteresis response of daytime net ecosystem exchange during drought. *Biogeosciences* 7, 1159-1170.

[125] Körner, C., 1995. Leaf diffusive conductances in the major vegetation types of globe, in: *Ecophysiology of photosynthesis*, edited by: Schulze, E. D. and Caldwell, M. M., Springer, Berlin, 463-490.

[126] Bahn, M., Rodeghiero, M., Anderson-Dunn, M., Dore, S., Gimeno, C., Drösler, M., Williams, M., Ammann, C., Berninger, F., Flechard, C., Jones, S., Balzarolo, M., Kumar, S., Newesely, C., Priwitzer, T., Raschi, A., Siegwolf, R., Susiluoto, S.,

Tenhunen, J., Wohlfahrt, G., and Cernusca, A., 2008. Soil respiration in european grasslands in relation to climate and assimilate supply. *Ecosystems* 11, 1352-1367.

[127] Gaumont-Guay, D., Black, T. A., Griffis, T. J., Barr, A. G., Jassal, R. S., and Nesic, Z., 2006. Interpreting the dependence of soil respiration on soil temperature and water content in a boreal aspen stand. *Agr. Forest Meteorol.* 140, 220-235.

[128] Graf, A., Weiherm¨uller, L., Huisman, J. A., Herbst, M., Bauer, J. and Vereecken, H., 2008. Measurement depth effects on the apparent temperature sensitivity of soil respiration in field studies. *Biogeosciences* 5, 1175-1188.

[129] Tang, J., Baldocchi, D. D. and Xu, L., 2005. Tree photosynthesis modulates soil respiration on a diurnal time scale. *Glob. Change Biol.* 11, 1298-1304.

[130] Boroujerdi, M., 2002. *Pharmacokinetics. Principle and Applications.* McGraw-Hill, New York, 420p.

[131] Shargel, L. and Yu A. B. C., 1999. *Applied Biopharmaceutics & Pharmacokinetics* (4th ed.). McGraw-Hill, New York, 768p.

[132] Chiang, C. W., Barnett G., 1984. Marijuana effect and delta-9-tetrahydrocannabinol plasma level. *Clin. Pharmacol. Ther.* 36, 234-238.

[133] Stanski, D. R., Hudson, R. J., Homer, T. D., Saidman, L. J., Meathe, E., 1984. Pharmacodynamic modeling thiopental anesthesia. *J. Pharmacokinet. Biopharm.* 1, 223-240.

[134] Ohlsson, A., Lindgren, J. E., Wahlen, A., Agurell, S., Hollister, L. E., Gillespie, H. K., 1980. Plasma delta-9-tetrahydrocannabinol concentrations and clinical effect after oral and intravenous administration and smoking. *Clin. Pharmacol. Ther.* 28, 409-416.

[135] Blundell, J. E., Campbell, D. B., Lesham, M., Tozer, R., 1975. Comparison of the time course of the anorectic of fenfluramine and amphetamine with drug levels in blood. *J. Pharm. Pharmacol.* 27, 187-192.

[136] Campbell, D. B., 1990. The use of kinetic-dynamic interactions in the evaluation of drugs. *Psychopharmacology* 100, 433-450.

[137] Nagashima, R., O'ReiUy, R. A., Levy, G., 1969. Kinetics of pharmacologic effects in man; the anticoagulant action of warfarin. *Clin. Pharmacol. Ther.* 10, 22-35.

[138] Shetty H. G. M, Fennerty, A. G, Routledge, P. A., 1989. Clinical pharmacokinetic considerations in the control of oral anticoagulant therapy. *Clin. Pharmacokinet.* 16, 238-253.

[139] Van Dyke, C., Jatlow, P., Ungerer, J., Barash, P.G., Byck R., 1978. Oral cocaine: plasma concentrations and central effects. *Science* 200, 211-213.

[140] Zahter, R., Wachtel, P., Jatlow, P., Byck, R., 1982. Kinetics of drug effect by distributed lags analysis: an application to cocaine. *Clin. Pharmacol. Ther.* 31, 775-782.

[141] Kalant, H.; LeBlanc, A. E.; Gibbons, R. T., 1971. Tolerance to and dependence on some non-opiate psychotropic drugs. *Pharmacol. Rev.* 23, 135-190.

[142] Jones, B. M.; Vega, A., 1972 Cognitive performance measured on the ascending and descending limb of the blood alcohol curve. *Psychopharmacologia* 23(2), 99-114.

[143] Rosenfeld, G. C., Burks, T. F., 1977. Single-dose tolerance to morphine hypothermia in the rat: differentiation of acute from long-term tolerance. *J. Pharmacol. Exp. Ther.* 202, 654-659.

[144] Schuttler, J., Schwilden, H., Stoeckel, H., 1985. Infusion strategies to investigate the pharmacokinetics and pharmacodynamics of hypnotic drugs: etomidate as an example. *Eur. J. Anaesthesiol.* 2 133-142.

[145] Ellinwood, E. H. Jr, Heatherly, D. G., Nikaido, A. M., Bjornsson, T. D., Kilts, C., 1985. Comparative pharmacokinetics and pharmacodynamics of lorazepam, alprazolam and diazepam. *Psychopharmacology* 86, 392-399.

[146] Sunzel, M., Paalzow, L., Berggren, L., Eriksson, I., 1988. Respiratory and cardiovascular effects in relation to plasma levels of midazolam and diazepam. *Br. J. Clin. Pharmacol.* 25, 561-569

[147] Semmar, N., 2010. A new mixture design-based approach to graphical screening of potential interconnections and variability processes in metabolic systems. *Chem. Biol. Drug Des.* 75, 91-105.

[148] Semmar N., 2013. Two Computational Simplex Approaches to Graphical Highlighting Metabolic Phenotypes and their Functional Origins: Correspondence Analysis and Weighted Metabolic Profiles Analysis. In: *Metabolomics Coming of Age with its Technological Diversity*. Advances in Botanical Research, Vol. 67, 1st Edition, D. Rolin (Ed.). Academic Press, Burlington, 441-492.

[149] Bonifati, V. and Meco, G., 1999. New selective cathecol-O-methyltransferase inhibitors as therapeutic agents in parkinson's disease. *Pharmacol. Ther.* 81, 1-36.

[150] Wade, L. A. and Katzman R., 1975. 3-O-Methyldopa uptake and inhibition of L-dopa at the blood-brain barrier. *Life Sci.* 17, 131-136.

[151] Gervas, J. J., Muradas, V., Bazan, E., Aguado, E. G., de Yébenes, J. G., 1983. Effects of 3-OM-dopa on monoamine metabolism in rat brain. *Neurology* 33, 278-282.

[152] Nutt, J. G., Fellman, J. H., 1984. Pharmacokinetics of levodopa. *Neuropharmacol.* 7, 35-49.

[153] Semmar N., 2011. *Computational Metabolomics*. Nova Science Publishers, New York, 238p.

[154] Semmar, N., Jay, M. and Nouira, S., 2007. A new approach to graphical and numerical analysis of links between plant chemotaxonomy and secondary metabolism from HPLC data smoothed by a simplex mixture design. *Chemoecology* 17, 139-156.

[155] Semmar, N., Jay, M., Farman, M. and Roux, M., 2008. A New Approach to Plant Diversity Assessment Combining HPLC Data, Simplex Mixture Design and Discriminant Analysis. *Environ. Model. Assess.* 13, 17-33.

[156] Hock, H. S., Schöner, G., Giese, M., 2003. The dynamical foundations of motion pattern formation: stability, selective adaptation, and perceptual continuity. *Perception and Psychophysics* 64, 429-457.

[157] Kelso, J. A. S., 1995. *Dynamic Patterns: Self-organization of Brain and Behavior*. MIT Press, Cambridge, MA, USA.

[158] Hock, H. S., Kelso, J. A. S., Schöner, G., 1993. Bistability and hysteresis in the organization of apparent motion patterns. *J. Exper. Psychology: Human Perception and Performance* 19, 63-80.

[159] Kruse, P., Stadler, M., Wehner, T., 1986. Direction and frequency specific processing in the perception of long-range apparent movement. *Vision Research* 26, 327-335.

[160] Ramachandran, V. S., Anstis, S. M., 1985. Perceptual organization in multistable apparent motion. *Perception* 14, 135-143.

[161] Von Schiller, P., 1933. Stroboskopische Alternativbewegungen. *Psychologische Forschung* 17, 179-214.

In: Hysteresis
Editor: Jos Carlos Dias

ISBN: 978-1-63321-336-4

Chapter 12

NEUROMECHANICAL MODELS REVEALING HYSTERESIS MECHANISM IN GAIT TRANSITIONS OF ANIMALS

Shinya Aoi [1,2*]
[1] Dept. of Aeronautics and Astronautics, Graduate School of Engineering, Kyoto University, Kyoto Daigaku-Katsura, Nishikyo-ku, Kyoto, Japan
[2] JST, CREST, 5 Sanbancho, Chiyoda-ku, Tokyo, Japan

Abstract

Animals use various gaits in their locomotion and hysteresis appears when the gaits vary in accordance with the locomotion speed. That is, the gaits change at different locomotion speeds depending on the speed change direction. Although hysteresis is a typical characteristic of nonlinear dynamic systems, the hysteresis mechanism in the gait transitions remains unclear. In this chapter, neuromechanical models of a quadruped and an insect are constructed and dynamic characteristics of gaits and gait transitions are investigated. The simulation results show that these animal models produce different gaits depending on the locomotion speed through dynamic interactions between the body mechanical and nervous systems. In addition, the animal models exhibit the gait transitions with hysteresis by changing the locomotion speed. This chapter examined the hysteresis mechanism of the gait transitions in quadruped and insect locomotion from a dynamic viewpoint.

Keywords: hysteresis, quadruped, insect, gait transition, neuromechanical model, central pattern generator, phase resetting

1. Introduction

Humans and animals change their gaits depending on the locomotion speed. Different animals have a different number of legs and the gaits are characterized by the number of legs used for locomotion. Humans walk bipedally and employ walking and running gaits.

*Correspondence: shinya_aoi@kuaero.kyoto-u.ac.jp

Quadruped animals with four legs use various gaits, such as walking, trotting, and galloping gaits. Insects (hexapods) have six legs and use metachronal and tripod gaits. These gaits are generated over a limited range of locomotion speeds and they are characterized by parameters that vary discontinuously at the transition [1]. In the walk–run transition of humans, the relative phase between the leg segments (that is, the intralimb (or intersegmental) coordination pattern) changes [17]. On the other hand, in the walk–trot–gallop transition of quadrupeds and the metachronal–tripod gait transition of insects, the relative phase between the legs (that is, the interlimb coordination pattern) varies [1, 58]. Despite such gait transitions having been investigated from various viewpoints including mechanics, energetics, kinematics, and kinetics [21, 28, 33, 49, 56], their underlying mechanism remains largely unclear.

Locomotion in humans and animals is a self-organizing phenomenon that emerges through dynamic interactions between the body and nervous systems and it has a number of nonlinear characteristics. In particular, hysteresis appears when the gaits change in accordance with the locomotion speed [17, 28, 30, 34, 41, 49, 52, 56]. In other words, the gait changes at different locomotion speeds depending on whether the speed is increasing or decreasing. Although hysteresis is a typical characteristic of nonlinear dynamic systems, the hysteresis mechanism in the gait transitions has not been clarified.

To data, biomechanical and physiological studies have been independently conducted to elucidate the locomotor mechanism in humans and animals. Biomechanical studies mainly examine the functional roles of the musculoskeletal system, whereas physiological studies generally investigate the configurations and activities of the neural system. However, locomotion is a well-organized motion generated by dynamic interactions between the body and nervous systems and it is thus difficult to fully analyze locomotion mechanisms solely from a single perspective. To overcome the limitations of studies based on a single approach, neuromechanical models for the locomotion of humans and animals have been developed by integrating the biomechanical and physiological findings [14, 20, 31, 39, 45, 48, 54, 55, 57].

This chapter focuses on the walk–trot transition in quadrupeds and the metachronal–tripod gait transition in insects and shows the neuromechanical models [2, 23] to reveal the hysteresis mechanism in their gait transitions from a dynamic viewpoint.

2. Neuromechanical Model

Neuromechanical models for a quadruped and an insect were developed by integrating the body mechanical model and nervous system model.

2.1. Body Mechanical Model

Figure 1**A** and **B** show the body mechanical models of a quadruped and an insect, respectively. The quadruped model consists of one rigid body and two pairs of legs (leg 1–4). Each leg is attached to the body through a pitch joint (joint 1) and is comprised of two rigid links connected by a pitch joint (joint 2). The insect model consists of one rigid body and three pairs of legs (leg 1–6). Each leg has three rigid links (coxa, femur, and tibia). The body-coxa jonit moves the leg tip forward or backward relative to the body and the coxa-femur and femur-tibia joints raise or lower the leg tip.

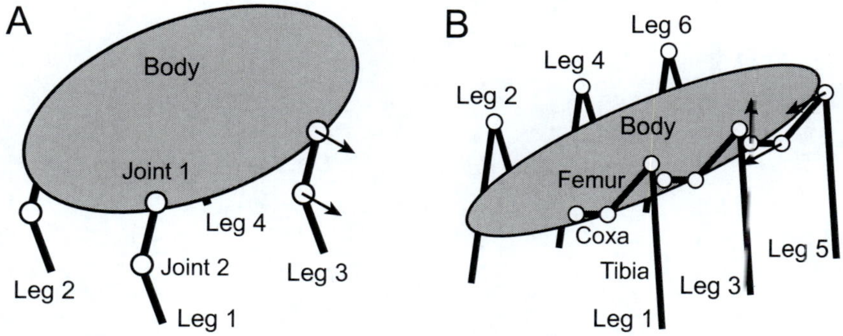

Figure 1. Body mechanical models of (**A**) a quadruped and (**B**) an insect.

The equations of motion of the body mechanical models were derived using a Lagrangian equation, where viscoelastic elements were used to model the contact between the leg tips and the ground and large values were used for the viscoelastic parameters so that the leg tips rarely slipped during foot contact. Forward dynamic simulations were performed by solving the equations of motion using a fourth-order Runge-Kutta method.

2.2. Nervous System Model

Physiological studies have shown that central pattern generators (CPGs) greatly contribute to producing motor commands for rhythmic leg movements in both vertebrates and invertebrates [6, 29, 42, 43, 53], which are located in the spinal cord of vertebrates and in the thoracic ganglia of invertebrates. Although the organization of CPGs remains unclear, physiological findings suggest that CPGs consist of hierarchical networks composed of rhythm generator (RG) and pattern formation (PF) networks [9, 40, 50]. The RG network produces the basic rhythm for locomotion and the PF network shapes the rhythm into spatiotemporal patterns of motor commands. In this study, simple oscillator network models were used for the RG and PF models of the nervous system model for the quadruped and insect models (Fig. 2).

2.2.1. Rhythm generator model

In this study, the RG model produces rhythm information for locomotion through dynamic interactions between the body mechanical and nervous systems. For the RG model, four phase oscillators were used for the quadruped model (oscillator 1–4) and six phase oscillators were used for the insect model (oscillator 1–6) to generate the basic rhythm and phase information for the corresponding leg based on commands related to the desired locomotion speed and gait. The phase of the oscillator i ($i = 1, \ldots, 4$ for the quadruped model, $i = 1, \ldots, 6$ for the insect model) is denoted by ϕ_i ($0 \leq \phi_i < 2\pi$) and they are governed by the following phase dynamics:

$$\dot{\phi}_i = \omega + g_{1i} + g_{2i} \tag{1}$$

where ω is the basic oscillator frequency, g_{1i} is the interaction between the oscillators based on the interlimb coordination (see Section 2.2.3.), and g_{2i} is the sensory regulation based on a phase resetting mechanism (see Section 2.2.4.).

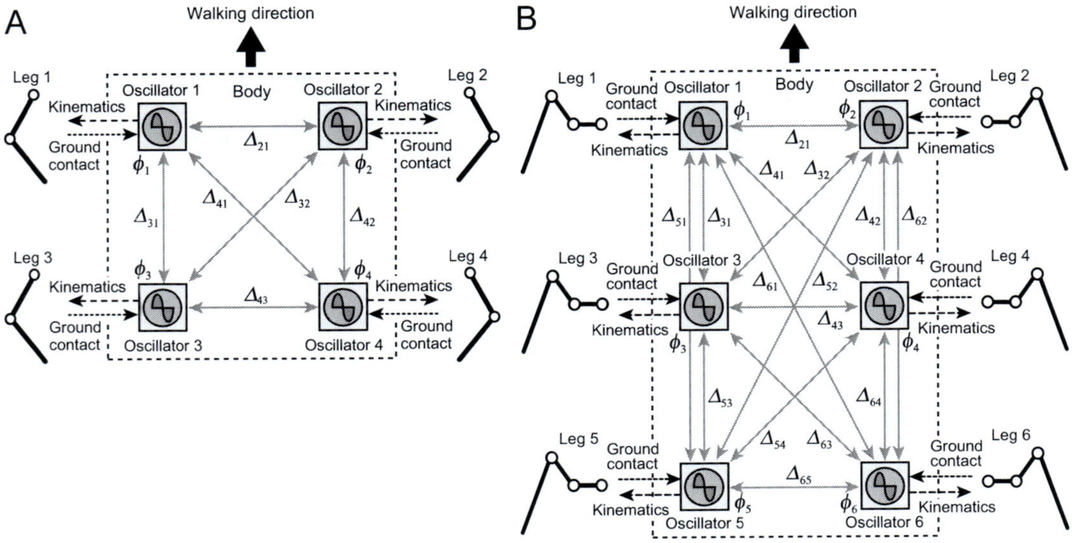

Figure 2. Nervous system models for (A) the quadruped model and (B) the insect model. Four phase oscillators are used for the quadruped model and six phase oscillators are used for the insect model. The oscillators interact with each other based on the phase relationship Δ_{ij}. The oscillator phases are modulated by tactile sensor information. The oscillator phases determine the leg joint kinematics.

2.2.2. Pattern formation model

Physiological studies have revealed that nervous systems receive sensory information and encode global parameters of the leg kinematics, such as the length and orientation of the limb axis (position of the leg tip relative to the root) [8, 11, 46, 47]. In this study, the PF model was used to determine the leg kinematics based on the oscillator phase ϕ_i from the RG model and to produce motor torques for generating the desired kinematics.

To put it simply, locomotion involves transporting the entire body against gravitational force by using the legs. This is achieved by moving the swing leg forward and by supporting the entire body and producing a propulsive force from the ground using the stance leg. For that purpose, a simple leg kinematics was used for both quadruped and insect models, which is composed of the swing and stance phases (Fig. 3). The swing phase consists of a simple closed curve for the leg tip that includes the anterior extreme position (AEP) and the posterior extreme position (PEP). It starts from the PEP and continues until the leg touches the ground. The line segment between the AEP and the PEP is parallel to the body. The stance phase is a straight line from the landing position (LP) to the PEP. During this phase, the leg tip moves in the opposite walking direction relative to the body. The body travels in the walking direction while the leg tips are in contact with the ground. These leg kinematics are common in the quadruped and insect models.

The distance between the AEP and the PEP is denoted by D. The swing and stance phase durations are denoted by T_{sw} and T_{st}, respectively, for the case when the leg tip contacts the ground at the AEP (LP $=$ AEP). The duty factor β (the ratio between the stance phase and the step cycle duration), the basic frequency ω in (1), the locomotion

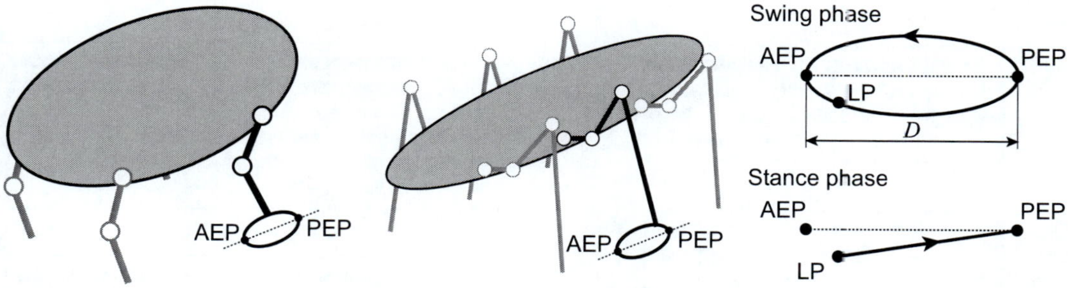

Figure 3. Leg joint kinematics composed of swing and stance phases. The swing phase for the leg tip is a closed curve that includes the anterior extreme position (AEP) and the posterior extreme position (PEP). The stance phase is a straight line from the landing position (LP) to the PEP. When the leg lands on the ground, it changes from the swing to the stance phase. When the leg tip reaches the PEP, it moves into the swing phase. These kinematics are common in the quadruped and insect models.

speed v, and the stride length S are respectively given by

$$\beta = \frac{T_{\rm st}}{T_{\rm sw} + T_{\rm st}}, \quad \omega = \frac{2\pi(1-\beta)}{T_{\rm sw}}, \quad v = \frac{(1-\beta)D}{\beta T_{\rm sw}}, \quad S = \frac{D}{\beta} \tag{2}$$

The locomotion speed v was varied by changing the duty factor β in the same manner as observed in the locomotion of humans and animals [15, 16, 32, 43, 44], where the frequency ω and the stride length S also vary with the duty factor β. The same values of these parameters were used for all the legs in each animal model.

These trajectories for the swing and stance phases are given as functions of the corresponding oscillator phase, where we used $\phi_i = 0$ at the PEP and $\phi_i = \phi_{\rm AEP}(= 2\pi(1-\beta))$ at the AEP. Therefore, the desired joint kinematics is given as a function of the oscillator phase and each joint is controlled by the joint torque based on proportional-derivative (PD) feedback control to produce the desired kinematics.

2.2.3. Gait pattern

In both quadruped and insect models, the leg kinematics is determined by the corresponding oscillator phase. Therefore, the interlimb coordination pattern is determined by the relative phase between the oscillators, which is denoted by the matrix $\Delta_{ij} = \phi_i - \phi_j$ ($i, j = 1, \ldots, 4$ for the quadruped model, $i, j = 1, \ldots, 6$ for the insect model, $0 \leq \Delta_{ij} < 2\pi$). Because the relationships $\Delta_{ij} = -\Delta_{ji}$, $\Delta_{ij} = \Delta_{ik} + \Delta_{kj}$, and $\Delta_{ii} = 0$ are satisfied ($i, j, k = 1, \ldots, 4$ for the quadruped model, $i, j, k = 1, \ldots, 6$ for the insect model), the gait is determined by three state variables for the quadruped model, such as $[\,\Delta_{21}\ \Delta_{31}\ \Delta_{43}\,]$, and by five state variables for the insect model, such as $[\,\Delta_{21}\ \Delta_{31}\ \Delta_{43}\ \Delta_{53}\ \Delta_{65}\,]$. For the quadruped model, $[\,\Delta_{21}\ \Delta_{31}\ \Delta_{43}\,] = [\,\pi\ \pi/2\ \pi\,]$ is satisfied for the walking gait and $[\,\Delta_{21}\ \Delta_{31}\ \Delta_{43}\,] = [\,\pi\ \pi\ \pi\,]$ is satisfied for the trotting gait (Fig. 4A). For the insect model, $[\,\Delta_{21}\ \Delta_{31}\ \Delta_{43}\ \Delta_{53}\ \Delta_{65}\,] = [\,\pi\ 2\pi/3\ \pi\ 2\pi/3\ \pi\,]$ is satisfied for the metachronal gait and $[\,\Delta_{21}\ \Delta_{31}\ \Delta_{43}\ \Delta_{53}\ \Delta_{65}\,] = [\,\pi\ \pi\ \pi\ \pi\ \pi\,]$ is satisfied for the tripod gait (Fig. 4B).

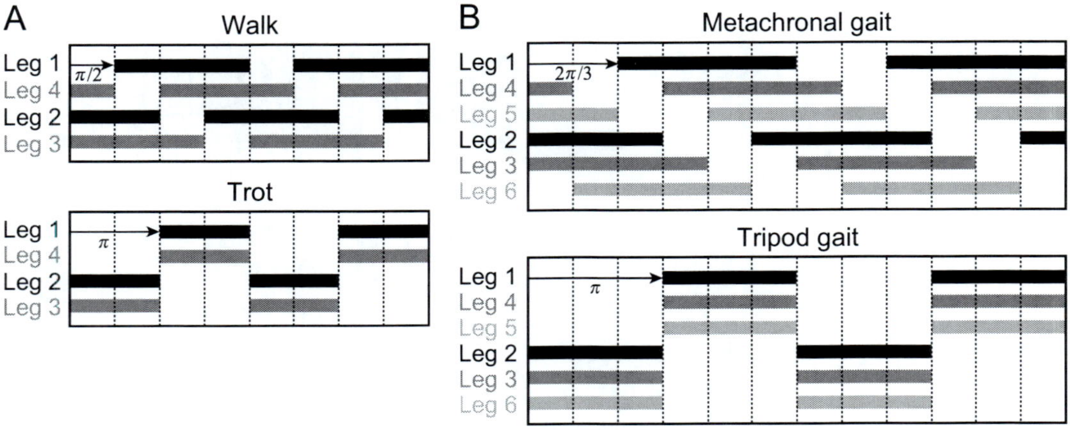

Figure 4. Footprint diagrams for (A) the quadruped model and (B) the insect model. The quadruped model has walking and trotting gaits and the insect model has metachronal and tripod gaits, where the right and left legs move in antiphase.

Function g_{1i} in (1) manipulates these relative phases, which is given by

$$g_{1i} = -\sum_{j=1}^{N} K_{ij} \sin(\Delta_{ij} - \Delta_{ij}^*) \tag{3}$$

where $N = 4$ for the quadruped model and $N = 6$ for the insect model, Δ_{ij}^* is the desired relative phase, and K_{ij} (≥ 0) is the gain constant ($i, j = 1, \ldots, N$). When a large value is used for K_{ij}, the relationship $\Delta_{ij} = \Delta_{ij}^*$ is satisfied.

2.2.4. Phase resetting

For adaptive and effective locomotor behaviors, adequate sensory regulation of motor commands is important. Physiological studies have shown that the locomotor rhythm and phase are modulated by producing phase shift and rhythm resetting based on sensory afferents and perturbations (phase resetting) [6, 13, 18, 50, 51]. In addition, neuromusculoskeletal models have demonstrated the contributions of phase resetting to the generation of adaptive locomotion [3, 4, 59, 60].

In this study, the phase resetting mechanism was incorporated to produce adaptive locomotion through dynamic interactions between the body mechanical and nervous systems. Because cutaneous afferents were observed to contribute to these resetting behaviors [18, 51], the phase ϕ_i of the oscillator i is reset to ϕ_{AEP} when the leg i lands on the ground. Function g_{2i} in (1) corresponds to this sensory regulation and is given by

$$g_{2i} = (\phi_{\mathrm{AEP}} - \phi_i)\delta(t - t_{\mathrm{land}}^i) \tag{4}$$

where t_{land}^i is the time when the leg i contacts the ground and $\delta(\cdot)$ denotes the Dirac delta function. The tactile sensor signals not only modulate the locomotion rhythm and phase but also switch the leg movements from the swing to the stance phase, as explained in Section 2.2.2..

2.2.5. Constraints for gait

The gait pattern is determined by the relative phase between the oscillators, which depends on the interactions among the oscillators (3) and the sensory regulation by phase resetting (4). Without (3) and (4), the relative phase remains in the initial state and the gait never changes. When all the elements of matrix Δ^*_{ij} are determined based on the desired gait and large values are used for the gain constants K_{ij} in (3), the desired gait is established when the gait is stable. On the other hand, when small values are used for K_{ij}, a different gait can be established from the desired one due to the sensory regulation by phase resetting (4).

Based on the relationship of the leg movements in the locomotion of animals [58], some constraints were used for the relative phase Δ_{ij}. For the walking and trotting gaits in quadrupeds and the metachronal and tripod gaits in insects, the right and left legs move in antiphase. Therefore, the quadruped model used

$$\Delta^*_{21} = \Delta^*_{43} = \pi \tag{5}$$

and a large value for K_{12}, K_{21}, K_{34}, and K_{43}. The insect model used

$$\Delta^*_{21} = \Delta^*_{43} = \Delta^*_{65} = \pi \tag{6}$$

and a large value for K_{12}, K_{21}, K_{34}, K_{43}, K_{56}, and K_{65}. In addition to (6), because the intervals between the steps of the fore leg and middle leg and between the middle leg and hind leg are identical (they vary with the locomotion speed), the insect model considered

$$\Delta_{31} = \Delta_{53} \; (\Delta_{42} = \Delta_{64}) \tag{7}$$

To satisfy this condition, the following desired phases were modified using the actual relative phases by

$$\Delta^*_{31} = \Delta_{53}, \quad \Delta^*_{35} = \Delta_{13}, \quad \Delta^*_{42} = \Delta_{64}, \quad \Delta^*_{46} = \Delta_{24} \tag{8}$$

and a large value was used for K_{31}, K_{35}, K_{42}, and K_{46}.

From the conditions (5), (6), and (7), $\Delta_{21} = \Delta_{43} = \pi$ is generally satisfied for the quadruped model and $\Delta_{21} = \Delta_{43} = \Delta_{65} = \pi$ and $\Delta_{31} = \Delta_{53}$ are satisfied for the insect model. Therefore, there are two constraints for the three state variables of the quadruped gait and four constraints for the five state variables of the insect gait. Because the other K_{ij} were set to zero, the gait is determined by a single phase relationship in both quadruped and insect models, such as Δ_{31}, which is obtained through the locomotion dynamics. For the quadruped model, a walking gait is achieved when $\Delta_{31} = \pi/2$ and a trotting gait is produced when $\Delta_{31} = \pi$ (Fig. 4A). For the insect model, a metachronal gait is established when $\Delta_{31} = 2\pi/3$ and a tripod gait is obtained when $\Delta_{31} = \pi$ (Fig. 4B).

3. Results

3.1. Dependence of Gaits on Locomotion Speed

First, what gaits the quadruped and insect models generate depending on the locomotion speed was investigated, where $\beta = 0.748$ and 0.751 were used for the quadruped model

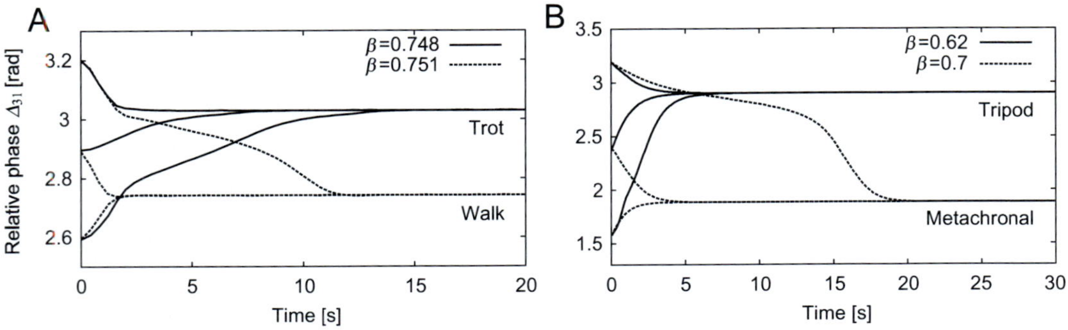

Figure 5. Relative phase Δ_{31} for (A) the quadruped model and (B) the insect model, plotted at contact of leg 1 for three initial values. The quadruped model used $\beta = 0.748$ and 0.751 and the insect model used $\beta = 0.62$ and 0.7. For the quadruped model, Δ_{31} converges to 3.0 rad (trot) for $\beta = 0.748$, while Δ_{31} converges to 2.75 rad (walk) for $\beta = 0.751$. For the insect model, Δ_{31} converges to 2.9 rad (tripod) for $\beta = 0.62$, while Δ_{31} converges to 1.9 rad (metachronal) for $\beta = 0.7$.

and $\beta = 0.62$ and 0.7 were used for the insect model. Three initial values for the relative phase Δ_{31} were used and where Δ_{31} converged was examined.

Figures 5A and B show the results of Δ_{31} for the quadruped and insect models, respectively, plotted when the leg 1 touches the ground. For the quadruped model, Δ_{31} converged to 3.0 rad for $\beta = 0.748$, which shows that the quadruped model established the trot at high speed. In contrast, Δ_{31} converged to 2.75 rad for $\beta = 0.751$, indicating that the quadruped model performed the walk at low speed (Although $\Delta_{31} = 2.75$ rad is also close to π, it was considered to indicate the walk to distinguish from the trot of 3.0 rad). For the insect model, Δ_{31} converged to 2.9 rad for $\beta = 0.62$, showing that the insect model established the tripod gait at high speed. On the other hand, Δ_{31} converged to 1.9 rad for $\beta = 0.7$, which indicates that the insect model performed the metachronal gait at low speed. Both quadruped and insect models produced different gaits depending on the locomotion speed.

3.2. Appearance of Hysteresis in Gait Transitions

After the animal models produced stable gait, the locomotion speed was slowly increased or decreased. Specifically, for the quadruped model, the locomotion speed was slowly increased by reducing the duty factor β from 0.751 to 0.748 or the locomotion speed was slowly decreased by increasing β from 0.748 to 0.751. For the insect model, the locomotion speed was slowly increased by reducing the duty factor β from 0.7 to 0.62 or the locomotion speed was slowly decreased by increasing β from 0.62 to 0.7. How the gait of the quadruped and insect models changed through locomotion dynamics was investigated.

Figures 6A and B show the results of Δ_{31} of the quadruped and insect models, respectively, for increasing and decreasing the locomotion speed. For the quadruped model, Δ_{31} varied between 3.0 and 2.75 rad, indicating that the gait changed between the walking and trotting gaits. When the locomotion speed was reduced, the trotting gait transitioned to the walking gait at about $\beta = 0.75$. In contrast, when the locomotion speed was increased, the walking gait changed to the trotting gait at about $\beta = 0.749$. For the insect model, Δ_{31}

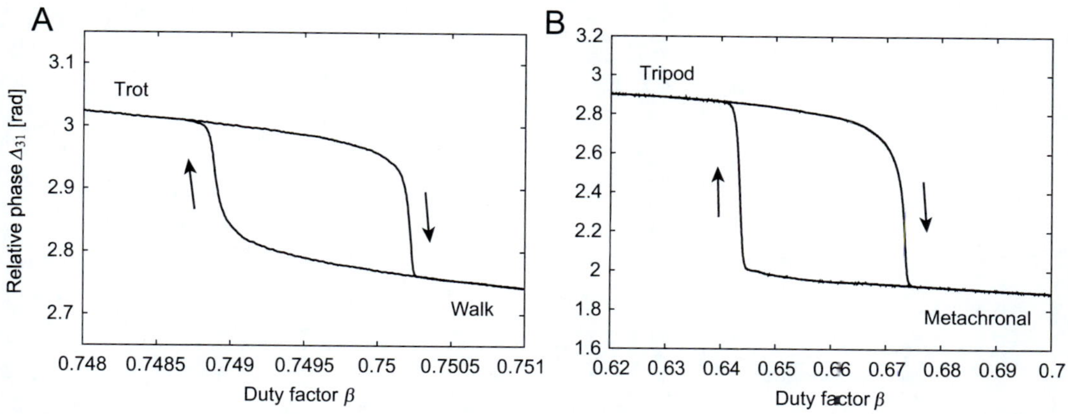

Figure 6. Gait transitions of (A) the quadruped model and (B) the insect model, induced by changing the locomotion speed through duty factor β. Walk-to-trot and trot-to-walk transitions in the quadruped model and metachronal-to-tripod and tripod-to-metachronal gait transitions in the insect model occur at different locomotion speeds in each animal model. Hysteresis appears in both animal models.

changed between 2.9 and 1.9 rad, showing that the gait varied between the metachronal and tripod gaits. When the locomotion speed was decreased, the tripod gait changed to the metachronal gait at about $\beta = 0.67$. On the other hand, when the locomotion speed was increased, the metachronal gait varied to the tripod gait at about $\beta = 0.645$. These results mean that the gait transitions occur at different locomotion speeds in both quadruped and insect models depending on the direction of the speed change, i.e., hysteresis appeared in their gait transitions.

3.3. Stability Characteristics in the Hysteresis

To clarify the stability structure of locomotion dynamics that induces the hysteresis in the gait transitions, the return map of Δ_{31} was used, where the relationship between Δ_{31n} at the foot contact of leg 1 for the nth gait cycle and Δ_{31n+4} for the $(n+4)$th gait cycle was plotted. By using the return map, possible gaits and their stabilities can be determined from the intersection with the diagonal line ($\Delta_{31n+4} = \Delta_{31n}$). More specifically, the intersection corresponds to the equilibrium point for the gait, and when the slope at the intersection is less than 1 and larger than -1, the equilibrium point is asymptotically stable. When the slope is larger than 1 or less than -1, the equilibrium point is unstable. The return map elucidates not only the local stability, but also the global stability for the gait dynamics.

Figures 7A and B show the results of the return map of Δ_{31} for the quadruped and insect models, respectively. For the quadruped model, when $\beta = 0.7487$, there is only one intersection with a diagonal line, and the trotting gait is the only attractor. For $\beta = 0.7495$, three intersections appear and there are two stable gaits (trot and walk) and one unstable gait between the stable gaits (indicated by the open dot). When $\beta = 0.7503$, the trotting gait disappears, which the loss of the two intersections reflects, and the walking gait becomes the only attractor. For the insect model, when $\beta = 0.64$, the tripod gait is the only attractor. When $\beta = 0.655$, there are two stable gaits (tripod and metachronal) and one unstable

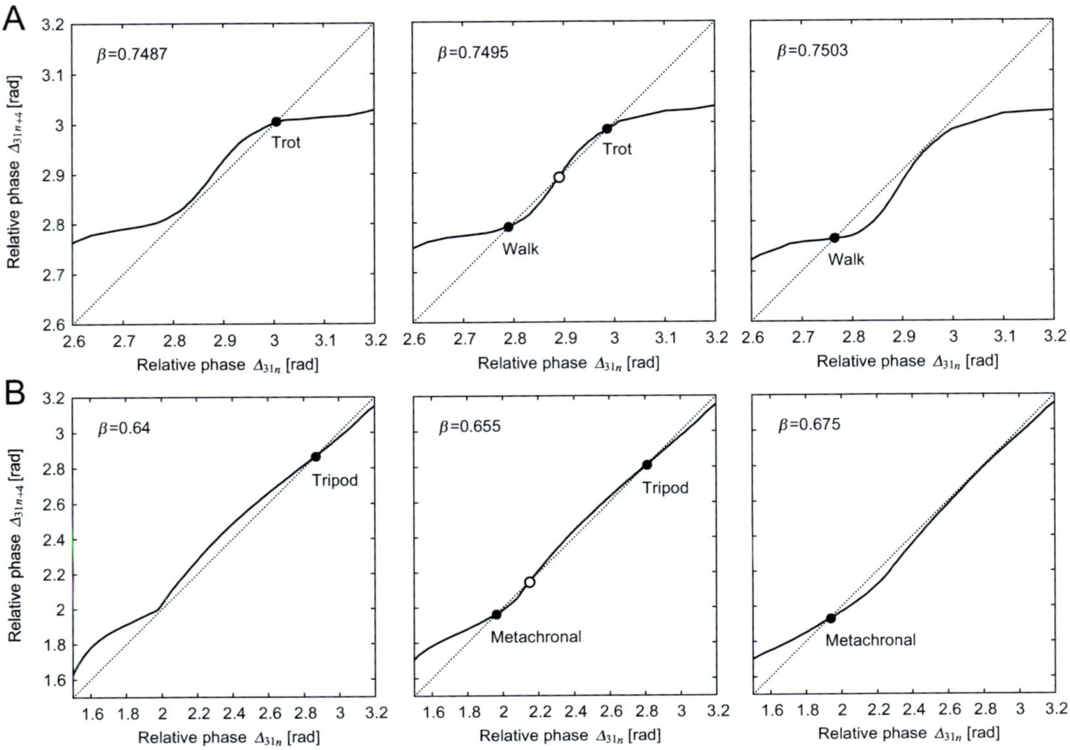

Figure 7. Return maps of the relative phase Δ_{31} for (A) the quadruped model and (B) the insect model, by plotting the relationship between the relative phase Δ_{31n} for the nth gait cycle and the relative phase Δ_{31n+4} for the $(n+4)$th gait cycle. Solid and open dots indicate stable and unstable gaits, respectively.

gait between the stable gaits. For $\beta = 0.675$, the tripod gait disappears due to the loss of the intersections and the metachronal gait is the only attractor. These results show that the gait stability passes through the saddle-node bifurcation twice in both quadruped and insect models.

Figures 8A and B show the stable and unstable gaits for the quadruped and insect models, respectively, obtained by calculating the return maps for each locomotion speed. For the quadruped model, the trotting gait is stable from $\beta = 0.748$ to 0.75, while the walking gait is stable from $\beta = 0.749$ to 0.751, indicating that two different stable gaits coexist from $\beta = 0.749$ to 0.75. For the insect model, the tripod gait is stable from $\beta = 0.62$ to 0.67 and the metachronal gait is stable from $\beta = 0.645$ to 0.7, showing the coexistence of two different stable gaits from $\beta = 0.645$ to 0.67. In both animal models, the stable gaits are connected by the unstable gaits (fold catastrophe). These stability structures induce a jump of the gait as indicated by the arrows, which give rise to the hysteresis.

4. Conclusion

So far, to elucidate the gait generation and transition mechanisms, oscillator network models have been developed that represent animal gaits in terms of the phase relationship between

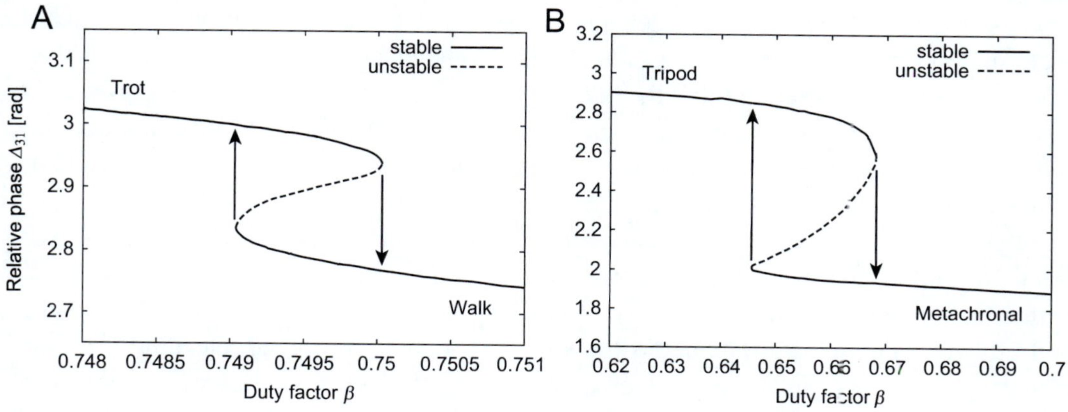

Figure 8. Stable and unstable gaits calculated from the return maps for (A) the quadruped model and (B) the insect model. Two stable gaits and one unstable gait coexist from $\beta = 0.748$ to 0.749 in the quadruped model and from $\beta = 0.645$ to 0.67 in the insect model. Fold catastrophe occurs at connections of the stable and unstable gaits and induces jump of gaits.

the leg movements [10, 12, 24, 25, 26, 27, 35, 37, 38]. However, these models did not incorporate the contribution of the body mechanical system. Because locomotion is a well-organized motion generated through dynamic interactions between the body and nervous systems, it was difficult to fully elucidate the mechanisms solely in terms of the nervous system. It is important to clarify the contributions of sensory-motor coordination in the gait generation and transition. In this study, neuromechanical models for a quadruped and an insect were constructed to emulate their dynamic locomotion. The quadruped model established the trotting gait at fast locomotion speeds and the walking gaits at slow locomotion speeds, and the insect model produced the tripod gait at fast locomotion speeds and the metachronal gait at slow locomotion speeds, as observed in quadruped and insect locomotion [1, 15, 44, 58]. In addition, these animal models exhibited gait transitions with hysteresis. These results were not because the leg movements and locomotion control systems were designed so that these models achieved gait transitions and hysteresis, but because the stability structure changed through dynamic interactions between the body mechanical and nervous systems.

Even without rhythmic input and proprioceptive feedback, CPGs in biological systems can produce oscillatory motor commands. However, adequate sensory regulation of motor commands is crucial to achieve adaptive and effective locomotion. The locomotor rhythm and phase have been reported to be modulated by producing phase shift and rhythm resetting based on sensory afferents and perturbations (phase resetting) [6, 13, 18, 50, 51]. Furthermore, spinal cats establish locomotion on treadmills and their gait changes in accordance with the belt speed [22, 43], which suggests that tactile sensory information influences the locomotor rhythm and phase generated by CPGs [19]. In this study, the phase resetting mechanism was used for the sensory regulation model during locomotion. Without this sensory regulation, all the values of the relative phase Δ_{31} are neutral and the body mechanical system makes no contribution to the gait. The sensory regulation based on phase resetting changed the stability structure so as to produce equilibrium points of Δ_{31}, where the two legs for the quadruped model and the three legs for the insect model are synchronized at high speeds whereas the legs are not synchronized at low speeds. Stability analysis using the return map revealed that the equilibrium points vary through the saddle-node bifurcation, which induces hysteresis in gait transitions. The neuromechanical models show the contribution of the interaction between the body mechanical and nervous systems to the gait generation and transition from a dynamic viewpoint.

To produce locomotion, humans and animals use a large number of degrees of freedom. To understand their locomotor mechanism, simple physical models constructed by extracting the fundamentals of their locomotion dynamics are useful. In particular, stability and bifurcation structures obtained using simple models will provide meaningful biological insights [5, 31]. In the quadruped and insect models in this chapter, the movement of one leg was denoted by the oscillator phase and the degrees of freedom of their gait was reduced by employing biologically adequate constraints, which allowed to clarify the stability and bifurcation structures of locomotion dynamics. To better understand the gait transition mechanisms in locomotion dynamics, more biomechanically and physiologically sophisticated models should be developed in the future.

Acknowledgments

This chapter is based on collaborative studies with Kazuo Tsuchiya, Tsuyoshi Yamashita, Soichiro Fujiki, Tetsuro Funato, Nozomi Tomita, and Kei Senda. This chapter is supported in part by a Grant-in-Aid for Scientific Research (B) No. 23360111 and a Grant-in-Aid for Challenging Exploratory Research No. 23656185 from the Ministry of Education, Culture, Sports, Science, and Technology of Japan.

References

[1] R.McN. Alexander, *Optimization and gaits in the locomotion of vertebrates*, Physiol. Rev., 69(4):1199-1127, 1989.

[2] S. Aoi, T. Yamashita, and K. Tsuchiya, *Hysteresis in the gait transition of a quadruped investigated using simple body mechanical and oscillator network models*, Phys. Rev. E, 83(6):061909, 2011.

[3] S. Aoi, N. Ogihara, T. Funato, Y. Sugimoto, and K. Tsuchiya, *Evaluating functional roles of phase resetting in generation of adaptive human bipedal walking with a physiologically based model of the spinal pattern generator*, Biol. Cybern., 102(5):373-387, 2010.

[4] S. Aoi, T. Kondo, N. Hayashi, D. Yanagihara, S. Aoki, H. Yamaura, N. Ogihara, T. Funato, N. Tomita, K. Senda, and K. Tsuchiya, *Contributions of phase resetting and interlimb coordination to the adaptive control of hindlimb obstacle avoidance during locomotion in rats: a simulation study*, Biol. Cybern., 107(2):201-216, 2013.

[5] S. Aoi, Y. Egi, and K. Tsuchiya, *Instability-based mechanism for body undulations in centipede locomotion*, Phys. Rev. E, 87(1):012717, 2013.

[6] U. Bässler and A. Büschges, *Pattern generation for stick insect walking movements–multisensory control of a locomotor program*, Brain Res. Rev., 27:65-88, 1998.

[7] J.A. Bender, E.M. Simpson, B.R. Tietz, K.A. Daltorio, R.D. Quinn, and R.E. Ritzmann, *Kinematic and behavioral evidence for a distinction between trotting and ambling gaits in the cockroach Blaberus discoidalis*, J. Exp. Biol., 214:2057-2064, 2011.

[8] G. Bosco and R.E. Poppele, *Proprioception from a spinocerebellar perspective*, Physiol. Rev., 81:539-568, 2001.

[9] R.E. Burke, A.M. Degtyarenko, and E.S. Simon, *Patterns of locomotor drive to motoneurons and last-order interneurons: Clues to the structure of the CPG*, J. Neurophysiol., 86:447-462, 2001.

[10] C.C. Canavier, R.J. Butera, R.O. Dror, D.A. Baxter, J.W. Clark, and J.H. Byrne, *Phase response characteristics of model neurons determine which patterns are expressed in a ring circuit model of gait generation*, Biol. Cybern., 77:367-380, 1997.

[11] A. Casabona, M.S. Valle, G. Bosco, A. Garifoli, S.A. Lombardo, and V. Perciavalle, *Anisotropic representation of forelimb position in the cerebellar cortex and nucleus interpositus of the rat*, Brain Res., 972:127-136, 2003.

[12] J.J. Collins and I. Stewart, *Hexapodal gaits and coupled nonlinear oscillator models*, Biol. Cybern., 68:287-298, 1993.

[13] B.A. Conway, H. Hultborn, and O. Kiehn, *Proprioceptive input resets central locomotor rhythm in the spinal cat*, Exp. Brain Res., 68:643-656, 1987.

[14] H. Cruse, T. Kindermann, M. Schumm, J. Dean, and J. Schmitz, *Walknet–a biologically inspired network to control six-legged walking*, Neural Netw., 11:1435-1447, 1998.

[15] F. Delcomyn, *The locomotion of the cockroach Periplaneta americana*, J. Exp. Biol., 54:443-452, 1971.

[16] F. Delcomyn and P.N.R. Usherwood, *Motor activity during walking in the cockroach Periplaneta americana I. Free walking*, J. Exp. Biol., 59:629-642, 1973.

[17] F.J. Diedrich and W.H. Warren Jr., *Why change gaits? Dynamics of the walk-run transition*, J. Exp. Psychol. Hum. Percept. Perform., 21:183-202, 1998.

[18] J. Duysens, *Fluctuations in sensitivity to rhythm resetting effects during the cat's step cycle*, Brain Res., 133(1):190-195, 1977.

[19] J. Duysens, F. Clarac, and H. Cruse, *Load-regulating mechanisms in gait and posture: Comparative aspects*, Physiol. Rev., 80:83-133, 2000.

[20] Ö. Ekeberg, M. Blümel, and A. Büschges, *Dynamic simulation of insect walking*, Arthropod Struct. Dev., 33:287-300, 2004.

[21] C.T. Farley and C.R. Taylor, *A mechanical trigger for the trot-gallop transition in horses*, Science, 253:306-308, 1991.

[22] H. Forssberg and S. Grillner, *The locomotion of the acute spinal cat injected with clonidine i.v.*, Brain Res., 50:184-186, 1973.

[23] S. Fujiki, S. Aoi, T. Funato, N. Tomita, K. Senda, and K. Tsuchiya, *Hysteresis in the metachronal-tripod gait transition of insects: A modeling study*, Phys. Rev. E, 88(1):012717, 2013.

[24] R.M. Ghigliazza and P. Holmes, *A minimal model of a central pattern generator and motoneurons for insect locomotion*, SIAM J. Appl. Dyn. Syst., 3(4):671-700, 2004.

[25] M. Golubitsky, I. Stewart, P.L. Buono, and J.J. Collins, *A modular network for legged locomotion*, Physica D, 115:56-72, 1998.

[26] M. Golubitsky, I. Stewart, P.L. Buono, and J.J. Collins, *Symmetry in locomotor central pattern generators and animal gaits*, Nature, 401:693-695, 1999.

[27] D. Graham, *Simulation of a model for the coordination of leg movement in free walking insects*, Biol. Cybern., 26:187-198, 1977.

[28] T.M. Griffin, R. Kram, S.J. Wickler, and D.F. Hoyt, *Biomechanical and energetic determinants of the walk-trot transition in horses*, J. Exp. Biol., 207:4215-4223, 2004.

[29] S. Grillner, *Locomotion in vertebrates: central mechanisms and reflex interaction*, Physiol. Rev., 55(2):247-304, 1975.

[30] N.C. Heglund and C.R. Taylor, *Speed, stride frequency and energy cost per stride: how do they change with body size and gait?*, J. Exp. Biol., 138:301-318, 1998.

[31] P. Holmes, R.J. Full, D. Koditschek, and J. Guckenheimer, *The dynamics of legged locomotion: Models, analyses, and challenges*, SIAM Rev., 48(2):207-304, 2006.

[32] M.G. Hoy and R.F. Zernicke, *Modulation of limb dynamics in the swing phase of locomotion*, J. Biomech., 18:49-60, 1985.

[33] D.F. Hoyt and C.R. Taylor, *Gait and the energetics of locomotion in horses*, Nature, 292:239-240, 1981.

[34] A. Hreljac, R. Imamura, R.F. Escamilla, and W.B. Edwards, *Effects of changing protocol, grade, and direction on the preferred gait transition speed during human locomotion*, Gait Posture, 25:419-424, 2007.

[35] S. Ito, H. Yuasa, Z. Luo, M. Ito, and D. Yanagihara, *A mathematical model of adaptive behavior in quadruped locomotion*, Biol. Cybern., 78:337-347, 1998.

[36] D.L. Jindrich and R.J. Full, *Many-legged maneuverability: dynamics of turning in hexapods*, J. Exp. Biol., 202:1603-1623, 1999.

[37] G. Schöner, W.Y. Jiang, and J.A.S. Kelso, *A synergetic theory of quadrupedal gaits and gait transitions*, J. Theor. Biol., 142:359-391, 1990.

[38] S. Kimura, M. Yano, and H. Shimizu, *A self-organizing model of walking patterns of insects*, Biol. Cybern., 69:183-193, 1993.

[39] R. Kukillaya, J. Proctor, and P. Holmes, *Neuromechanical models for insect locomotion: Stability, maneuverability, and proprioceptive feedback*, Chaos, 19:026107, 2009.

[40] M. Lafreniere-Roula and D.A. McCrea, *Deletions of rhythmic motoneuron activity during fictive locomotion and scratch provide clues to the organization of the mammalian central pattern generator*, J. Neurophysiol., 94:1120-1132, 2005.

[41] C.J.C. Lamoth, A. Daffertshofer, R. Huys, and P.J. Beek, *Steady and transient coordination structures of walking and running*, Hum. Mov. Sci., 28:371-386, 2009.

[42] E. Marder and D. Bucher, *Central pattern generators and the control of rhythmic movements*, Curr. Biol., 11(23):R986-R996, 2001.

[43] G.N. Orlovsky, T. Deliagina, and S. Grillner, Neuronal control of locomotion: from mollusc to man, *Oxford University Press*, 1999.

[44] K. Pearson, *The control of walking*, Sci. Am., 235:72-86, 1976.

[45] K. Pearson, Ö. Ekeberg, and A. Büschges, *Assessing sensory function in locomotor systems using neuro-mechanical simulations*, Trends Neurosci., 29:625-631, 2006.

[46] R.E. Poppele, G. Bosco, and A.M. Rankin, *Independent representations of limb axis length and orientation in spinocerebellar response components*, J. Neurophysiol., 87:409-422, 2002.

[47] R.E. Poppele and G. Bosco, *Sophisticated spinal contributions to motor control*, Trends. Neurosci., 26:269-276, 2003.

[48] J. Proctor, R.P. Kukillaya, and P. Holmes, *A phase-reduced neuro-mechanical model for insect locomotion: feed-forward stability and proprioceptive feedback*, Phil. Trans. R. Soc. A, 368:5087-5104, 2010.

[49] A.J. Raynor, C.J. Yi, B. Abernethy, and Q.J. Jong, *Are transitions in human gait determined by mechanical, kinetic or energetic factors?*, Hum. Mov. Sci., 21:785-805, 2002.

[50] I.A. Rybak, N.A. Shevtsova, M. Lafreniere-Roula, and D.A. McCrea, *Modelling spinal circuitry involved in locomotor pattern generation: insights from deletions during fictive locomotion*, J. Physiol., 577(2):617-639, 2006.

[51] E.D. Schomburg, N. Petersen, I. Barajon, and H. Hultborn, *Flexor reflex afferents reset the step cycle during fictive locomotion in the cat*, Exp. Brain Res., 122(3):339-350, 1998.

[52] V. Segers, P. Aerts, M. Lenoir, and D. De Clercq, *Spatiotemporal characteristics of the walk-to-run and run-to-walk transition when gradually changing speed*, Gait Posture, 24:247-254, 2006.

[53] M.L. Shik and G.N. Orlovsky, *Neurophysiology of locomotor automatism*, Physiol. Rev., 56(3):465-501, 1976.

[54] G. Taga, Y. Yamaguchi, and H. Shimizu, *Self-organized control of bipedal locomotion by neural oscillators in unpredictable environment*, Biol. Cybern., 65:147-159, 1991.

[55] T.I. Tóth, S. Knops, and S. Daun-Gruhn, *A neuromechanical model explaining forward and backward stepping in the stick insect*, J. Neurophysiol., 107:3267-3280, 2012.

[56] M.T. Turvey, K.G. Holt, M.E. LaFlandra, and S.T. Fonseca, *Can the transition to and from running and the metabolic cost of running be determined from the kinetic energy of running?*, J. Mot. Behav., 31:265-278, 1999.

[57] T. Wadden and Ö. Ekeberg, *A neuro-mechanical model of legged locomotion: single leg control*, Biol. Cybern., 79:161-173, 1998.

[58] D.M. Wilson, *Insect walking*, Annu. Rev. Entomol., 11:103-122, 1966.

[59] S. Yakovenko, V. Gritsenko, and A. Prochazka, *Contribution of stretch reflexes to locomotor control: A modeling study*, Biol. Cybern., 90:146-155, 2004.

[60] T. Yamasaki, T. Nomura, and S. Sato, *Possible functional roles of phase resetting during walking*, Biol. Cybern., 88:468-496, 2003.

INDEX

A

absorbance, 248
acid, 226, 228, 230, 248, 249, 250, 251, 252, 267, 276, 279
activation, 51, 230, 265
adenine, 248, 250
aerodynamics, v, vii, 107, 108, 136, 137, 138, 139, 140, 141
aeroelastic, 107, 108, 119, 136, 137, 138, 139, 140
airways, 243, 244, 245, 279
alveoli, 243, 245, 246
angle of attack, 107, 108, 109, 110, 111, 112, 113, 114, 115, 116, 117, 118, 119, 120, 121, 122, 124, 125, 126, 127, 128, 129, 130, 131, 132, 133, 135, 136, 137
anti-clockwise, 109, 230, 243, 258, 261, 262, 263, 267, 271
aspect ratio, 272, 274
atelectasis, 245, 246, 279
attractor, 31, 232, 233, 293, 294

B

band of inaction, 5, 20, 22, 23, 27, 34, 36
base, 98, 103, 105, 129, 131, 136, 144, 149, 153, 155, 161, 163, 228, 230, 248, 249, 250, 251, 252, 276
bifurcation diagram, 229, 237, 240
bistability, 46, 57, 229, 230, 257, 274, 281, 283
boundary layer, 108, 109, 110, 113, 114, 116, 117, 118, 119, 137, 138
boundary layer separation, 108, 113, 114, 118

C

cell membrane, 230, 253, 280
channel, 189, 218, 230, 253, 254, 255, 256, 257, 280, 281
circular magnetic field, 44, 45, 46, 48, 49, 50, 52, 53, 54, 55, 56
clockwise, 109, 116, 230, 241, 242, 253, 254, 261, 262, 264, 267, 271
cocaine, 264, 282
coercivity, 50, 74, 79, 81, 83, 87, 94, 96, 97
conductance hysteresis, 253, 254, 256
conformation, 228, 248, 252, 253, 254, 265, 267, 277
conformational change, 249, 251, 252, 253, 254
cooling, 230, 240, 241, 242, 243, 276, 277, 278
critical nucleus size, 49, 50
crocodile, 241, 242, 277, 278
crustacean, 242, 243, 277, 278
current, 2, 14, 19, 20, 22, 24, 25, 26, 30, 44, 45, 46, 47, 48, 49, 57, 98, 99, 144, 187, 189, 190, 197, 198, 199, 200, 202, 203, 204, 205, 206, 208, 211, 212, 213, 214, 215, 216, 218, 219, 220, 223, 224, 229, 231, 253, 254, 255, 256, 257, 258, 280

D

damping, 108, 110, 117, 118, 119, 120, 121, 123, 124, 125, 126, 133, 137, 139, 140, 145, 146, 149, 150, 154, 157, 158, 161, 163
deflation, 230, 243, 244, 245, 248
depletion zone, 102
diazepam, 264, 265, 283
diurnal cycle, 228, 259
domain nucleation, 46, 47, 48, 49, 50, 51, 53, 55, 56
domain suppression effect, 53
domain walls propagation, 46, 48

dopamine, 268, 272
drainage, vii, 165, 166, 167, 171, 178, 179, 181, 183
drift of surface domains, 54
drug, 85, 228, 229, 260, 261, 262, 263, 264, 265, 268, 269, 276, 282, 283
dynamic stall, 107, 108, 109, 114, 116, 118, 137, 138, 141
dynamical system, 239, 272

E

ecosystem exchange, 258, 281
effect compartment, 262
electron mobility, 101, 102
energy barrie, 49, 50, 51, 55, 60, 76, 77, 195, 217, 228, 250
energy harvesting, 107, 137
enzyme, 230, 265, 266, 267, 276
enzyme conformation, 265, 266, 267
equilibrium state, 31, 121, 227, 232, 235, 237, 239, 248
euphorbia, 264

F

fitting, 103, 104, 105, 150

G

galloping, 107, 108, 109, 110, 119, 120, 121, 122, 123, 124, 125, 126, 127, 128, 129, 130, 133, 136, 137, 138, 139, 140, 141
galloping instability, 109, 120, 121, 129, 130, 136, 137, 138, 139, 140
gaussian distribution, 90, 91

H

heart rate, 228, 230, 240, 241, 242, 277, 278
heating, 46, 219, 230, 240, 241, 242, 243, 277, 278
horizontal, 25, 27, 91, 92, 104, 108, 149, 152, 153, 160, 161, 199, 272, 274
hyperbolic distribution, 91
hysteresis, v, vi, vii, viii, 1, 2, 3, 5, 6, 7, 9, 10, 11, 13, 15, 17, 19, 20, 21, 22, 23, 24, 25, 27, 28, 29, 30, 32, 33, 34, 36, 39, 40, 41, 42, 43, 46, 52, 53, 54, 55, 59, 60, 63, 64, 65, 67, 69, 70, 72, 74, 77, 79, 80, 81, 82, 83, 84, 89, 90, 91, 92, 93,94, 95, 96, 97, 98, 99, 100, 101, 102, 103, 104, 105, 107, 108, 110, 112, 114, 115, 116, 117, 118, 122, 123, 125, 126, 127, 128, 130, 131, 133, 134, 135, 136, 137, 139, 143, 144, 145, 146, 147, 148, 149, 150, 154, 155, 156, 159, 160, 162, 163, 165, 166, 167, 183, 185, 186, 187, 188, 189, 191, 196, 197, 198, 199, 200, 203, 204, 205, 206, 207, 208, 209, 210, 211, 214, 215, 216, 218, 219, 220, 221, 222, 224, 225, 226, 227, 228, 229, 230, 231, 232, 237, 239, 240, 241, 242, 243, 244, 245, 246, 247, 248, 249, 250, 251, 252, 253, 254, 256, 257, 258, 259, 260, 261, 262, 263, 264, 265, 266, 267, 272, 274, 275, 276, 277, 278, 279, 280, 281, 283, 285, 286, 287, 289, 291, 292, 293, 294, 295, 296, 297, 298, 299

I

imbibition, vii, 165, 166, 167, 170, 171, 172, 174, 175, 178, 179, 181, 183
inactivation, 230, 253, 254, 255, 256
inflation, 42, 230, 243, 244, 245, 246, 248, 278
iteration, 103, 104, 203, 273

L

labor market, 20, 23, 27, 34
laminar, 108, 110, 113, 114, 117, 119, 136, 137, 141, 142
L-dopa, 268, 269, 283
lift, 109, 113, 114, 115, 116, 117, 119, 120, 121, 122, 124, 140
ligand, 265, 266, 267
lung pressure, 243, 244, 245
lung volume, 228, 230, 243, 244, 245
lysenin, 253, 254, 256, 257, 280

M

magnetic insulator, 96, 98
magnetic microwires, 56, 57
magnetic transistor, 98
market entry/exit, 20
mathematica, 103
mathematical model, 90, 94, 105, 110, 122, 298
Matlab, 103
Mayergoyz, 20, 21, 27, 41, 90, 105
memory, v, vii, 28, 31, 40, 57, 187, 188, 189, 190, 191, 192, 193, 194, 195, 196, 197, 198, 200, 203, 204, 205, 206, 207, 208, 209, 210, 211, 212, 213, 214, 216, 225, 228, 229, 252, 253, 254, 256, 267, 276, 277, 280, 281
metabolic trend, 268, 269
metabolite, 230, 262, 268
metastability, 227, 228, 237, 250
MOKE magnetometer, 44, 46

MOKE microscope, 44, 46
MOS, 1, 2, 10, 22, 34, 60, 90, 91, 103, 108, 110, 137, 143, 144, 145, 149, 153, 166, 194, 195, 199, 203, 215, 275

N

natural frequency, 109, 119, 123, 125, 133, 136, 146
non-ideal relay, 19, 20, 22, 23, 24, 25, 27, 29, 34
non-planar vortex-type structure, 54
normalization, 89, 103, 105, 145, 223

O

opening probability, 257
option (value), 20

P

Parkinson, 121, 123, 128, 139, 267, 268, 269, 271, 283
path-dependence, 20, 21, 30, 31, 32, 217, 225
PDF, 162, 213, 214
peanut, 258, 259
percept, 272, 273, 274, 277, 297
percolation theory, vii, 165, 166, 185
pH, 11, 162, 163
pharmacodynamic model, 260, 261, 282
pharmacological hysteresis, 261
photosynthetically active radiation, 258, 259
plasma, 260, 261, 262, 268, 282, 283
play-hysteresis, 20, 24, 25, 35, 36
precursor, 149, 230, 262, 267, 271
preisach model, 20, 27, 90, 91
protein channel, 253, 255
proteresis, 261, 264
protonation, 230, 248, 249, 250

Q

quantum, 59, 60, 61, 79, 85, 94, 212

R

radiation, 230, 258
receptor, 260, 261
recirculation bubble, 113, 117
reduced frequency, 108, 110, 111, 112, 114, 115, 116, 119, 137, 141
relative permeability, vii, 165, 166, 168, 169, 170, 172, 176, 178, 179, 180, 181, 182, 183, 184, 185, 186
repelor, 228
reptile, 242

S

saddle point, 233, 236, 237
scanning loop, 249
sensor, 56, 89, 280, 281, 288, 290
Si, 61, 69, 85, 101, 102, 187, 189, 190, 191, 192, 193, 195, 201, 206, 207, 208, 210, 212, 213, 214
sigmoid, 90, 91
simplex, 269, 283
site of action, 260, 261, 262
spin, vii, 60, 61, 63, 65, 67, 69, 74, 79, 80, 84, 86, 89, 91, 94, 95, 96, 98, 99, 100, 105, 106
spin transistor, 98
spin-valve, vii, 89, 94, 96, 98, 99, 100
stall, 108, 109, 113, 114, 116, 118, 137, 138, 141, 142
structure, vii, 2, 17, 44, 45, 46, 47, 48, 49, 52, 53, 54, 55, 56, 57, 59, 60, 69, 84, 86, 96, 105, 106, 107, 109, 120, 122, 123, 125, 126, 136, 137, 138, 154, 161, 164, 179, 187, 188, 189, 190, 191, 192, 193, 194, 195, 196, 199, 200, 203, 206, 207, 208, 210, 211, 213, 216, 224, 225, 226, 256, 280, 293, 295, 297
sunk (adjustment) costs, 20
surface magnetic domain, 43, 44, 48, 52, 53, 54
surface physics, vii, 89, 105
surface reconstruction, 99, 102
switch, 2, 4, 22, 28, 33, 253, 254, 276, 290

T

temperature, 51, 60, 61, 63, 66, 67, 69, 70, 71, 72, 74, 75, 76, 77, 78, 79, 80, 81, 83, 84, 85, 87, 96, 97, 98, 99, 101, 106, 149, 216, 219, 220, 221, 230, 240, 241, 242, 243, 258, 277, 278, 279, 281, 282
titration, 230, 248, 249, 250, 251, 252, 276, 279
tolerance, 230, 262, 264, 265, 282
transverse galloping, 107, 110, 125, 136, 139, 140
turbulent, 108, 110, 113, 114, 116, 117, 119, 136, 137

U

uncertainty, vii, 1, 2, 6, 17, 20, 25, 26, 34, 36, 40, 42
unit root dynamics, 20, 30, 31

uptake, 230, 249, 258, 262, 283

V

vectors, 94, 96
vertical, 27, 104, 123, 131, 132, 153, 272, 274
visual perception, 231
voltage, v, 94, 102, 187, 188, 189, 190, 191, 192, 193, 196, 197, 198, 199, 200, 203, 204, 205, 206, 210, 211, 212, 214, 215, 216, 218, 219, 220, 221, 222, 223, 224, 225, 226, 230, 253, 254, 255, 256, 257, 280, 281

W

wait-and-see strategy, 20, 26
wind, 107, 108, 109, 110, 111, 112, 121, 128, 129, 130, 131, 132, 133, 136, 137, 138, 139, 140, 141, 142, 144, 155, 161, 164
wind tunnel, 107, 108, 110, 111, 112, 128, 130, 131, 132, 133, 136, 137, 139, 140, 141